Renaissance Art and Science @ Florence

Habent sua fata libelli

Early Modern Studies Series

Renaissance Art & Science @ Florence

Susan B. Puett and J. David Puett

Early Modern Studies 17
Truman State University Press
Kirksville, Missouri

tsup.truman.edu

Cover art: Egnazio Danti, Equinoctial armillary (1574), façade of the Church of Santa Maria Novella (photograph by Sailko, retrieved from Wikimedia Commons, GNU Free Documentation License.)
Cover design: Lisa Ahrens (artistic modifications of photo by Sailko).

Library of Congress Cataloging-in-Publication Data

Names: Puett, Susan B., 1939- author. | Puett, J. David, 1939- author.
Title: Renaissance art and science @ Florence / Susan B. Puett and J. David Puett.
Other titles: Renaissance art and science at Florence
Description: Kirksville, Missouri : Truman State University Press, 2016. |
Series: Early modern studies ; 16 | Includes bibliographical references.
Identifiers: LCCN 2016033641 (print) | LCCN 2016033805 (ebook) | ISBN 9781612481852 (pbk.) | ISBN 9781612481388
Subjects: LCSH: Art and science—Italy—Florence—History. | Renaissance--Italy--Florence. | Florence (Italy)--Civilization.
Classification: LCC N72.S3 P84 2016 (print) | LCC N72.S3 (ebook) | DDC 709.02/4--dc23
LC record available at https://lccn.loc.gov/2016033641

The paper in this publication meets or exceeds the minimum requirements of the American National Standard for Information Sciences—Permanence of Paper for Printed Library Materials, ANSI Z39.48–1992.

To our family, David, Michael, Mary, Brannon, Connor, and Meg, with whom we have shared Italy and found, much to our delight, that they too enjoy that magnificent country.

Contents

Figures and Illustrations | xv

Acknowledgments | xxi

Preface | xxiii

Introduction: The Art in the Science, the Science in the Art | 1

Chapter 1: The Evolution of Art, Science, and Polity in Renaissance Florence | 3
- **1.1** Geography and the Establishment of Florence | 3
- **1.2** The Emergence of Humanism | 5
- **1.3** Rise of the Florentine City-State | 7
- **1.4** From Medieval *Commune* to Modern Republic | 9
- **1.5** Why Florence and Why Then? | 13

Chapter 2: The Science of Art | 15
- **2.1** Color Production and Visualization | 15
  - **2.1.1** Pigments and Their Compositions | 15
    - **2.1.1.1** Blue | 16
    - **2.1.1.2** Red | 17
    - **2.1.1.3** Yellow | 18
    - **2.1.1.4** Green | 19
    - **2.1.1.5** Other Colors | 19
  - **2.1.2** Visualization of Colors | 20
- **2.2** Painting | 21
  - **2.2.1** Fresco | 21
    - **2.2.1.1** Chemistry and Methods | 21

2.2.1.2 Examples of Florentine Frescoes | 23
a. Santa Maria Novella | 23
b. Florence Cathedral (Duomo), Santa Maria del Fiore | 24
c. Ognissanti | 24
d. Santa Trinita | 26
e. Sant'Apollonia | 26
f. Santa Maria del Carmine | 26
g. San Marco | 27
h. Santa Croce | 27
2.2.1.3 Intersection of Renaissance Art and Modern Science | 27
2.2.2 Tempera | 29
2.2.2.1 Chemistry and Applications | 29
2.2.2.2 The Painting Process | 30
2.2.2.3 Tempera Masterpieces in Florentine Art | 31
a. Santa Maria Novella | 31
b. Uffizi Gallery | 31
c. San Marco | 31
2.2.3 Oil Paints | 31
2.2.3.1 Composition and Applications | 31
2.2.3.2 The Painting Process | 33
2.2.3.3 Florentine Oil Painting | 34
a. Uffizi Gallery | 34
2.3 Sculpture | 36
2.3.1 Marble | 36
2.3.1.1 Formation, Composition, and the Sculpting Process | 36
2.3.1.2 Examples of Florentine Marble Sculpture | 37
a. The Church of Orsanmichele | 37
b. Duomo Museum | 38
c. Galleria dell'Accademia | 38
2.3.2 Porphyry | 41
2.3.2.1 Formation, Composition, and Historical Perspective | 41
2.3.2.2 Examples of Porphyry in Florence | 42
a. San Lorenzo | 42
b. Santa Maria Novella | 42
c. Palazzo Vecchio | 42
d. Piazza Santa Trinita | 43

2.3.3 Wood | 43
2.3.3.1 Representative Florentine Works | 45
a. Santa Maria Novella and Santa Croce | 45
b. Duomo Museum | 45
c. Bargello Museum | 45
2.3.4 Florentine Terracotta | 47
2.3.4.1 The della Robbia Workshop | 47
a. Duomo Museum | 47
b. Bargello Museum | 48
2.3.4.2 Contemporary Analyses | 49
2.3.5 Bronze | 50
2.3.5.1 The Lost-Wax Method of Bronze Casting | 51
2.3.5.2 Illustrative Florentine Bronzes | 52
a. Baptistery | 52
b. San Lorenzo | 55
c. Orsanmichele | 55
d. Bargello Museum | 56
c. Loggia dei Lanzia (Loggia della Piazza Signoria) | 56
2.4 Glass | 60
2.4.1 Composition | 60
2.4.2 Glassmaking in Florence | 61
2.4.3 Mosaics | 62
2.4.3.1 An Illustrative Florentine Work | 63
2.4.3.2 Contemporary Scientific Studies | 63
2.4.4 Stained Glass | 65
2.4.4.1 The Creative Process | 66
2.4.4.2 Examples of Florentine Stained Glass | 66
a. Duomo | 66
b. Santa Maria Novella | 66
2.5 Semi-Precious Hardstone | 68
2.5.1 The Technology | 68
2.5.1.1 Opificio delle Pietre Dure | 68
2.5.1.2 San Lorenzo (Chapel of the Princes) | 70
2.5.2 Stones Utilized | 70

2.6 Ceramics | 71
2.6.1 Pottery in Italy | 71
2.6.1.1 Bargello Museum | 73
2.6.2 Porcelain | 73
2.6.2.1 Bargello Museum | 75
2.6.3 Decorative Italian Technique of *Sgraffito* | 75
2.7 Textiles | 76
2.7.1 Florentine Tapestries | 76
2.8 Printmaking and Engraving | 78
2.8.1 Illustrative Florentine Works | 78
Chapter 3: Applying Mathematics to Art and Cartography | 80
3.1 Use of Geometric Perspective in Art | 81
3.1.1 The Origins of Modern Geometry | 81
3.1.2 Mathematical Perspective in Renaissance Art | 82
3.1.2.1 Establishment of the Principle | 82
3.1.2.2 Seminal Treatises on Perspective | 84
3.1.2.3 Applying the Principles of Perspective | 90
a. Masaccio | 90
b. Lorenzo Ghiberti | 96
c. Donatello | 97
d. Paolo Uccello | 99
e. Fra Angelico | 101
f. Continuing the Process | 101
3.2 Cartography: Florentine Maps, a Blending of Art and Science | 104
3.2.1 Early Maps | 104
3.2.2 Florentine Mapmaking | 107
3.2.2.1 Palazzo Vecchio | 108
Chapter 4: Renaissance Medicine: Physicians, Hospitals, Apothecaries, and the Artistic Dimension | 115
4.1 Foundations of Western Medicine | 116
4.2 Doctors and the Practice of Medicine | 118
4.2.1 Training, Health Care, and Regulation | 118
4.2.2 Understanding the Human Body for Medicine and Art | 123
4.2.3 The Role of the Divine in the Healing Process | 125

4.3 Florentine Hospitals | 126
4.3.1 Santa Maria Nuova and Sant'Egidio | 127
4.3.1.1 History and Mission | 127
4.3.1.2 Healing through Art | 128
4.3.2 Ospedale degli Innocenti | 130
4.3.2.1 History and Mission | 130
4.3.2.2 Importance of Art | 132
4.4 Apothecaries | 136
4.4.1 Monastery and Convent Apothecaries | 136
4.4.2 Ingredients and Recipes | 137
4.4.3 Standardization of Recipes | 139
4.4.4 Vital Role of Apothecaries in Art | 140
4.5 Medicine Depicted in Art | 143
4.5.1 Campanile (Duomo) | 143
4.5.2 Bargello Museum | 144
4.5.3 Santa Felicitá and the Palazzo Vecchio | 144
4.5.4 Oratorio Buonomini di San Martino | 145

Chapter 5: Astronomy and Time Reckoning | 147
5.1 Ancient Origins of Renaissance Astronomy | 147
5.2 Solar Time Measurement: The Sites, the Scientists, and the Art | 151
5.2.1 The Baptistery | 152
5.2.2 The Florence Cathedral (Duomo), Santa Maria del Fiore | 153
5.2.3 Santa Maria Novella | 157
5.2.4 The Uffizi | 162
5.2.5 The Pitti Palace | 162
5.3 Measuring the Hours in Florence | 163
5.3.1 Ponte Vecchio | 163
5.3.2 Palazzo Vecchio | 163
5.3.3 Duomo | 165
5.3.4 Galileo Museum | 165
5.4 Understanding the Solar System: Galileo Galilei and the European Scientific Revolution | 166
5.5 Astronomical Instruments and Their Artistic Dimensions | 171

5.6 Astronomy Depicted in Art | 172
5.6.1 Campanile | 172
5.6.2 Santa Croce | 175
5.6.3 San Lorenzo | 176
5.6.4 Duomo, Uffizi, and Ognissanti | 177
5.7 Astronomy in Post-Renaissance Florence: The Sites and Scientists | 178
5.7.1 The Ximenes Observatory (Osservatorio Ximeniano) | 178
5.7.2 La Specola | 180
5.7.3 Piazza dei Giudici | 181

Chapter 6: Art and Technology | 182
6.1 Rediscovery and Refinement of Ancient Technology | 182
6.2 Transport of Marble for Sculpting and Building | 186
6.2.1 Marble Quarries | 186
6.2.2 Selection and Extraction | 187
6.2.3 Transportation | 188
6.3 Technology Developed for Construction | 191
6.3.1 The Florence Cathedral (Duomo), Santa Maria del Fiore | 191
6.3.1.1 Building the Cathedral | 193
6.3.1.2 The Dome Rises | 196
a. Enter Brunelleschi | 196
b. Specifications for the Shells | 199
c. Structural Specifications | 201
d. Building Methods | 202
e. The Timeframe of Construction | 204
6.3.1.3 The Lantern Completes the Construction | 205
6.3.1.4 Materials for the Dome and the Lantern | 207
6.3.1.5 Machines Used in the Construction of the Dome | 207
a. The Great Hoist | 208
b. The Great Crane | 209
6.3.1.6 Additional Machines Used in the Construction of the Lantern | 210
6.3.1.7 Art and the Cupola | 211
6.3.1.8 Intersection of Art and Contemporary Science | 214

6.3.2 Renovation of the Palazzo Vecchio | 214
6.3.2.1 Salone dei Cinquecento | 215
a. The Technological Challenge | 215
b. Enhancement with Art | 216
6.3.2.2 Studiolo | 217
6.4 Military Engineering and Fortification | 219
6.4.1 Leonardo da Vinci | 220
6.4.2 Michelangelo | 220
6.5 Performing Arts and Technology | 221
6.5.1 Background | 221
6.5.2 Sacred Enactments | 221
6.5.3 Secular Theatrical Presentations | 224
6.5.3.1 Intermezzi | 224
6.5.3.2 Opera | 225

Concluding Remarks: Connections between Science and Art | 227

Site Links | 231

Works Cited | 237

About the Authors | 247

Figures and Illustrations

Chapter 2

Figure 2.1: Domenico Ghirlandaio, *Last Supper* (ca. 1480), fresco, Church of Ognissanti (Alinari Archives, Florence, photograph by Serge Domingie, reproduced with the permission of Ministero per i beni e le Attivita Culturali [AGC-F-002068–0000]). | 25

Figure 2.2: Sandro Botticelli, *Primavera* (ca. 1482), tempera on panel, Uffizi Gallery (IBERFOTO/Alinari Archives, Florence [AIS-F-066036–0000]). | 32

Figure 2.3: Leonardo da Vinci, *Adoration of the Magi* (ca. 1481), oil on wood, Uffizi Gallery (Alinari Archives, Florence, photograph by Nicola Lorusso, reproduced with the permission of Ministero per i beni e le Attivita Culturali [AGC-F-000514–0000]). | 35

Figure 2.4: Luca della Robbia, *Cantoria* (1430–38), sculpture, Museo dell'Opera del Duomo (photograph by Jastrow, retrieved from Wikimedia Commons). | 38

Figure 2.5: Michelangelo, *David* (1501–4), marble sculpture, Galleria dell'Accademia (photograph by Marcus Obal, retrieved from Wikimedia Commons, GNU Free Documentation License). | 40

Figure 2.6: Tadda, *La colonna della Giustizia* (ca. 1569–79), porphyry sculpture, Piazza Santa Trinita (photograph by Marie-Lan Nguyen, retrieved from Wikimedia Commons, Creative Commons License). | 44

Figure 2.7: Donatello, *Saint Mary Magdalene* (ca. 1430s), wood sculpture, Museo dell'Opera del Duomo (photograph by Sailko, retrieved from Wikimedia Commons, GNU Free Documentation License). | 46

Figure 2.8: Luca della Robbia, *Madonna and Child in the Rose Garden* (late 1450s), terracotta sculpture, Bargello Museum (Alinari Archives, Florence, photograph by Daniela Cammilli, reproduced with the permission of Ministero per i beni e le Attivita Culturali [CAL-F-003744–0000]). | 48

Figure 2.9: Lorenzo Ghiberti, *Sacrifice of Isaac* (1401–ca. 1403), cast bronze panel for the competition, Bargello Museum (Web Gallery of Art, retrieved from Wikimedia Commons). | 53

Figure 2.10: Filippo Brunelleschi, *Sacrifice of Isaac* (1401–ca. 1403), cast bronze panel for the competition, Bargello Museum (photograph by Richard Fabi, retrieved from Wikimedia Commons, GNU Free Documentation License). | 53

Figure 2.11: Donatello, *David* (ca. mid-1440s), bronze sculpture, Bargello Museum (photograph by Rufus46, retrieved from Wikimedia Commons, Creative Commons License). | 57

Figure 2.12: Giovanni Cellini, *Perseus with the Head of Medusa* (1545–54), bronze sculpture, Loggia della Signoria (photograph by Marie-Lan Nguyen, retrieved from Wikimedia Commons, Creative Commons License). | 59

Figure 2.13: Mosaic ceiling (ca. 1225–1300s), Baptistery (photograph by Jebulon, retrieved from Wikimedia Commons, Creative Commons License). | 64

Figure 2.14: Lorenzo Ghiberti and Niccolò di Pietro, *Assumption of the Virgin* (1405–15), stained glass window, Duomo (Alinari Archives, Florence, photograph by Nicolò Orsi Battaglini, reproduced with the permission of Ministero per i beni e le Attivita Culturali [OBN-F-000349–0000]). | 67

Chapter 3

Figure 3.1: Depiction of Alberti's perspective method (drawing by Justine E. Stevens commissioned by the authors). | 86

Figure 3.2: Depiction of foreshortening (drawing by Justine E. Stevens commissioned by the authors). | 89

Figure 3.3: Masaccio, *Trinity* (ca. 1425–27), fresco, Santa Maria Novella (Alinari Archives, Florence, photograph by George Tatge [CAL-F-009620–0000]). | 91

Figure 3.4: Masaccio, *Payment of the Tribute Money* (ca. 1425–27), fresco, Brancacci Chapel, Santa Maria del Carmine (Alinari Archives, Florence, photograph by Serge Domingie [AGC-F-000550–0000]). | 94

Figure 3.5: Lorenzo Ghiberti, *Story of Jacob and Esau* (ca. 1429–52), cast bronze, east Baptistery door (Nicolò Orsi Battaglini/Alinari Archives, Florence, photograph by Nicolò Orsi Battaglini [OBN-F-000343–0000]). | 98

Figure 3.6: Paolo Uccello, *Battle of San Romano* (ca. 1456), fresco, Uffizi Gallery (Alinari Archives, Florence, photograph by Nicola Lorusso, reproduced with the permission of Ministero per i beni e le Attivita Culturali [AGC-F-000112–0000]). | 100

Figure 3.7: Fra Angelico, *Annunciation* (ca. 1438–45), fresco, Convent of San Marco (Nicolò Orsi Battaglini/Alinari Archives, Florence, photograph by Nicolò Orsi Battaglini [OBN-F-000057–0000]). | 102

Figure 3.8: Egnazio Danti, Map of the British Isles (1565), Guardaroba of the Palazzo Vecchio (photograph by QwHdenMWIhNP8g, Google Cultural Institute, retrieved from Wikimedia Commons). | 110

Figure 3.9: Egnazio Danti, Terrestrial Globe (ca. 1563–75), Guardaroba of the Palazzo Vecchio (photograph by Sailko, retrieved from Wikimedia Commons, GNU Free Documentation License). | 112

Chapter 4

Figure 4.1: Leonardo da Vinci, Anatomical sketch: Profile of Man and Two Riders (1490–ca. 1504), red chalk, pen, and ink on paper, Accademia of Venice (Web Gallery of Art, retrieved from Wikimedia Commons). | 124

Figure 4.2: Filippo Brunelleschi, Ospedale degli Innocenti façade arches (photo by Warburg, retrieved from Wikimedia Commons, GNU Free Documentation License). | 133

Figure 4.3: Filippo Brunelleschi, Arches at Ospedale degli Innocenti (photograph by authors). | 133

Figure 4.4: Andrea della Robbia, Roundels added to Ospedale degli Innocenti (1487) (photograph by Bruno Barral, retrieved from Wikimedia Commons, Creative Commons License). | 134

Figure 4.5: Title page, *Ricettario fiorentino di nuovo illustrato* [Fiorenz]: [Marescotti], 1597 (ETH-Bibliothek, Zurich, Alte und Seltene Drucke; EDIT 16 CNCE 3199 [http://dx.doi.org/10.3931/e-rara-9313]). | 141

Figure 4.6: Andrea Pisano, Hexagonal tile depicting patients and physician examining urine (ca. 1340s), Campanile (photograph by authors). | 145

Chapter 5

Figure 5.1: Leonardo Ximenes's depiction of beam of light admitted by the gnomon in Santa Maria del Fiore (Duomo) at the summer solstice. Meridian designed by Paolo dal Pozzo Toscanelli (ca. 1475), drawing from Ximenes, *Del vecchio e nuovo gnomone fiorentino e delle osservazioni astronomiche* . . . , 1757 (MED 1809, tav. VIII, photograph by Renato Idra, inv. # 68400, courtesy of Museo Galileo, Florence). | 155

Figure 5.2: Paolo dal Pozzo Toscanelli, Marble disc (ca. 1475), modified by Leonardo Ximenes (mid-18th century), in the floor of the Duomo (photograph by Sailko, retrieved from Wikimedia Commons, GNU Free Documentation License). | 156

Figure 5.3: Egnazio Danti, Astronomical quadrant (1572) on façade of the Church of Santa Maria Novella (photograph by Sailko, retrieved from Wikimedia Commons, GNU Free Documentation License). | 158

Figure 5.4: Egnazio Danti, Equinoctial armillary (1574), façade of the Church of Santa Maria Novella (photograph by Sailko, retrieved from Wikimedia Commons, GNU Free Documentation License). |159

Figure 5.5: Sundial (mid-1300s) on the Ponte Vecchio (photograph by Sailko, retrieved from Wikimedia Commons, GNU Free Documentation License). | 164

Figure 5.6: *Left*: Galileo Galilei, Drawing of the moon surface (1609–10), published in *Sidereus nuncius* (1610). *Right*: Photographic image of the same view (photograph by ECeDee, composition by ECeDee, retrieved from Wikimedia Commons). | 168

Figure 5.7: Egnazio Danti (attr.), Astrolabe (ca. 1570), Galileo Museum (photograph by Sailko, retrieved from Wikimedia Commons, Creative Commons Attribution 3.0 Imported License). | 173

Figure 5.8: Antonio Santucci, Armillary sphere (1588–93), Galileo Museum (photograph by Sailko, retrieved from Wikimedia Commons, Creative Commons Attribution 3.0 Imported License). | 174

Figure 5.9: Andrea Pisano, *Gionitus* tile (ca. 1334–36) on the Campanile, Museo dell'Opera del Duomo (photograph by Sailko, retrieved from Wikimedia Commons, GNU Free Documentation License). | 175

Figure 5.10: Pesello (attr.), interior of cupola depicting the sky over Florence, July 4, 1442 (1442–43), Old Sacristy, Church of San Lorenzo (Alinari Archives, Florence, photograph by George Tatge [CAL-F-001256–0000]). | 177

Figure 5.11: Sandro Botticelli, *Saint Augustine in His Study* (ca. 1480), fresco, Church of Ognissanti, Florence (Archive/Alinari Archives, photograph by George Tatge [SEA-S-F12000–0006]). | 179

Figure 5.12: Juxtaposition of the Ximenes Observatory with the Duomo used by Paolo dal Pozzo Toscanelli for his astronomical measurements (photograph by authors). | 180

Chapter 6

Figure 6.1: The dome (1420–36) of the Florence Cathedral (Duomo), Santa Maria del Fiore (photograph by Sailko, retrieved from Wikimedia Commons, GNU Free Documentation License). | 192

Figure 6.2: Sketch showing partial structural components of the dome, Santa Maria del Fiore (drawing by Justine E. Stevens [modified from illustration by George Rich], commissioned by the authors). | 200

Figure 6.3: Herringbone brickwork designed by Brunelleschi for the dome of Santa Maria del Fiore, Florence (photograph by authors). | 204

Figure 6.4: Il Taccola, Sketch of great hoist (1420) designed by Brunelleschi (retrieved from Wikimedia Commons, public domain). | 208

Figure 6.5: Giorgio Vasari and Federico Zuccari, *Last Judgment* (1571–79), fresco, interior of the dome of Santa Maria del Fiore, Florence (Nicolò Orsi Battaglini/Alinari Archives, Florence, photograph by Nicolò Orsi Battaglini, reproduced with the permission of Ministero per i beni e le Attivita Culturali [OBN-F-000043–0000]). | 213

Figure 6.6: View of ceiling structural support (1559–63) designed by Giorgio Vasari in the Palazzo Vecchio (photograph by authors). | 216

Figure 6.7: Giorgio Vasari, Studiolo (1570–73) in the Palazzo Vecchio (Web Gallery of Art, retrieved from Wikimedia Commons). | 218

Figure 6.8: Giovanni Standano, *The Alchemist's Studio* (1570), painting, Studiolo, Palazzo Vecchio (paleopatologia.it, retrieved from Wikimedia Commons). | 219

Acknowledgments

Over the years we have benefited greatly from the expertise of numerous friends and colleagues. We wish to acknowledge in particular Professors Shelley Zuraw, Donatella Lippi, Christopher Robinson, Thomas B. Settle, and Thomas Polk for their enlightening artistic and scientific discussions and suggestions. We are most indebted to the University of Georgia Foundation Fellows program of the Honors College, the Franklin College of Arts and Sciences, and the Lamar Dodd School of Art for supporting the unique study and travel abroad programs in which we were fortunate to participate. In particular, we thank President Jere Morehead and Professors Richard Johnson, R. G. Brown, Benjamin Reynolds, and David Williams. Lectures by and conversations with Professor Betty Jean Craige on the integration of humanities and sciences were sources of inspiration. We thank the enthusiastic students with whom we have interacted in Florence and in the classroom. Observing them share their perspectives and acquire a more tangible comprehension of the interwoven forces at play during the Renaissance was an important impetus for us. The helpful interactions with our two sons, Dr. David W. Puett and Professor Michael J. Puett, have provided valuable insights for which we are most grateful. In addition, our grandson Connor Puett gave expert assistance with the figures, three of which were designed and drawn by Justine E. Stevens; to them we are most appreciative. Mary, Meg, and Brannon Puett have our thanks for cheering us on through this lengthy process. We are indebted to Professor Michael Wolfe, Queens College, CUNY, general editor of the Early Modern Studies series for Truman State University Press, for his astute comments and suggestions, as well as two anonymous reviewers who provided cogent ideas that enhanced the manuscript. Lastly, we extend our deepest gratitude to Barbara Smith-Mandell, Director and Editor-in-Chief of Truman State University Press, for her extraordinarily careful editing and insightful recommendations. The authors are of course responsible for any errors, misinterpretations, or omissions.

Preface

Enticing visitors for centuries, Florence has continued as a source of inspiration to countless generations of artists, writers, and others who are interested in this important site of the Renaissance, arguably one that had a major impact on the continued development of Western civilization. The advent of humanism, with its ramifications on many aspects of intellectual endeavors, ushered a new era that redefined the role of scholarly and artistic pursuits in man's quest for the essence of nature.

While teaching honors seminars, accompanying students (many in the Honors College) and faculty from the University of Georgia on study and travel abroad programs in Italy, it became obvious to the authors that those in the humanities/fine arts and those concentrating on the sciences had diverse views of Renaissance masterworks. For them, the art and architecture were inspiring on different levels: that of the artistic works themselves and the scientific explanations of the processes involved. Excitement was generated by the synergy of the resulting discussions and our interactions with the students and faculty from various disciplines. It was these experiences, vividly contrasting today's dualistic mind-set with that of the early Florentines and our fervent love for the Renaissance, that stimulated us to write this book.

We posit that the effective integration of art and science was a major catalytic force for the development and fluorescence of the Florentine Renaissance. This book is based on that premise, and herein we are using the term "science" in the broadest sense to include all aspects such as experimental observations and the development and testing of hypotheses, mathematics, cartography, engineering, technology, medicine, and astronomy.

S.B.P. and J.D.P.
2016

Introduction

The Art in the Science, the Science in the Art

Although art and science in the twenty-first century intersect on many levels—witness for example the impact of the computer on art and design—the practitioners of science are perceived as operating in a realm quite different from the one inhabited by those pursuing artistic endeavors. To apply that same dualistic cultural assumption to the period of the Florentine Renaissance is to assume a prevailing view that did not exist during that relatively short period that followed the Middle Ages and preceded the Scientific Revolution and Age of Enlightenment in Europe.

In the Florentine Renaissance,[1] science, particularly mathematics and technology, was often used to enhance the exquisiteness of art, and in turn art frequently stimulated scientific inventiveness and innovation. Scientific and technological advances usually occurred in the workshops of artists, in the design and construction of churches and public buildings, in apothecaries, and in hospitals, as well as in the laboratories of alchemists. Science became a means of extending artistic dimensions with the use of mathematical perspective in painting and the creation of new materials and techniques. In hospitals, art was used to reinvigorate the soul while the physicians were focused on healing the body. Technology enabled artists to expand the possibilities of their painting and sculpture. Glassware and maps, for example, were transformed from utilitarian items into objects to be admired for their beauty. Art enabled the depiction of scientific observations, and advances in construction technology made buildings canvasses for artistic grandeur. Visitors to the Florence of today are enthralled by the celebrated art in the Uffizi Gallery, the sculpture in the Bargello National Museum, the magnificence of Michelangelo's *David* in the Galleria dell'Accademia, the

1. "Renaissance" is a French word and was first introduced in reference to Florence by the French historian Jules Michelet in 1858. Shortly thereafter, Jacob Burckhardt, a Swiss historian, cemented this term into popular usage in his seminal book, *The Civilization of the Renaissance in Italy*, published in 1860. Thus, the contemporary usage of "Renaissance Florence" refers to a term introduced centuries after the time period addressed in this book.

depth of the frescoes in the Brancacci Chapel, Santa Maria Novella, and other churches, and the majesty of the Florence Cathedral, to name but a few of the treasures from the Renaissance. These sites, as well as many more, also include examples of remarkable scientific achievements.

This volume explores the accomplishments of the Florentine Renaissance in the fields of color production and artistic methodology, geometric applications in painting and cartography, medicine and medicinals, astronomy, solar reckoning, and technology and construction. Viewed in conjunction with the art, an understanding of the scientific endeavors gives the splendor that is Florence an even deeper meaning.

Chapter 1

The Evolution of Art, Science, and Polity in Renaissance Florence

> "That wondrous Florentine spirit, at once keenly critical and artistically creative, was incessantly transforming the social and political condition of the State, and as incessantly describing and judging the change."
>
> —Jacob Burckhardt, *The Civilization of the Renaissance in Italy*

The elements of Florentine society coalesced to create an environment within which the Renaissance emerged. The Florentine social and political structure provided the essential components for the observational approach to exploration, creativity, and innovation across the arts and sciences, expanding the possibilities of each. A journey through the panorama of the Florentine Renaissance can begin at many levels and proceed along many paths. Because art became integrated with science and technology—indeed they were synergistic and catalytic one with the other—it is useful to start this particular journey with a synopsis of Florentine geography and history in order to understand the prevailing forces that resulted in this magnificent period in Western culture.

1.1 Geography and the Establishment of Florence

The mountains and hills of Italy are dotted with lush valleys where settlements grew up along navigable rivers. As early as the Neolithic Age through the tenth century BCE, the region now known as Florence was inhabited by Ligurian and later Italic tribes who settled in an isolated valley along the Arno River that provided fertile land and abundant water. They were gradually outnumbered in about 900 BCE by the Etruscans who migrated to Italy from Lydia, located in present-day northwestern Turkey. By 600 BCE the culturally advanced Etruscans, possessing a writing system and a societal infrastructure, had established twelve settlements, one of which was Fiesole in the hills north of present-day Florence. As is true at the many Etruscan settlements in Italy, artifacts including exquisite pottery and jewelry, characterized by a high level of artistic development, have been uncovered mainly from funerary excavations. Those from Fiesole are on exhibit in the local museum.

Eventually the Etruscans were defeated by the Romans and by 205 BCE all of what is now Tuscany was under Roman rule. The Romans established the colony of Florentia on the banks of the Arno River to control the route between northern Italy and Rome. While there is some controversy over the exact date of Roman settlement, most scholars accept the year 59 BCE. The colonization of the area by the Romans lasted for several hundred years, and they left a distinctly Roman mark on the city that became Florence. Present-day streets in Florence basically follow the Roman grid with Via Roma as the main north–south artery and Via del Corso and Via Strozzi defining the east–west axis. The modern-day Piazza della Repubblica is the site of the original Roman forum, and traces of the outline of the Roman amphitheater also remain. It can be argued that the possession of a Roman legacy contributed to the personae of the citizens of Florence. Indeed, that classical heritage remained a part of who they were as a people and the ideals they aspired to emulate and advance.

In cities like Florence, cultures developed in relative isolation. Situated at a latitude and longitude of 43°46'N and 11°15'E, present-day Florence is located in the valley of the Arno River and surrounded on three sides by tree-covered hills. To the north are the hills of Careggi and Fiesole; to the south those of Arcetri, Poggo Imperiale, Scandicci, and Bellosguardo; and to the east lies the range of Settignano. The Apennine and other mountain ranges tended to isolate towns and villages, offering some degree of protection from invading armies, which allowed those areas of Italy to develop distinctly regional cultures, including local dialects. The Arno, with its origins on Mount Falterona in the Casentino area of the Apennine Mountains, flows westward through Florence into the Tyrrheanian Sea at Pisa, about forty miles from Florence. Although the Arno provided a vital link to the Mediterranean, thus enabling the later Florentines to engage in international shipping and banking enterprises, its propensity for unpredictable flooding proved to be a continuing problem for the developing city-state. Due in part to the geographical features, in spite of the Arno's periodic flooding, the area was ripe for growth and prosperity.

After the collapse of the Roman Empire, the Goths, Lombards, and Byzantines fought for control of the Italian peninsula. The Lombards occupied the Tuscan region in the latter part of the sixth century until Charlemagne invaded in 774 and put Tuscany under Carolingian rule. Beset by attacks on Italy from various invaders, the Carolingian hold faltered in the ninth century. Italian towns, freed from external control, began to consolidate their authority over the surrounding countryside, thus enabling city-states to emerge. In 962 Otto I was crowned emperor, reconsolidating the Holy Roman Empire. Assuming control of the Florentine region, he provided the necessary stability for many fledgling city-states to mature and gain autonomy. One of these new city-states was Florence. Although Florence was not dominant in those early years, by the latter part of the eleventh century, when the city was ruled by the powerful Countess Matilda who controlled present-day Tuscany, Emilia-Romagna, and Lombardy, Florence gained in population and stature and began its quest for hegemony in the area. When Pope Gregory VII struggled with Holy Roman Emperor Henry IV over investiture, Countess Matilda supported the papacy, recognizing the pope as her overlord in opposition to Henry's claims, thereby helping to break Henry's domination of the region. For Florence this alliance with the papacy led to a period of prosperity and growth that began an era of relative independence and increased city development. After defeating its rival Fiesole in 1125, Florence gradually achieved the status that was to last for centuries. Between the twelfth and fourteenth centuries, Florence continued to expand in Tuscany and became a major international economic center.

The question as to why Florence became the nucleus of the glorious era to follow, blessed with its extraordinarily talented and insightful people, may never be fully answered, but some clues can be gleaned from an examination of the Florentine society that materialized in the late medieval period.

1.2 The Emergence of Humanism

In parts of Europe, a determined effort to return to that which made ancient classical Greece and Rome the intellectual and artistic centers of the Western world contributed to the emergence from the Middle Ages. An interest in works of classical antiquity had not been altogether lost during the medieval period, but had been limited. Religious communities continued to study and use Latin, legal scholars examined Roman law and history, and scholars studied Latin literature, albeit from a medieval perspective. The church in Italy, where people took pride in their Roman ancestry, was especially active in classical scholarship.

In the late thirteenth century, a Florentine writer who used the Tuscan vernacular began to be recognized as a major figure in literature. Dante Alighieri (ca. 1265–1321) published his early works in the Italian of Florence: *Vita nuova* (*New Life*), a compilation of poetry and verse, exalted Dante's love, albeit from a distance, of Beatrice Portinari; *De vulgari eloquentia* (*On Eloquence in the Vernacular*) argued for the preference of writing poetry in the vernacular rather than Latin; and *Convivio* (*Banquet*) was a celebration of the long lyrical poem or song (*canzone*). Dante is now best known for *Commedia*, written while in exile and much later referred to as *The Divine Comedy* (*Divina commedia*). In *The Divine Comedy*, Dante begins his journey through the realms of the dead as a tormented soul and travels through classical, Christian, medieval Tuscan, and Florentine history and culture, encountering figures from each period before ending his journey in a state of happiness. Dante's influence on literature was profound as he planted the seed for the coming humanist movement.

Francesco Petrarch (1304–74), born to a Florentine family exiled in Arezzo, was also a catalyst for the humanistic mode of thinking. Embracing the classical writers, he promoted secular interests in his own writings. In addition to his works in Latin, the purview of the intellectual few, Petrarch wrote in the Tuscan vernacular advanced by Dante, and in doing so popularized the language. Petrarch served as a mentor to his close friend Giovanni Boccaccio (1313–75) and encouraged him to pursue a career in literature embracing humanist views. Boccaccio took Petrarch's advice most seriously and wrote numerous works, the most famous being the *Decameron*.

These and other writers were gradually redirecting their focus away from an imposed religious perspective toward the secular interests of man and nature. This movement, leading to the emergence of a better-educated citizenry more capable of engaging in civic life, became known as "humanism," a term coined in the beginning of the nineteenth century and derived from the Italian word *umanista* prevalent in the fifteenth century. Championed in the arenas of literature and thought, this humanistic outlook gained momentum with the arrival in 1375 in Florence of Coluccio Salutati (1331–1406). In a short time he became a respected and influential member of society. As an intellectual interested in philosophical discussion, Salutati was a driving force in a group of similarly minded individuals who met regularly in the convent of Santo Spirito to share ideas and discuss their scholarly writing. Although some of the ideas espoused by Salutati are filled with contradictions, the seeds of humanist thought are evident. Other writers and thinkers of the many who contributed significantly to the growing humanistic movement in Florence include the following.

Leonardo Bruni (1370–1444), historian and statesman who penned twelve Florentine history books, *Histories of the Florentine People* (*Historiarum Florentini populi libri XII*).

Leon Battista Alberti (1404–72), artist, architect, priest, and author of numerous books, including *On Painting* (*Della pittura*) and *On the Art of Building* (*De re aedificatoria*), as well as on topics such as cryptography, education, marriage, and family.

Marsilio Ficino (1433–99), philosopher, teacher, translator, priest, and scholar who served as the head of the Florentine Platonic Academy.

Angelo Ambrogini, known as Poliziano (1454–94), poet, philosopher, educator, and translator of the *Iliad* (from Greek to Latin), as well as others.

Giovanni Pico della Mirandola (1463–94), philosopher, scholar, and writer whose best known work, *Oration on the Dignity of Man* (*De hominis dignitate*), became known as the manifesto of the Renaissance.

Michelangelo Buonarroti (1475–1564), artist, architect, poet, and sculptor acclaimed for the *David*, *Pieta*, and the Sistine Chapel frescoes, to name but a few of his extraordinary contributions.

Francesco Guicciardini (1483–1540), statesman, scholar, and writer of the epic *The History of Italy* (*Storia d'Italia*).

Early humanists like Petrarch and his contemporaries focused their attention on classical literature, but the humanist approach soon expanded to encompass mathematics, architecture, art, music, and the sciences. Classical artistic achievements, some hidden and others largely ignored, were rediscovered and examined, as were classical texts, architectural treatises, and scientific manuscripts. Humanism provided the freedom to think and act in a different manner, critically posing questions about the natural world, and the idea of relating art, architecture, medicine, mathematics, and astronomy to physical phenomena and nature laid a basis for Florentine Renaissance thought.

1.3 Rise of the Florentine City-State

An integral factor in the economic success of Florence was the rise of a third class of citizenry, the wealthy merchants. The merchant class, along with the aristocracy and the clergy, formed a triumvirate of power, albeit an often-factious one, that contributed to the city-state's strength. Many of these merchants searched aggressively for new markets—both inland and through seaports—and their success in this venture was an important component of Florence's early economic achievements.[1] During that same time period, individuals from different economic backgrounds who were involved in trades and crafts began to form guilds (*arti*) that became the basic organs of municipal governing structures, thus allowing the town to evolve into a *comune*, or city-state. This process began in the late twelfth century. The recognized guilds provided manufacturers, artisans, shopkeepers, and merchants with a group identity and with political power that had previously been limited to the elite. The guilds were important in maintaining the economic integrity of their trade, setting standards for quality of workmanship and qualifications for membership, and enforcing rules governing misconduct and fraud in their given areas. In addition, guilds were vital contributors to Florentine society, as they often assumed financial responsibility for the building and decorating of major churches and city hospitals, including commissioning important works of art for both venues. Their patronage, as well as that of the wealthy elite, played a significant part in the flowering of the Renaissance.

By the thirteenth century, the government of Florence formally recognized seven major guilds (the Arti Maggiori) and fourteen lesser guilds (the Arti Minori). The Arti Maggiori included those of lawyers (Arte dei Giudici e Notai); wool merchants (Arte della Lana); silk merchants (Arte di Por Santa Maria); cloth merchants (Arte di Calimala); bankers (Arte del Cambio); doctors, apothecaries, shopkeepers, and some painters (Arte dei Medici, Speziali e Merciai); and dealers in animal fur and skins (Arte dei Vaccai e Pellicciai). The Arti Minori comprised, in general terms, butchers, smiths, armorers, bakers/cooks, stonemasons, innkeepers, vintners, leatherworkers and tanners, timber merchants, cobblers, strap makers, oil merchants, locksmiths, and tailors, and the number of lesser guilds increased to more than seventy over time.[2] In 1293, guild members comprised 28 to 30 percent of the adult male population of Florence.[3] In addition, many other smaller guilds were regularly forming, merging, disbanding, or at times joining the established guilds. Even with the expanding numbers, many segments of the working population operated outside of the guild organization.

The wool and silk industries were major contributors to Florence's early wealth. The business of wool production was a conglomeration of many trades that involved citizens from all levels of society. In the first third of the trecento (1300s), wool manufacturing and its satellite businesses supported approximately 30,000 people of an estimated population of

1. Goldthwaite, *Economy of Renaissance Florence*, 32.

2. Hibbert, *House of Medici*, 25.

3. Najemy, *History of Florence*, 43.

100,000.[4] The entire wool manufacturing process involved importers who brought the wool from other areas of Europe, mainly England and Spain; supervisors who oversaw large numbers of relatively unskilled workers in plants engaged in the preparation of the wool; skilled spinners and weavers working in small shops or homes; dyers, shearers, and washers; and brokers who delivered and collected the wool at the various stages and paid the workers. Subsidiary trades were established to supply products required for the process. By the time of the quattrocento (1400s), the silk industry, formerly based primarily in Lucca, began to flourish in Florence. The silk produced in Florence was of such high quality that it was soon in demand throughout Europe. Silk manufacturing required far fewer workers than the wool industry, which by that time was in serious decline due to the low profit margins, largely attributed to the decimation of the population from the Black Death in the mid-fourteenth century.

The other important facet of Florentine economy was the wealth and power of the bankers whose influence spread across Europe, encouraging trade on a wide scale. First minted in Florence in 1252, the gold florin became the most valued coin in Europe and was accepted as an international currency. Members of the banking community loaned large sums of money both locally and throughout Europe, giving the Florentines far-reaching control and authority. The families of the Pazzi, Pitti, Strozzi, Bardi, Alberti, and of course the Medici, were synonymous with this influential sector. Although usury (charging interest for money loaned) had long been forbidden by the Catholic Church, religious leaders mostly ignored the practice in daily transactions. Those offending bankers could find a penitential avenue in the support of the arts, resulting in the financing of church construction, chapels, and works of art commissioned for the glory of God and, not incidentally, as a show of status and influence.

Other wealthy citizens joined the bankers and became important contributors to the development of artistic expression in Florence. Their patronage, motivated by either selfish or altruistic aims, became vital to Renaissance innovation, elevating the artists from mere craftsmen to respected and prestigious members of society. The influential role of Florentine patrons has become the focus of interesting and in-depth discussions by authors such as Jill Burke in her treatise on the topic.[5] Martin Kemp has extensively explored the motivation behind the creation of many of the Florentine masterpieces that have, through the years, been analyzed and reanalyzed by numerous art historians.[6]

As in all the city-states of Italy, Florence also had a large laboring class who had few rights and were very poor. Living in the least desirable areas of Florence, they were most susceptible to floods, famine, and plagues that often ravished the city. A devastating blow to the Florentines was the bubonic plague (Black Death), resulting from the gram-negative bacterium *Yersinia pestis*, that initially struck in 1348 and returned repeatedly in subsequent years. The population at the time of the first outbreak, some 100,000, was reduced by more than half. Many of the wealthy families were able to escape to the countryside where hygienic

4. Tognetti, "Development of the Florentine Silk Industry," 59.

5. Burke, *Changing Patrons*.

6. Kemp, *Behind the Picture*.

conditions were better, and thus were more likely to avoid exposure to the Black Death. The poor, on the other hand, suffered disproportionately due to their overcrowded living conditions and lack of proper hygiene. Yet even during this challenging period, the Florentines were only slowed, but not halted, in their quest for continued improvement of their city.

Superimposed on the economic and social composition of Florence were the rivalries between the city-states and among the various groups within the city itself. Two thirteenth-century factions in particular were in a fairly constant state of acrimony: the Guelfs, who supported the pope and were primarily from the merchant class, and the Ghibellines, who backed the Holy Roman Emperor and were most often noblemen. Each group formed alliances with like-minded factions or with other city-states as they fought for governmental control. The Guelfs were defeated in 1260, but regained power in 1266. Later they became divided into the Black and White Guelfs with the strife between the two often resulting in the banishment of prominent citizens.

Another segment of society, equally important and at times rather turbulent, was composed of the religious orders and confraternal associations of laypersons. Major religious orders in Florence during this formative period were the Franciscans, Dominicans, Augustinians, Servites, Friars of the Sack, Carmelites, and Umiliati. Pious laypeople could join a confraternal association to engage in prayer and philanthropic projects. These confraternities had their own rules governing eligibility for membership, as well as different levels of observance, discipline, and activity. The Catholic religion permeated all levels of society, and in many cases became an important factor in decisions, both political and social.

1.4 From Medieval *Comune* to Modern Republic

In the 1100s, the noble families of Florence began to consolidate their power into a central government. The first governing body was a committee designated as the Comune that was composed of up to twelve consuls elected annually from the elite class by the nobles. From this early beginning until the Medicis exerted their influence in the early part of the fifteenth century, the governmental structure underwent numerous changes in its organization. For example, the first mention of a chief executive (*podestà*) who governed with the advice of the elected consuls is found in a document dated to 1192/93. As the guilds were formed, recognized, and increased in size, both guild members and citizens from specific neighborhoods began to act in concert to voice issues that affected them. From this humble beginning, the pressure increased for the Comune to become a more inclusive political body. As a result, in 1250 Florence established a new governmental entity, the Popolo (the People), which consisted of a chief office of twelve elders, elected twice a year by leaders of the neighborhood organizations and guilds, whose function was to attend to the daily affairs of government. The Popolo sent proposed legislation for approval to the Capitano del Popolo and his council, which had twenty-four regular members and one hundred residents from those same neighborhood organizations and guilds from which the elders were elected. Although power did shift somewhat from the nobility, the twelve elders were typically wealthy, influential men, usually members of the major guilds, and included prominent bankers and merchants, often from the noble class.[7] By the mid-thirteenth century, the Popolo was sufficiently established that Florence constructed the Palazzo del Podestà (now the site of the Bargello Museum), with its distinctive tower and façade, as the seat of government.

Over time, the *comune* system led by the Popolo was unable to stop a small number of noble families from gaining and maintaining power, often resorting to force. In addition, it became increasingly difficult to maintain order and suppress violence between the warring Guelfs and Ghibellines as they struggled for political dominance. The development of governing bodies and changes in the makeup and systems of selecting members for those bodies continually evolved in reaction to public sentiment. Despite efforts to create a governmental structure that represented a cross-section of the population, some influential families came to dominate political offices.

In 1293 the Ordinances of Justice were enacted by the Popolo, disenfranchising those powerful families that had become violent, authoritarian, and lawless. Although the provisions remained in effect over the ensuing years, they were modified as attitudes and opinions changed. A new governing body, the Signoria, was formed, drastically altering the previous Comune. It was composed of nine members who were elected for two-month terms. They in turn were advised by two groups, the twelve *buonuomini* elected for three-month terms and the sixteen *gonfalonieri* elected for terms lasting four months. Civil servants were appointed in non-elected positions to ensure continuity and proper functioning of the governmental process. Two reformed legislative assemblies, the Council of the Popolo and the Council of the Comune, consisted of a total of five hundred members who served two-month terms. The Signoria submitted legislative proposals to those two bodies, and the decisions on whether to approve the provisions were in their hands. The legislators, known as *priori*, were selected by drawing from leather bags (*borse)* names of eligible guild members aged thirty and older.

The city began construction of a new structure to house the government in 1299, and the resulting Palazzo della Signoria (later also known as the Palazzo Vecchio) became one of the largest such civic buildings in the region. Those selected to serve moved to the Palazzo della Signoria and lived there for their two-month terms. Because names were screened before being placed in the leather bags and only members of recognized guilds were eligible to hold office, the system could not be classified as truly democratic. It was, however, more inclusive than those of most other city-states, a fact of which Florentines were extremely proud.

With minor changes, this basic political format remained in effect until 1382 when the conservative patricians in control established a new governmental system. Theirs was clearly a regime dedicated to preserving the security and status of the wealthy, and the poor were discouraged from active dissent by a government that was able to gain wide support by maintaining order and stability during extremely turbulent times in the city-states. Although the guilds had only token representation in the new system, members who served in the legislative bodies were, for the most part, content with the arrangement. In the main, Florentines accepted their responsibility to pay taxes and to contribute as much as their

7. Najemy, *History of Florence*, 66–67.

abilities allowed. That aspect of the Florentine societal belief system helped the elite sustain a balance between the needs of the citizenry and their own interests.

After Cosimo de' Medici (1389–1464), known as Cosimo the Elder, returned from a year in exile imposed by rival factions, he assumed political control in 1434. Under Cosimo, power was concentrated in the hands of a few influential people, of whom he was the dominant figure. Guilds continued to send elected representatives to serve in the legislature, but many were preselected to ensure loyalty. Cosimo's vast wealth, inherited from his father, Giovanni di Bicci, and augmented by his own interests and financial success in banking, commerce, and industry, provided the resources to buy support both locally and abroad. To consolidate his base, Cosimo filled important elected offices with supporters; those who openly opposed him were often sent into exile. In order to appeal to the populace, Cosimo financed elaborate festivals and tournaments. Central to his success and political clout was his astute handling of dissident factions in a manner that was neither obvious nor offensive, while resisting the temptation to flaunt his wealth or sense of entitlement. In addition, Cosimo had a talent for mastering details and political nuances, was focused on foreign policy, and appreciated intellectual qualities in others.[8]

Political acumen and the resulting stability were not Cosimo's only legacy. An earnest encouragement and substantive support of the arts and literature were landmarks of his time in power, and his circle included Donato di Niccolò di Betto Bardi (known as Donatello, ca. 1386–1466), Filippo Brunelleschi (1377–1446), Luca della Robbia (ca. 1399–1482), Paolo Uccello (1397–1475), Guido di Pietro (known as Fra Angelico, ca. 1400–1455), Filippo Lippi (ca. 1406–69), and others. By establishing the Platonic Academy, founding the still-famous Laurentian Library, and commissioning many high-profile construction projects throughout the city, he managed to leave the citizenry with the impression that without the power of the Medici, Florence would fail to maintain the stature it enjoyed. Under Cosimo the Elder, the line of Medici in the political forefront was firmly established.

The son of Cosimo the Elder, Piero the Gouty (1416–69), was very different. Sickly and not blessed with either political savvy or a good head for business, he died five years after his father and added little to the wealth or aura of the Medici dynasty. In contrast, Piero's son Lorenzo (1449–92), known as Lorenzo the Magnificent (Il Magnifico), was highly intelligent and adept at diplomacy. His grandfather Cosimo chose Lorenzo and trained him to assume the role of the leader of Florence, although serving without portfolio, as Cosimo had managed for years. In contrast to the more circumspect Cosimo, Lorenzo was a leader who did not shy from exploiting his personal power. He was far more interested in affairs of state and his own personal passions than in the family enterprises, and the Medici banking business suffered under his years of control. Importantly, Lorenzo was also a lover of the arts and literature. He wrote poetry and sonnets, and encouraged and supported writers, humanists, painters, and sculptors, including the young Michelangelo, who was invited to live in Lorenzo's *palazzo* where the maturing sculptor and painter refined his artistic skills. The period of Lorenzo's political control was fraught with intrigues led by those wishing

8. Brucker, *Renaissance Florence*, 159–60.

to usurp power. Most notable was the conspiracy mounted by the Pazzi family who, in an effort to rid the city of Medici domination, attempted in 1478 to assassinate Lorenzo and his brother Giuliano. The twenty-five-year-old Giuliano was killed during a service at the Florence Cathedral, but Lorenzo and several friends were able to escape into the New Sacristy. His reprisals were swift and brutal.

At the time of Lorenzo's death in 1492, foreign interests were threatening, the government was less stable, and the Dominican prior of San Marco, Girolamo Savonarola (1452–98), was preaching against the Medici rule, the decadence of society, and the corruption and excesses of the papacy in Rome. Lorenzo's son Piero II (1472–1503) assumed the position of power vacated by his father. A weak and disinterested ruler, Piero quickly faced a military crisis when Charles VIII invaded the Italian peninsula in 1494 and threatened Florence's independent existence. To prevent military occupation by the French forces, Piero, who originally opposed Charles VIII, capitulated and surrendered Pisa and other strategic areas under Florentine dominion. This was a crushing blow to the Florentines, who responded by forcing Piero and his family into exile from Florence. Taking advantage of the power vacuum, Savonarola assumed control of the government and reestablished a republic.

In 1497, under Savonarola's direction, youths went from house to house in Florence demanding that the wealthy rid themselves of their frivolous art, jewelry, books, and finery. These items were taken to the Piazza della Signoria and burned in what became known as the Bonfire of the Vanities. Savonarola's message was that of a moral crusader urging a return to a simple life dictated by Christian ideals. After he denounced the egregious behavior of the papacy, particularly in the personage of Pope Alexander VI, Savonarola was excommunicated, and in 1498 the church, in collaboration with the government of Florence, had him arrested and charged with heresy and tyranny. Savonarola and two followers were tortured and then hanged and burned in the Piazza della Signoria on the site of the Bonfire of the Vanities. A plaque in the pavement marks the location of this public execution.

The Florentines called upon Piero di Tommaso Soderini (1450–1522) to lead the city as it attempted to regain stability after this tumultuous period. Soderini had been appointed *prior* of Florence in 1481 and served as ambassador to France; he was elected *gonfaloniere* for life in 1502. When the Medici returned to Florence in 1512, however, Soderini was sent into exile. Soderini was later called to Rome by Pope Leo X, and there he continued to work on Florentine affairs, but he never returned to his native city.

Under Soderini, Niccolò Machiavelli (1469–1527) held the position of second chancellor and secretary to the Ten of War, the council given responsibility for warfare and diplomacy, as well as ambassador to several courts. Machiavelli too, upon the return of the Medicis to power, was expelled from government in 1512, and afterwards was allowed to hold only a few temporary appointments. Machiavelli maintained his innocence throughout, but was nonetheless implicated in a plot to overthrow the Medicis in 1513. For this he was sentenced to life in prison, where he underwent torture for his purported actions. After Giovanni de' Medici was elected pope as Leo X in 1513, he issued a general pardon and Machiavelli retired to his farm near Florence. From there, during this period of exile, he wrote the works for which he is best known: *The Prince*, the *History of Florence*, and others.

After Soderini was exiled, the republic, greatly weakened, continued until 1537 under several Medicis, including Lorenzo's second son, Giovanni (Pope Leo X), who governed Florence via various relatives until his death in 1521. After uprisings and some futile attempts to rule, the Medici returned to full power in 1530 when Alessandro (1510–37), a descendant of Lorenzo the Magnificent, took the title of Duke of Florence. Following Alessandro's death at the hands of his cousin, a member of the cadet branch of the Medici family, Cosimo I (1519–74), also of the cadet branch, finally gained control of a unified Tuscany and proclaimed himself grand duke in 1569. He was an authoritarian ruler with the ability to govern with competence, intelligence, and a strong will. Although the government could no longer be viewed as democratic, with the power of the Signoria and the political structure appropriated by the grand dukes, Florence did regain its prosperity and experienced a revival in artistic and scientific achievement. As part of Cosimo's attempt to stimulate economic growth, he encouraged the introduction of new industries in Florence, including tapestry production (see chapter 2), and the expansion of existing industries such as printing and engraving. Printing had been introduced in Florence in 1471,[9] but was never able to achieve the level of output experienced in Venice.

A total of seven Medicis from the cadet branch ruled as grand dukes, the last being Gian Gastone (1671–1737).[10] Upon her death in 1743, Gastone's sister, Anna Maria Ludovica, bequeathed to Florence the priceless art and treasures that had been in Medici possession for centuries. Anna Maria's gift ensured that Florence has continued as a living museum, reflecting the breadth and depth of the stellar accomplishments during that remarkable period from the thirteenth to seventeenth centuries.

1.5 Why Florence and Why Then?

In a relatively brief period of time, this region of Tuscany experienced major changes in social and economic development, the rise of political institutions with a form of government that attempted to be more inclusive, and the amassing of wealth and power by various competing families and individuals. Strong and diverse guilds, a philosophical and emotional connection with the Roman past sparking a return to study of the classics and the development of humanism, the Florentine ethos of a responsibility to support public institutions and the arts, and the personal interest in promoting patronage of the arts converged to create an atmosphere in which the historical period we call the Florentine Renaissance could thrive. While major civilizations around the world—like those in China, India, Egypt, Greece, and the Middle East—have experienced periods of renaissance, it is the Florentine Renaissance that seems to have retained a special place in the minds of Western culture. That Renaissance shaped much of how European civilization would progress, and scholars and laypeople alike continue to be fascinated by the innovations and innovators who came out of that Renaissance.

9. Richardson, *Printing, Writers, and Readers*, 4.

10. Medici rulers of the cadet branch who served as grand dukes were Cosimo I (1519–74), Francesco I (1541–87), Ferdinando I (1549–1609), Cosimo II (1590–1621), Ferdinando II (1610–70), Cosimo III (1642–1723), and Gian Gastone (1671–1737).

We may never be able to fully answer the question, why Florence and why then?, but one thing is clear. The circumstances and forces that allowed people to explore the arts, sciences, and technology in new ways coalesced in Florence at a time when resources and incentives were optimal to allow such creativity to flourish. The following chapters attest to an inventiveness that was palpable. In portraying Florence as unique, one can only describe the people who graced this book as extraordinary individuals nurturing and defining an extraordinary period.

Chapter 2

The Science of Art

> "It is the impulse of a noble mind which moves some towards this art, pleasing to them through their natural love. The intellect delights in invention; and nature alone draws them."
>
> Cennino Cennini, *Craftsman's Handbook*

The glory that is Florence rests with the passionately visual culture that the Renaissance so uniquely generated. Based on careful observation and a keen knowledge of the properties of materials and techniques utilized, Florentine artists undertook innovative design and manipulations using media ranging from fresco to tempera to oils, from stone and wood sculpting to terracotta, from bronze casting to glass, and others. As artists and scholars studied the science behind art and used their knowledge to pioneer new techniques, the former craftsman's workplace evolved into an artist's studio that was a center of experimentation and learning.[1] The story of that evolution begins with the science of color and the technology behind the art forms.

2.1 Color Production and Visualization

2.1.1 Pigments and Their Compositions

Renaissance artists were skilled at mixing pigments to obtain the myriad colors and shades they desired.[2] Indeed, they greatly advanced existing techniques and experimented with new formulations and mixtures of various pigments, some of which had been in use for thousands of years.[3] Three of the prominent masters who wrote about the practice of art were

1. Cole and Pardo, "Origins of the Studio," 4.
2. Cennini, *Book of Art*, 31–50.
3. Bomford et al., *Art in the Making*, 31–43.

Cennino d'Andrea Cennini (ca. 1370–ca. 1440), author of *Craftsman's Handbook* (*Il Libro dell'arte*), which was probably penned sometime during the 1390s; Leon Battista Alberti (1404–72), who wrote *On Painting* (*Della pittura*), first in Latin and then translated into Italian in 1436; and Leonardo da Vinci (1452–1519), who filled numerous notebooks with text and illustrations concerning the discipline of art. We are indebted to them and others for recording the materials and methods involved in color production, utilizing primarily pigments obtained from plant, mineral, and synthetic sources. Those naturally occurring are generally of two major types: lake pigments and earth pigments. The former, referring to organic structures (with no relationship to a body of water), denotes pigments derived from plants; the latter denotes pigments derived from inorganic or mineral sources that contain one or more metals. In chapter 4 the role of apothecaries and their relationship to art are discussed, notably their services as purveyors of numerous pigments available to artists. By and large, the Renaissance artists chose pigments that had previously been used by the Romans, Greeks, and Egyptians, although several new and expensive ones were introduced in Florence. The sources of some of the pigments are delineated below, and their use during the Renaissance has been documented by recent scientific studies in which the compositions of paintings have been analyzed.

2.1.1.1 Blue

Lapis lazuli, a rock containing the intense blue mineral component composed of 25 to 40 percent lazurite ($[Na,Ca]_8[Al,Si,O_4]_6[S,SO_4,Cl]_{1-2}$), as well as several other compounds, including white calcite, blue sodalite, yellow pyrite, and other minor constituents, entered Florence from northern Afghanistan via Venice. Containing sodium, aluminum, silicon, sulfur, and oxygen, lapis lazuli was ground into a powder to obtain the color of the pigment ultramarine. Extracting ultramarine was an intricate, time-consuming procedure, and a much improved method developed by Cennini became widely used. The rock was ground into a powder and then mixed with resins, wax, and oil. The mixture that resulted was then wrapped in cloth and kneaded in a lye solution, causing the particles of blue to amass at the bottom of the cloth. Additional extractions result in lower-quality ultramarine. Those lesser-quality extractions from lapis lazuli yielded a gray-blue fraction called ultramarine ash, which was also utilized by the artists. Due to the difficulties associated with importing lapis lazuli from the Middle East and the rather detailed extraction procedure required to obtain ultramarine, the color was expensive and used only in works commissioned by wealthy patrons.

A deep blue smalt, made from silica (silicon dioxide, SiO_2, obtained from quartz or sand), potash (K_2CO_3, prepared from leaching wood ashes), and cobalt oxide (CoO, one source being smalite), was a lower-priced option. A cobalt ore such as smalite was roasted and the resulting cobalt oxide was then added to molten glass (silica and potash). When the mixture was poured into water, it formed particles that could be collected and ground to the desired size. The resulting smalt was generally about 65 to 70 percent silica, 2 to 20 percent cobalt oxide, and 10 to 20 percent potassium, and contained a number of impurities, generally oxides of magnesium, sodium, copper, nickel, manganese, and others.

Another less costly source of a blue pigment was azurite, a copper-containing ore ($2CuCO_3 \cdot Cu[OH]_2$) mined in Germany, Spain, and France, that was known to the early Greek and Roman artists. The pigment was obtained by crushing and grinding the naturally blue azurite.

Cennini also discussed the preparation of a blue pigment obtained from lazulite (*azzuro della magna*, $[Mg,Fe^{2+}]Al_2[PO_4]_2[OH]_2$). Lazulite is found adjacent to veins of silver near Siena as well as in Germany and other locales. The lazulite is ground in water to form a powder and was recommended for use on walls and wood panels.

2.1.1.2 Red

Cinnabar (mercuric sulfide, HgS) is a common ore of mercury found in volcanic eruptions and mined in Greece and in Italy since Roman times. The color vermilion was generally obtained from cinnabar using a technique developed by alchemists in the eighth century. The procedure involved heating metallic mercury and sulfur and collecting the resulting vapor on a colder surface, yielding the solid deposits. This process of sublimation (formation of solids from the vapor), subsequent to the initial chemical reaction catalyzed by heating, speaks well for advances made by the early alchemists in developing a less-than-obvious procedure. Because of the difficulty of the process, Cennini advised artists to purchase the vermilion from an apothecary rather than attempt its preparation in their workshops.

Lead oxide (red lead, Pb_3O_4, also $2PbO \cdot PbO_2$) could be used to obtain the color red. Mined in northwest Spain as an ore, it was heated in air at a high temperature to produce the pigment that was used both in painting and in the illustration of manuscripts.

The madder plant (*Rubia tinctorum*), in the genus *Rubia*, grows wild in Italy and contains several polyphenolic compounds, one of which is alizarin ($C_{14}H_8O_4$), a source of a red lake pigment. Madder had been widely cultivated since antiquity, and the red pigment was obtained by drying, crushing, and hulling the roots, then boiling them in acid. Potash was added to the extracted dye to form the pigment. According to Cennini, another method involved grinding the hulled root in urine, although he considered it unpleasant.

Other possible sources for red colors were naturally occurring red earth pigments such as hematite (ferric oxide, Fe_2O_3), from which the pigment *sinopia* was derived. During the Renaissance, *sinopia* was mined in Cappadocia and shipped to Italy from the Greek colony Sinop on the Black Sea. Prepared by grinding to the finest consistency possible, it was used for fresco or panel painting and was especially good for creating preliminary drawings. Interestingly, a pale red pigment, *cinabrese*, derived from a light-colored *sinopia* mixed with a lime white pigment, was utilized mainly in Florence for enhancing the flesh color of skin tones.

Lastly, a bright red resin known as dragon's blood was generally obtained from trees of the genera *Croton* and *Draceana*, which were imported from Asia. The resinous compound extracted from the trunks and branches of these trees consists of 50 to 70 percent benzoic acid and benzoyl-acetic acid, as well as numerous other organic and inorganic constituents.

2.1.1.3 Yellow

Artists achieved various shades of yellow using both plant and mineral sources. The yellow pigment gamboge was extracted from the evergreen trees of the genus *Garcinia*, found in Asia, Africa, and other locales. The gum-resin removed from the tree by tapping was about 72 to 74 percent resin and 22 to 24 percent gum, and contained moisture and traces of starch and woody fiber (lignin). After the gum was extracted, the resulting yellow resin, which was chemically rather complicated, contained a high concentration of gambogic acid. It was allowed to congeal until it became hard and brittle, and was then ground to a powder.

Several shades of yellow pigment came from mineral sources. Lead-tin yellow type I, lead stannate (Pb_2SnO_4), and type II ($Pb[Sn,Si]O_3$ or $PbSnO_3$) were produced for use in the glass-making industries, notably in southern Italy and northern Europe. The color was obtained by fusing tin, lead, and quartz at a high temperature, which resulted in a lead-glass pigment that was then finely ground. The last step was to screen the ground pigment through a mesh.

Naples yellow, a lead (II) antimonate ($Pb[SbO_3]_2/Pb_3[Sb_3O_4]_2$), was widely used in frescoes and on panels painted with tempera. Found in mountainous volcanic areas, lead antimonate was ground to a powder in clear water.

A yellow to reddish-yellow hue could be achieved using a lead monoxide (PbO) called massicot. The technique for extracting pigment from massicot was known from antiquity and practiced again from the 1300s in Europe. Massicot was mined from many areas but found in only small quantities, and it required a long period of heating.

Arsenic sulfide (As_2S_3), a poisonous compound, was occasionally used to create the yellow pigment orpiment. Arsenic sulfide is found in areas of volcanic activity and hot springs, including parts of the Italian peninsula, and the pigment orpiment has been available to artists since Roman times. Extremely hard and difficult to grind, arsenic sulfide was pulverized to a fine powder to create a yellow pigment. The pigment's toxicity and relatively poor stability in light prevented it from gaining widespread acceptance. In spite of its disadvantages, artists utilized its bright yellow color in painting.

Italian buckthorn berries (*Rhamnus saxatalis*) and weld (*Reseda luteola* L.) were local sources for yellow lake pigments. To obtain the pigments, the berries were crushed and boiled in water, and the extract was mixed with alum, a sulfate salt known to the Romans. These yellow pigments were either applied alone or combined with other pigments to produce a different shade of another color.

Hydrated iron oxide ($Fe_2O_3 \cdot H_2O$, with several impurities) was the source of yellow ochre. Mined near Florence, it is a natural pigment produced by grinding in clean water. Another yellow ochre, which came from goethite (FeOOH), was also readily accessible. Cennini recommended yellow ochre for its versatility, especially in fresco painting.

The saffron (*zafferano*) plant (*Crocus sativus*) was another source for a yellow pigment. Found in Asia, the Middle East, and the Mediterranean, the saffron plant was readily accessible to the Florentines. Extracting the pigment from the stigmas of the saffron flowers involved soaking the stigmas in glaire (egg whites) overnight. Artists applied the saffron pigment for gold leafing and on parchment, but in painting it was usually mixed with blues to create a green shade, as the saffron pigment alone tended to lose its color when exposed to air.

2.1.1.4 Green

Most Renaissance artists derived their various shades of green by mixing yellows and blues, and sometimes adding black and white. Green pigments, however, were also obtained from mineral sources.

The hydrated iron potassium silicate glauconite ($[K,Na][Fe^{3+},Al,Mg]_2[Si,Al]_4O_{10}[OH]_2$) was used to produce the green earth pigment (*terra verde*). Glauconite was mined in Italy and the pigment was prepared by grinding and sifting. This pigment was favored by artists as a foundation for gold leaf and for flesh tones in frescoes and panels. Additionally, the green earth pigments were frequently mixed with white silicates or with colored components to produce a variety of green tones. Other green earths included a complex group of iron silicates and aluminosilicate minerals containing magnesium, potassium, oxygen, and other elements.

A copper-containing malachite ($CuCO_3 \cdot Cu[OH]_2$) was used to produce the green pigment *verde azzurro*. It was readily available to Florentine artists as it was associated with secondary copper ore deposits found in Italy. Prepared by grinding and sifting, this pigment was unaffected by light and better used in tempera than in oil because it is coarse and mixes well with egg yolk. However, its pale green tone made it a less popular choice than some other options.

A green pigment called Spanish or Greek green (*verdigris*) was obtained by applying hot vinegar to copper strips, then scraping off the green crust that formed on the copper strips due to an oxidation reaction and collecting the scraped powder. This scraping was generally soaked in vinegar or wine, heated, and then mixed with a combination of egg yolk, vinegar, and gum water that served as a medium. This *verdegris* pigment was, as noted by Cennini, an ideal shade for depicting grass.

2.1.1.5 Other Colors

Black pigments came from a number of sources, including lamp black (soot from oil-burning lamps), charcoal from incomplete burning of various woods and nut shells, and black earth (iron oxides). Another plant source for black pigments was young shoots of vines that were burned, then immersed in water to quench the fire, and finally ground. This process resulted in what Cennini referred to as one of the most usable black pigments.

Lead white ($2PbCO_3 \cdot Pb[OH]_2$) has been accessible to artists throughout history with documented use as early as 400 BCE. This white pigment was readily made by allowing acetic acid vapors to contact metallic lead in the presence of carbon dioxide, and collecting the white crust that formed on the lead. This crust could easily be scraped off to obtain a lead white of excellent consistency. Artists applied lead white to lighten colors when painting tempera on panels.

Artists used a natural white pigment called lime-white (*bianco sangiovanni*), which was prepared by grinding the lime to a powder that was then placed in water for eight days before forming it into small cakes that were dried in the sun. Despite the steps required for preparation, lime-white was considered important for fresco because it was inexpensive, easily obtained, and durable. Chalk and gypsum were also employed to obtain white pigments.

Brown pigments of different hues were obtained from various types of clay. Another source of brown was a burned resinous wood that was boiled in lye to produce a brown pigment (*bistre*). Artists either mixed *bistre* with other pigments or used it separately.

Gold could be hammered into thin sheets called gold leaf. If the patron provided funds to cover the added expense, an artist could apply gold leaf to paintings for gilding.

2.1.2 Visualization of Colors

Florentine artists excelled in their choice and use of various colors and had a keen appreciation of the eye's ability to discern a wide variety of colors, shadings, and intensities, as evidenced by the magnificent paintings of this period. They, of course, did not have an understanding of the biochemical and physiological processes that enabled viewers to see and assimilate innumerable colors and compositions.

The human visual system is sensitive to wavelengths in a range of approximately 400 to 780 nanometers (nm, one billionth of a meter), although some individuals are unable to detect light at the extreme ends of the visible spectrum. While the breadth of human vision is rather narrow compared to the full range of frequencies or wavelengths in the electromagnetic spectrum, it is nonetheless exquisitely sensitive enough to distinguish many subtle shades between violet (lower wavelengths) and red (higher wavelengths). The human eye also has a remarkable ability to detect intensities ranging from a single photon of light to some 100 million photons. The molecular nature of the photoreceptors in our retina is that of a protein (rhodopsin) with a covalently attached vitamin A derivative, 11-*cis*-retinal. When incident photons are absorbed, 11-*cis*-retinal undergoes a conformational change to the *trans* form; this alteration, in turn, leads to a structural change in the protein, beginning a cascade of events resulting in vision.

Our light sensors are of two types, rods and cones, that are located in the retina to capture photons of light of different frequencies (or wavelengths). Rods are capable of detecting a single photon of light; however, they become saturated in moderate light and thus are not used extensively during daylight hours. Cones, on the other hand, are far less sensitive to light intensity and are primarily used in moderate and bright light. Present in three types of cells, cones are sensitive to short, medium, and long wavelength light, permitting us to discriminate a multitude of colors. In addition to rods and cones, the human eye has another type of photoreceptor that uses melanopsin as the photopigment. These photoreceptors, unlike those associated with rods and cones, are photosensitive retinal ganglion neurons that project to non-visual regions of the brain.

When light impinges on an object, that object either absorbs, transmits, or reflects the incident light. The process of absorption is attributed to chromophores, entities within molecules that undergo energetic shifts in their electrons due to electronic energies and the frequency (or wavelength) of the light. The wavelength that is not absorbed is reflected to the eye, and that wavelength (or wavelength range) corresponds to the color observed. The wavelength reflected from an opaque object is detected by the eye as a specific color. Translucent and transparent objects, in contrast to opaque or partially opaque objects, will either transmit the light or reflect some of it, perhaps giving the object a tinted appearance.

While Aristotle and other early scientists and philosophers studied vision and colors, it was not until the seventeenth century that Isaac Newton (1642–1726/27) provided a physical interpretation of white light as a mixture of all colors of the spectrum. He theorized that the retina contained receptors sensitive to red, green, and violet light and that pure spectral colors can be listed as red, orange, yellow, green, blue, and violet, each corresponding to a wavelength range. Red, for example, is typically in the range of about 635 to 700 nm. In his innovative studies, Newton included a seventh color—indigo—located between violet and blue. Several decades after Newton's pioneering investigations on color, Thomas Young (1773–1829) proposed what is now referred to as the trichromatic theory, namely, that the primary colors associated with human vision are red, green, and blue. Like modern artists, those in the Renaissance used red, yellow, and blue as their three basic colors, achieving enormous variations in the numbers of hues they were able to produce.

2.2 Painting

2.2.1 Fresco

In the Western world, artists have used a technique for painting murals known as fresco (*affresco*) for several millennia, with examples being found in many areas of the ancient world. In Crete frescoes have been dated to about 1500 BCE, and the early Egyptians also used a form of fresco to decorate their tombs. In Europe and the Mediterranean basin, Greek and Roman artists were creating frescoes by applying pigments to freshly laid or wet lime plaster to create wall paintings of noteworthy durability. However, it was in Florence in the fourteenth and fifteenth centuries that the art of fresco reached its zenith.

2.2.1.1 Chemistry and Methods

The process of fresco began with preparing calcium oxide (CaO), or porous lime, by heating limestone (calcium carbonate, $CaCO_3$) at high temperatures (800°C to 1,000°C), which gave the following chemical reaction, producing carbon dioxide (CO_2) as a by-product.

$$CaCO_3 \rightarrow CaO + CO_2$$

This procedure, referred to as "calcination," was followed by another process, termed "slaking." In this step, the plaster was prepared to receive pigment by adding additional water to the lime, which yielded calcium hydroxide ($Ca[OH]_2$) crystals in the form of a paste or putty.

$$CaO + H_2O \rightarrow Ca(OH)_2$$

It could take months or even years to get the putty to the consistency that artists preferred in terms of water content and particle size. Once the desired consistency was obtained, the final step involved mixing the lime putty with sand (or occasionally marble dust). The assorted textures favored by the various artists were achieved by using different proportions of sand to putty and by the choice of particle size of the sand. Usually frescoes were created by applying several layers of fresco plaster to the wall. The first layer contained larger sand particles and was mixed in a ratio of two to one with the slaked lime; for subsequent layers, finer grain sand was used and the ratio of sand to putty was decreased to perhaps one to one. As this mixture dried on the wall, water was lost by evaporation and absorption into the underlying structure, and the calcium hydroxide reacted with carbon dioxide to yield the starting chemical, calcium carbonate, but now of uniform consistency.

$$Ca(OH)_2 + CO_2 \rightarrow CaCO_3 + H_20$$

When drying was complete, the base layer (*arriccio*) was ready. After determining the amount of painting one expected to complete in a single session (*giornata*), the artist would mix fine particle sand with slaked lime to prepare what is known as the *intonaco*, which was then applied to the *arriccio*, spreading it evenly. Because the *intonaco* dried rapidly, it was important to develop the design and prepare the needed paint colors in advance. Speed in application was a necessary component of fresco. With the preliminary drawing complete, pigments were mixed with water to form a dispersion that was quickly applied to the *intonaco*. As the *intonaco* and paint dried, another chemical reaction occurred; this one permanently bonded the paint and the plaster. The fresco could, nevertheless, deteriorate due to humidity, salts (from the mortar or via pollution associated with ambient moisture), microbial/fungal action, excessive heat, poor plaster, and of course inappropriate restorations.

There were three techniques painters could employ to diagram the intended work before executing the fresco. In some cases, the artist drew a preliminary sketch (*sinopia*) of the final composition on the *arriccio*, or underlayer of plaster, on the wall before applying the final layers with pigments over the sketch. Another technique was to draw a smaller sketch onto a grid and transfer the sketch to a mathematically proportional grid on the wall to be frescoed. A third method was to transcribe the sketch to the wall area through the use of cartoons (from the Italian word *cartone*). In this process, the preliminary drawing was sketched on heavy paper, then one of two procedures was used to transfer the outline of the cartoon to the wall. In the first method, the paper was placed over the plaster, and an imprint of the outline was lightly pressed onto the plaster with a pointed instrument. As the artist applied the pigment the indentations were filled with paint. In the second method, holes were punctured along the outline of the drawing. After punching holes in the cartoon, the artist would rub charcoal through the holes to mark the outline on the plaster; the resulting charcoal dots were then covered during the painting process.

Overall, the finished fresco was very durable and the colors were especially intense. However, since the rapid drying process allowed the artist to complete only small sections of a wall in each session, small cracks could develop over time between areas painted on successive days. On occasion, the artist would add paint after the plaster had dried in order to enhance a small detail, make a correction, or add a color that had the potential of changing shades when applied to the wet *intonaco* because of a chemical reaction with the lime. Alterations introduced as a corrective step are referred to as *pentimento*, although the term *fresco secco* is also used. Another technique was utilized as a means of fresco and simply referred to the application of the pigments to dry plaster (*a secco*). Although not widely employed, a third method was to apply pigments to *intonaco* that was nearly dry (*mezzo-fresco*). In this method the pigments penetrated only a relatively short distance into the *intonaco*.

An artist had to possess considerable technical expertise to produce successful frescoes. Achieving the correct consistency of the color pigments and the plaster were of prime importance to ensure proper bonding. Also, the wetness of the plaster had to be equal in each section in order for the colors to dry in a manner that resulted in uniform tones throughout the fresco. Because colors could change during the drying process, artists had to understand how the original colors might change as the result of drying and make necessary adjustments in the pigments to accommodate those alterations. Each *giornata*, or area of wet plaster, represented a technological as well as an artistic challenge for the painter and associates.

2.2.1.2 Examples of Florentine Frescoes

Florence is blessed with a large number of extraordinary frescoes and a selection of those works is presented below.

a. Santa Maria Novella

Masaccio (1401–ca. 1428; see 3.1.2.3a for biographical details) was the earliest Florentine artist to use perspective. The *Trinity* in the church of Santa Maria Novella (elaborated on in detail in 3.1.2.3a) is creative in its use of perspective and innovative in its technology. The *Trinity* is significant from the standpoint of artistic technology because it contains markings, visible using certain lighting effects, made by a stylus on the wet plaster from a cartoon, one of the preparatory procedures. The stylus markings outlining the head of God and also indicating the placement of the vaulted ceiling ribs are probably discernable because of paint deterioration, but they document a fresco using a cartoon preliminary sketch.

Santa Maria Novella has other magnificent frescoes. A noteworthy cycle adorning the walls of the chancel was frescoed between 1485 and 1490 by Domenico Ghirlandaio (1449–94), an extremely gifted Florentine painter. Ghirlandaio, born Domenico di Tommasso di Currado di Doffo Bigordi, was nicknamed by his father, a goldsmith. After apprenticing in his father's workshop, Ghirlandaio worked with Alesso Baldovinetti studying painting and mosaics, finally apprenticing in the Florentine workshop of Andrea del Verrocchio. Ghirlandaio and his brothers eventually opened their own workshop, in which the young Michelangelo apprenticed. Commissioned by Giovanni Tornabuoni, their monumental Santa Maria Novella fresco cycle depicts the lives of the Virgin Mary and Saint John the Baptist. Each of the scenes contains a religious setting and also includes numerous contemporary figures shown as observers, many modeled after Tornabuoni family members and friends. Interestingly, the first fresco on the left depicts what many consider to be a self-portrait of the artist, standing with a hand on his hip. Extraordinary in its detail, this cycle may be Ghirlandaio's finest. It is unquestionably an excellent example of the amount of expertise needed to fresco a large space with such finely illustrated elements, recalling that frescoes require painting in small sections each day.

b. Florence Cathedral (Duomo), Santa Maria del Fiore

Paolo Uccello (1397–1475), born to an aristocratic Florentine mother and a father who was a surgeon and a barber, spent his apprentice years in the workshop of the illustrious Lorenzo Ghiberti. In 1415 he was admitted to the Arte dei Medici e degli Speziali, the Florentine painters guild. He was a master of the grid technique for transferring preliminary sketches to the wall area to be frescoed. In 1433, the Opera del Duomo asked painters to submit designs for a fresco honoring Sir John Hawkwood (ca. 1320–94) for a section of wall in the Duomo. Hawkwood, an English mercenary (*condottieri*), was to be so recognized for his contribution in defending Florence against Milan. The commission was awarded to Uccello, and the resulting *Sir John Hawkwood* fresco remains today in the Duomo, although it has since been framed and moved to canvas. Uccello's preparatory grid, a smaller version of the intended fresco, used mathematical precision to execute the work in a larger format. The grid has been preserved and can be seen in the Uffizi Gallery.

c. Ognissanti

One can view early fresco compositions by Ghirlandaio in the church of Ognissanti. There, in about 1472, Ghirlandaio painted *Descent from the Cross* and *Virgin Mary of Mercy* (created for the family of the Florentine navigator Amerigo Vespucci, who is interred at the foot of the altar). The church also houses a pair of complementary portraits by Sandro Botticelli (1445–1510) and Ghirlandaio. One of the geniuses in Renaissance art, Botticelli was a master colorist, outstanding in his understanding of perspective and anatomy, and innovative in his artistic manipulations and presentations. Born in Florence, Botticelli is thought to have been trained as a goldsmith by his brother and is believed to have apprenticed under Fra Filippo Lippi, but little is known of his early life. Botticelli's *Saint Augustine* and Ghirlandaio's *Saint Jerome*, both executed in about 1480, are now located opposite one another in the nave. The *Saint Jerome* fresco is an excellent example of the way an artist's understanding of color and composition in relation to the fresco process can produce a desired effect; in this case, Ghirlandaio achieved a warmth usually found only in oil painting. The *Saint Augustine* fresco (see 5.6.4 and fig. 5.11) demonstrates Botticelli's ability to incorporate light and shadow utilizing particular pigments.

In the Old Refectory at the church of Ognissanti, we see the *Last Supper* (fig. 2.1), created in 1480 by Ghirlandaio. This extremely moving fresco is notable for the way in which it gives the illusion of being a continuation of the physical space of the refectory. The fresco was damaged by the flooding of the Arno River in 1966, and during restoration, the fresco was detached from the wall using a technique known as *strappo*. This procedure involves attaching a canvas to the fresco, then gradually removing the canvas, thereby detaching from the wall a thin layer of plaster containing the pigments and remounting it at another location. After removal of the fresco the underlying preparatory *sinopia* created by Ghirlandaio was discovered; it was then moved and remounted on a separate wall. Interestingly, there are differences between the *sinopia* and the final fresco. For example, there are variations in the postures of some of the apostles, a figure drawn on the far left of the *sinopia* is absent in the

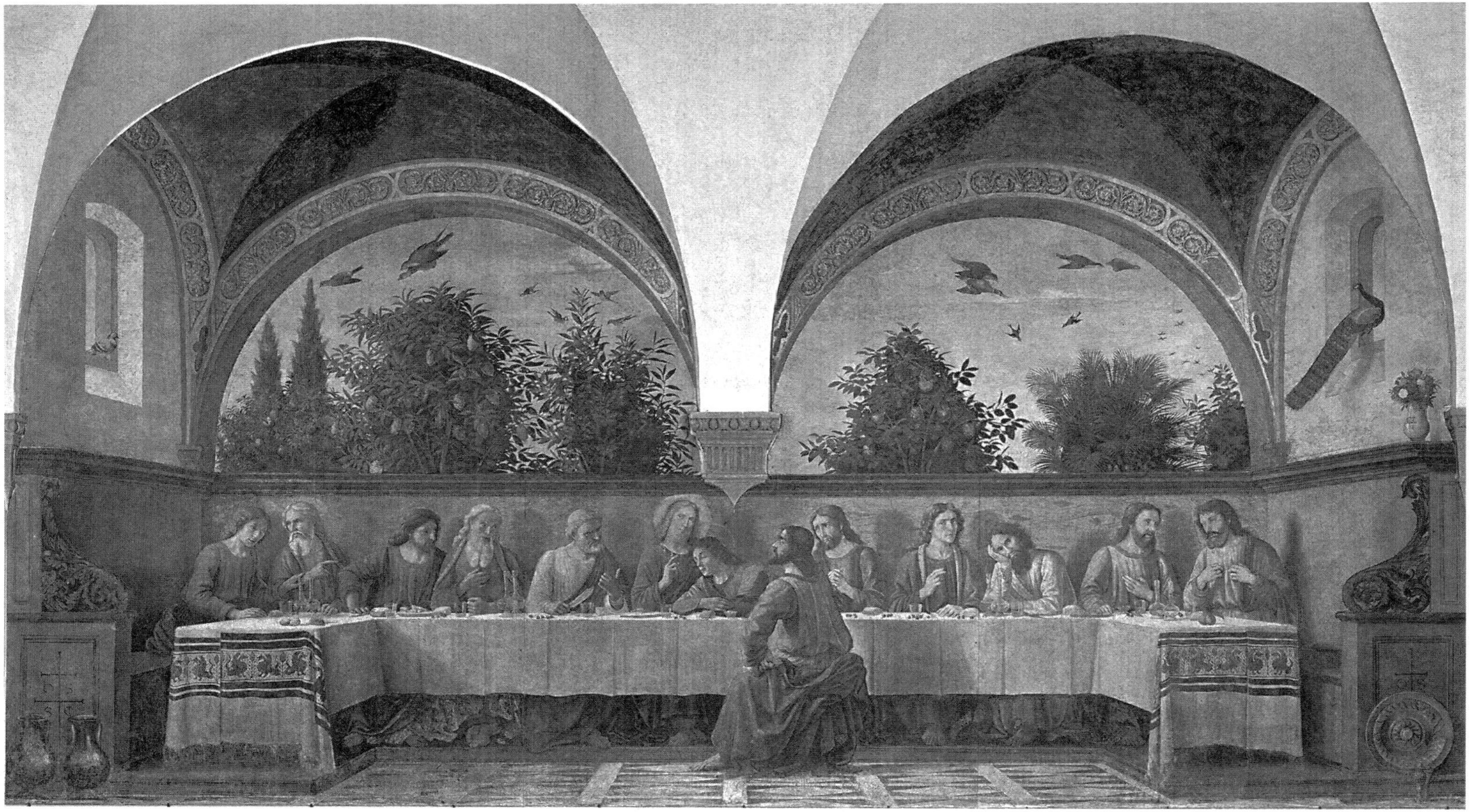

Figure 2.1: Domenico Ghirlandaio, *Last Supper* (ca. 1480), fresco, Church of Ognissanti (Alinari Archives, Florence, photograph by Serge Domingie, reproduced with the permission of Ministero per i beni e le Attivita Culturali [AGC-F-002068–0000]).

fresco, and the *sinopia* lacks all trees, birds, fruits, glasses, and bottles present in the fresco. These differences demonstrate Ghirlandaio's ability to continue his creative process even during the delicate and rapid execution of the final painting.

d. Santa Trinita

The church of Santa Trinita, constructed in the latter half of the fourteenth century, is the site of Ghirlandaio's frescoes in the Sassetti Chapel. It is arguably one of his important works because of his technique of utilizing pigments in such a way as to maximize the impact of their color within the limitations of fresco painting. Begun around 1482, this cycle, *Scenes from the Life of Saint Francis*, contains portraits of many prominent Florentines. An altarpiece (ca. 1485) and the resulting frescoes, like his others, are remarkable for the amount of detail and the depiction of life in the fifteenth century. For a discussion of Ghirlandaio's use of depth and perspective, see chapter 3 (3.1.2.3f).

e. Sant'Apollonia

Another fine example of the fresco technique can be viewed in the refectory of Sant'Apollonia, where Andrea del Castagno (ca. 1419–57) completed a *Last Supper* in 1450. Little is known about Castagno's early training, although he may have apprenticed in the workshops of Fra Filippo Lippi and Paolo Uccello. His Sant'Apollonia fresco is noteworthy for the geometric and linear perspective created by the flooring tiles, wall panels, and ceiling decoration. The higher section of the wall contains a representation of the *Resurrection, Crucifixion, and Entombment*, innovative for its three scenes in a single space unified by not only the landscape, but also by the use of perspective. As with Ghirlandaio's *Last Supper* in Ognissanti, the fresco in Sant'Apollonia provides the viewer with insight into this artist's preparatory process. Small sections of the *sinopia* were discovered when scenes were detached from the wall during a 1953 restoration. They are now mounted on a separate wall in the refectory.

A *sinopia* from Castagno's fresco *Christ with Angels*, originally commissioned for San Marco, is also on view at Sant'Apollonia, as is the actual fresco. Additionally, the refectory at Sant'Apollonia houses a small fragment of a *sinopia* from an artistically noteworthy fresco cycle depicting the *Life of the Virgin*, originally painted between 1439 and 1470 for the choir of the church of Sant'Egidio. This small portion of the *sinopia* is particularly important because the fresco, painted by Domenico Veneziano, Andrea del Castagno, Piero della Francesca, and Alessio Baldovinetti, was destroyed during reconstruction of the church in the sixteenth century.

f. Santa Maria del Carmine

An excellent example of fresco is found on the walls of the Brancacci Chapel in the church of Santa Maria del Carmine. This fresco depicts the *Life of Saint Peter* painted by Masaccio with Masolino from 1425 to 1427, and completed by Lippi in the early 1480s (see 3.1.2.3a). Masaccio's understanding of the alterations of color and light in relation to fresco and its location are evident throughout, but particularly in his consistent incorporation of the use of frescoed light sources that match the light emanating from actual chapel windows.

g. San Marco

Fra Angelico (Guido di Pietro, ca. 1400–1455; discussed in depth in 3.1.2.3e) executed several notable frescoes in San Marco. *Coronation of the Virgin* is especially interesting in illustrating his technical expertise in the art of fresco, with his remarkable and subtle use of whites, where purity is prominent in one area and the gradation of tones in others. In an art form where speed and knowledge of color changes are of constant concern when working with pigments and plaster, Fra Angelico was an adept master.

h. Santa Croce

Some of the earliest Florentine frescoes can be found in the Franciscan church of Santa Croce. The Baroncelli Chapel is frescoed with scenes containing dramatic lighting effects that recount the story of *Joachim and Anne* and the *Life of the Virgin Mary.* The Baroncelli Chapel was created from about 1332 to 1338 by Taddeo Gaddi (ca. 1290–1366), a pupil of Giotto.

In the Peruzzi Chapel, Giotto di Bondone (ca. 1266/67–1337) painted scenes from the lives of *Saint John the Baptist* and *Saint John the Evangelist.* This chapel was created in *fresco al secco* in which the pigment was applied to dry plaster and thus did not chemically adhere to the *intonaco.* Lacking the impregnation of the pigments, the frescoes have lost much of their color due to flaking of the paint, which adhered to the surface rather than bonding with the plaster.

Giotto, while in his youth, is believed to have been encouraged to become a painter by the artist Cimabue. Because only legends exist about his early life, he is known to us through his work. The Bardi Chapel, painted by Giotto in circa 1320, is artistically notable for the manner in which he illustrated the *Life of Saint Francis* using subjects similar to those in famous frescoes in Assisi. Unlike Giotto's frescoes in Padua, which evoke an emotional strength, these derive their power from the ordered logic and composed presentation. Giotto's technique represented a transition between the ornate and rigid Byzantine style and the more fluid frescoes discussed earlier.

2.2.1.3 Intersection of Renaissance Art and Modern Science

The use of modern non-invasive technology to study early painting can yield interesting information. The restoration process in current use is a prime example. Upon illuminating Giotto's Bardi Chapel frescoes with ultraviolet and infrared light, restorers were able to see an exceptional level of detail, not observed with visible light because of deterioration over time.[4] Such illumination is also being used in the restoration of other Florentine works under the aegis of the Opificio delle Pietre Dure. Restoration of Giotto's *Crucifix* was completed in 2011, and is now displayed in its original location, the church of Ognissanti.

4. Bellucci et al., "Integration of Image Data."

Reflectance spectroscopy in the wavelength region of 400 to 800 nm has been used to examine several frescoes in the Brancacci Chapel in Santa Maria del Carmine. Such a study enables researchers to identify the specific pigments and to determine any detrimental effects of prior restoration. For example, in the early twentieth century, restorers working on the Brancacci frescoes used a protein-rich egg white base mixture to repaint portions that had faded. Unfortunately the chemical reaction that resulted from applying this mixture to the fresco caused the colors to change. The paintings later studied using reflectance spectroscopy were *Raising of the Son of Theophilus*, *Saint Peter in the Pulpit*, and *Saint Peter Healing and His Shadow*. The researchers spread pure pigments on a tile containing about one centimeter of plaster. They then compared reflectance peaks from the spectra of the frescoes with those of the pure pigments on the tile and were able to determine the pigments that were used in painting the fresco. They found that the red pigments were likely derived from hematite, red bole, and/or Mars red, while malachite appears to be present in the green pigment.[5]

Researchers also analyzed the frescoes decorating the Saturn Room of the Palatine Gallery in the Pitti Palace, created toward the end of the seventeenth century by Ciro Ferri, who was a pupil of Pietro da Cortona, using Fourier-transform infrared (FT-IR) and Mössbauer spectroscopy, X-ray diffraction, and optical and scanning electron microscopy to determine the nature of the pigments. The results revealed that ultramarine blue and smalt were applied for blue coloring in the form of a potash glass containing cobalt, iron, potassium, arsenic, and other elements in lower amounts. Interestingly, the analysis also identified a synthetic form of ultramarine used in a nineteenth-century restoration. Other pigments identified were red, yellow, burnt ochres, raw sienna, green earth, carbon black, charcoal black, vine black, and raw white, all coloring agents commonly used throughout the Florentine Renaissance.[6]

The modern technique of unilateral nuclear magnetic resonance (NMR) provides important information to restorers regarding the causes of degradation, be it moisture, salts, heat, microbial/fungal action, or other factors. This technique was applied to three sets of frescoes from the house of Giorgio Vasari (1511–74; see 3.1.2.3a), two of which are in poor condition and characterized by crumbling and detachment. The authors found that the deterioration was in part caused by the outcropping of soluble salts.[7]

Recent scientific methods of investigation not only provide basic chemical knowledge of some of the pigments, but also verify historical documentation regarding the sources and origins of the pigments. This research also provides insight into the complexity of Florentine Renaissance pigments, many of which are found to contain varying quantities of impurities.

5. Bacci et al., "A Color Analysis."

6. Ajo et al., "Ciro Ferri's Frescoes."

7. Proietti et al., "Fresco Paintings Studied by Unilateral NMR."

2.2.2 Tempera

2.2.2.1 Chemistry and Applications

Tempera refers to the type of paint created by grinding color pigments and mixing them with egg yolks. Why egg yolk? Rich in lipids (fats), cholesterol, protein, small amounts of carbohydrate, and a number of vitamins, iron, calcium, magnesium, phosphorus, zinc, and potassium, egg yolk provides the major source of energy and structural lipids for the developing chick embryo. Many different components comprise the yolk lipid, including neutral lipids and phospholipids, with lecithin (phosphatidylcholine) being an important biomolecule.[8] The lipid constituents also give the yolk emulsifying and binding properties that make it particularly durable when mixed with pigments.

Tempera painting was usually done on wooden panels intended for altarpieces, crucifixes, and other religious works. Tempera was also used for decorative household pieces such as small devotional paintings and large wedding chests (*cassoni*), discussed further in the section on wood (see 2.3.3.1c). Preparation of the panels was an extremely important part of the process. Because few artists' workshops had the necessary woodworking tools, including the specialized saws and planes required to create the panels to be decorated, the wooden sections were frequently obtained from woodworkers who cut them to the size and shape ordered by the painter. Poplar trees, which are usually white and sufficiently large to be cut into planks of a good width, were most often used. Poplar, however, has the disadvantage of being a soft wood predisposed to warping, and so curing the wood was an essential step in the preparation procedure. After the planks were aged, they were glued together. In some cases, rods were inserted to reinforce the glued planks, and strips of wood were attached to the back for added stability. When a frame was required, it was also created by the woodworker, who was usually an expert carpenter.

After the panels were delivered to the artist's workshop, a ground (coating) was applied to the panels in order to render them suitable for painting. Cennini described the materials needed and a method of applying the ground in his treatise on art. Panels were sealed with a glutinous material, sanded, and then strengthened by adding linen strips soaked in strong glue, referred to as "size," that was spread by hand over the panel. After the panel was thoroughly dry, a mixture of glue from animal skins and ground plaster or chalk (*gesso grosso*) was spread over the surface of the panel with a large spatula (*stecca*) followed by other layers of a finer gesso (*gesso sottile*) that were added several times without waiting for each layer to dry completely. Cennini allocated several pages to the preparation of the *gesso sottile*, including a detailed recipe for achieving a successful product. To attain a smooth surface, he advised scraping the fully dried panels, then brushing off the loose *gesso* with hen or goose feathers.[9] Although not every panel was prepared in exactly the same manner as described by Cennini, researchers using scanning electron microscopy to examine a number of fourteenth-century

8. Kuksis, "Yolk Lipids."

9. Cennini, *Book of Art*, 94–101.

Florentine panels ascertained that indeed both *gesso grosso* and *gesso sottile* were used in all of those particular paintings. It was not until the fifteenth century that artists began to deviate from those protocols.[10]

Tempera painting was also done on canvas, mainly during the time artists were transitioning from tempera to oils as the medium of choice. Canvas, made from cotton, hemp, or linen, is composed primarily of cellulose, a polysaccharide with repeating units of $C_6H_{10}O_5$ linked to each other in a certain manner (β[1→4] glycosidic bonds), thus forming a long linear chain. Cellulose, a high molecular weight polymer, is responsible for canvas's strength and durability, which is necessary for maintaining the integrity of the painted surface. Before it could be used, the canvas was stretched across a wooden frame and coated with *gesso* and lead white.

2.2.2.2 The Painting Process

Once the surface had been prepared, the artist would either incise an outline of the picture to be created into the *gesso* or, more commonly, draw it on the panel, first with charcoal and then, when the artist was satisfied with the drawing, possibly with ink. For a painting that would include gilding, the gold leaf was applied prior to painting. The gold leaf, often obtained from hammered coins, was pressed into a reddish clay that was easily pulverized (*bole*) and layered on the panel in an overlapping fashion. If the artist's rendering called for portions of the panel to be raised above the surface in order to provide an area of low relief, small amounts of plaster were positioned in the desired form to give, for example, a halo or a decorative embellishment.

After these steps were completed, the artist could finally begin the meticulous process of painting with tempera. According to Cennini, the pigments were mixed with egg yolk in about equal amounts of yolk to the pigment paste, yielding an acceptable medium, albeit one that was difficult with which to work. Once the paint had been applied, the water in the mixture evaporated rapidly, while the egg proteins took longer to set. Because of this, tempera painting required that the artist make delicate, small, and repetitive strokes of the brush, applying the paint in numerous thin layers to avoid flaking or cracking. With the prepared paint drying rapidly, the artist had to determine exactly how much to mix for a single painting session to avoid wasting pigments, some of which were quite costly. Artists had various methods of organizing the execution of their paintings in order to maximize the life of the tempera. After the painting had thoroughly dried, a finishing varnish could be added. Although the process was exacting, the use of tempera had the advantage that colors did not change over time, either from a darkening effect or from the loss of brilliance.[11]

10. Dunkerton et al., *Giotto to Durer*, 163.

11. Adams, *Italian Renaissance Art*, 11; Paoletti and Radke, *Art in Renaissance Italy*, 26–27.

2.2.2.3 Tempera Masterpieces in Florentine Art

a. Santa Maria Novella

The church of Santa Maria Novella has a magnificent crucifix painted by Giotto, probably between 1288 and 1289. This example of tempera on wood was originally designed for the main altar, but was relocated to the inside wall of the façade in 1488. The work was restored in the 1980s and is now placed high above the central aisle. In this depiction of Christ, Giotto moved from the Byzantine idealized symbolism to a rendering of a realistic body almost drained of life, thereby setting an artistic path toward greater correspondence to more worldly representation.

b. Uffizi Gallery

Early works in tempera can be seen in the Uffizi Gallery (Galleria degli Uffizi), including Cimabue's *Madonna Enthroned* (1280–90), Duccio's *Madonna Enthroned* (ca. 1285), and Giotto's *Madonna d'Ognissanti* (ca. 1310). The Uffizi also displays tempera paintings by Ambrogio Lorenzetti (*Presentation at the Temple*, 1342), Taddeo Gaddi (*Madonna and Child Enthroned with Angels and Saints*, 1355), Gentile da Fabriano (*Adoration of the Magi*, 1434), Uccello (*The Battle of San Romano*, 1435–60; see 3.1.2.3d), Fra Filippo Lippi (*Coronation of the Virgin*, ca. 1439–47), Botticelli (*Primavera*, ca. 1482 [fig. 2.2] and *The Birth of Venus*, ca. 1484, both lyrical paintings), Leonardo da Vinci (whose mixture of tempera and oils is discussed in the section on oils), Raphael (*Self-Portrait*, 1506), Michelangelo (*Holy Family with Young Saint John*, ca. 1507), Vasari (*Portrait of Lorenzo il Magnifico*, 1534), and Bronzino (portraits of various Medici family members painted during the 1540s). The Uffizi collection contains multiple works by these artists, as well as many others, and this is only a partial representation of its many tempera masterpieces.

c. San Marco

The Museum of San Marco has some truly beautiful and unique panel pieces created by Fra Angelico (ca. 1387/1400–1455) and his workshop. Notable among these is the *San Marco Altarpiece* (1438–40) commissioned by Cosimo de' Medici. Although the condition of the panel painting is poor due to an eighteenth-century attempt to clean the surface with an abrasive that corroded the colors, it is technically and aesthetically a most significant work.

2.2.3 Oil Paints

2.2.3.1 Composition and Applications

Painting with oils involves suspending pigments in a medium such as linseed oil, which is obtained from the flax plant (*Linum usitatissium*, Linaceae). Linseed oil, derived from dried ripe flax seeds, is rich in the lipid triacylglycerol that contains a relatively high content

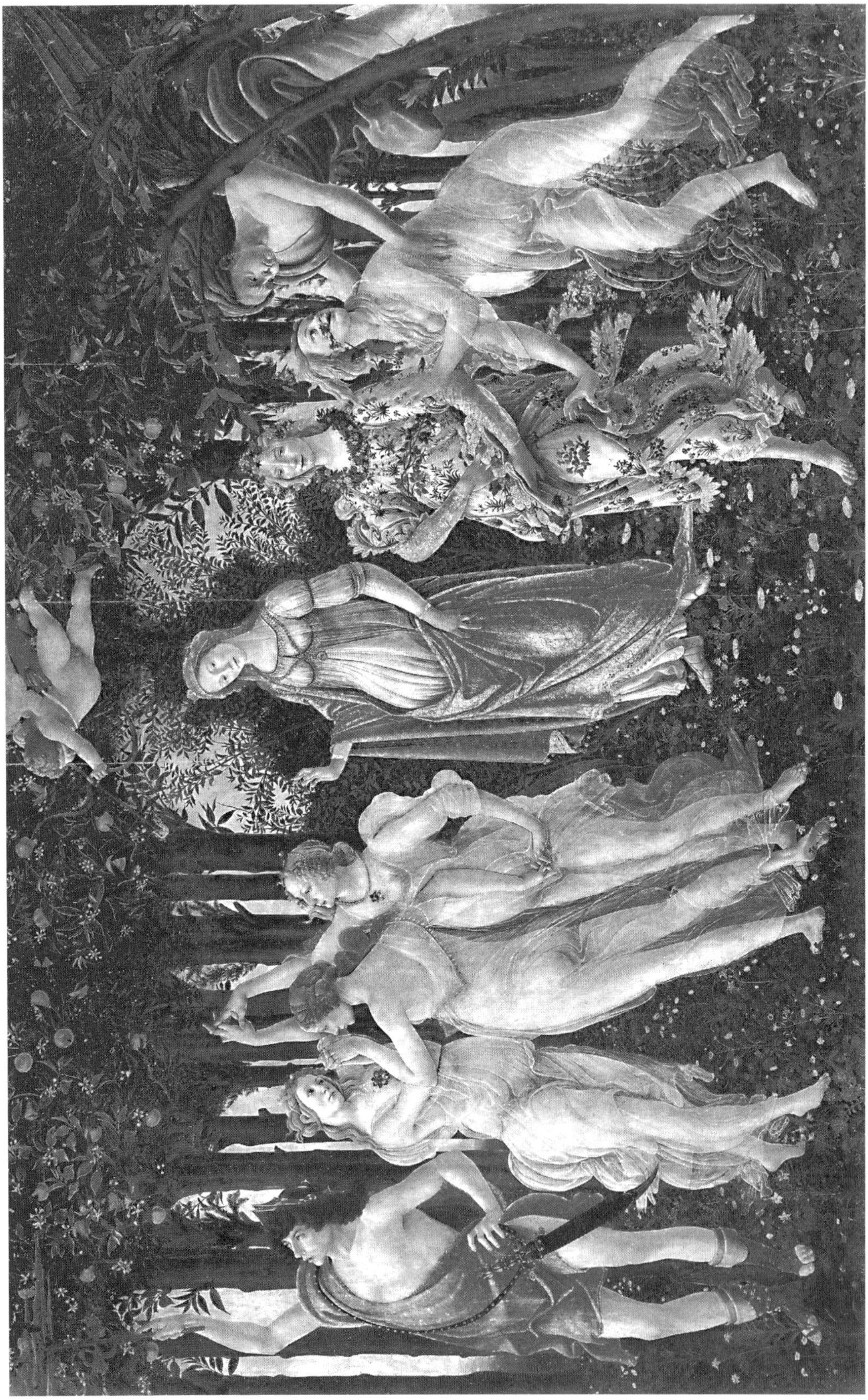

Figure 2.2: Sandro Botticelli, *Primavera* (ca. 1482), tempera on panel, Uffizi Gallery (IBERFOTO/Alinari Archives, Florence [AIS-F-066036-0000]).

of α-linolenic acid (an eighteen-carbon fatty acid with three double bonds). Upon being exposed to air, the linseed oil undergoes polymerization and forms a durable and protective base when mixed with pigments. Other oils were also used, such as safflower oil, walnut oil, and later, poppy seed oil. Drying time, the amount of luster, and the consistency of the paint varied depending upon the type of oil selected. As early as the sixth to ninth century, artists in Asia had been mixing pigments; the practice moved westward during the Middle Ages and eventually found acceptance in northern Europe. By the thirteenth century, artists in Norway and England were using oil paints. In *Lives of the Most Eminent Painters, Sculptors, and Architects* (ca. 1550), Giorgio Vasari (1511–74) erroneously credited Jan van Eyck of The Netherlands with the invention of oil painting. Although Vasari was not always accurate in his historical details, it was in fact the Netherlandish methods of preparing the oils—using a binder of mineral compounds, thereby optimizing the paint's ability to dry and set—that made the use of oil paints more palatable.

It was not until the mid-fifteenth century, however, that some Florentine artists began to work with oil paints. They were not inclined to simply adopt the Netherlandish techniques of mixing the paints. From a perusal of the numerous formulas for both the mixing of oil with pigments and the preparation of glazes reported by Italian painters, it is clear that major concerns were the amount of particulate matter in the oils, the drying process, and glazing/varnishing. Florentine artists already knew much about the durability of colors and the preparation of pigments from their experience working with other types of paint. Based on trial and error, they used their expertise to improve the integration of pigments with oil. During the first years of its use in Florence, oil paints were applied to wood panels using techniques similar to those for tempera paint; on occasion, artists mixed oil paints with tempera. Later, artists began using cloth (canvas base made from linen) as a painting surface, and oil-based paint supplanted tempera as the paint of choice.

2.2.3.2 The Painting Process

The process for applying oil paints is very different from applying tempera. Oil paint is more fluid than tempera and can be applied in layers, allowing the artist to change colors, shading, and the arrangement of the picture itself. By layering and partially covering areas previously painted, the artist could achieve a texture, depth, and intensity not possible with earlier techniques such as tempera and fresco. Although the artists from northern Europe usually incorporated a glaze or varnish into the oil paints, resulting in a lustrous finish, they and Italian painters experimented with methods of applying different colored glazes over one another to produce a diversity of hues. In Italy, Renaissance artists developed a glaze composed of an essential oil and a balsam, with the occasional addition of a resin. The glaze was thin and when warm could be evenly spread over the entire composition. Leonardo da Vinci spent time formulating methods for oil paint preparation, in particular with the use of walnut oil. He described an extraction technique that he found best for producing an oil with a degree of clarity that would not alter the color composition. Separating the oil from unwanted extraneous particles was key to maintaining the integrity of the colors. Pigments that were available for oils were the same as those used in earlier painting methods, but some, for example verdigris, were found to bind less easily with oil than with egg yolk used in tempera. In addition, malachite and ultramarine, which exhibit a low refractive index, appeared much darker when mixed with oil and thus became less popular choices for pigments.[12]

2.2.3.3 Florentine Oil Painting

With the time constraints of fresco and tempera no longer a consideration, oil painting provided artists the freedom to be more innovative during the painting process. The transition from fresco and tempera, however, occurred gradually. There were some settings for which painters felt tempera was better suited. With several media from which to choose, artists began experimenting with innovative techniques, achieving varying degrees of success. One method utilized by Florentine artists was to apply an underpainting of quickly drying tempera, then add layers of oil paints and glaze. This technique enhanced the intensity of the colors. Sandro Botticelli's *Adoration of the Magi* (ca. 1467), on view in the National Gallery in London, is an example of such a transitional work in which there appears to be an underpainting of tempera overlaid in oils.

a. Uffizi Gallery

Leonardo da Vinci (1452–1519), the quintessential Renaissance man, was born in Vinci near Florence. At fifteen years of age he was apprenticed to the Florentine artist and bronze sculptor Andrea del Verrocchio (1435–88), then remained in his workshop as an assistant; he eventually established his own studio in Florence. In 1482 Leonardo left Florence, having accepted a position as court artist and military tactician to the Duke of Milan, and did not return to Florence until 1500. As his reputation spread, Leonardo returned to Milan, then went to Rome, and was later called by Francis I to the French court, where he worked until his death. His influence in the world of art, anatomy, geography, and engineering is legendary. Like Botticelli, Leonardo also began to incorporate tempera and oils in a single work. An interesting example of such a mixture is *The Baptism of Christ* (1470–75), which was started by Andrea del Verrocchio and completed by Leonardo. Verrocchio executed the first part of the painting using tempera; when he was unable to complete the project, his young assistant Leonardo finished the work with oil paints. While the colors comprising the body of John the Baptist (done by Verrocchio) are very clear and exact, those of the figure of Christ (Leonardo's work) have a cloudy and indistinct quality achieved by subtle changes in the shading (*sfumato*). Underpainting with a dark pigment accounts for the muted colors achieved by Leonardo, who was also known to have used his fingertips in some instances to enhance the *sfumato* effect. *The Baptism of Christ* is exhibited in the Uffizi Gallery, as are Leonardo's *Annunciation* (ca. 1472) and *Adoration of the Magi* (ca. 1481, fig. 2.3). The latter, with its underpainting still in view and the oil glazing left unfinished, dramatically illustrates an evolving process.

12. Hall, *Color and Meaning*, 54.

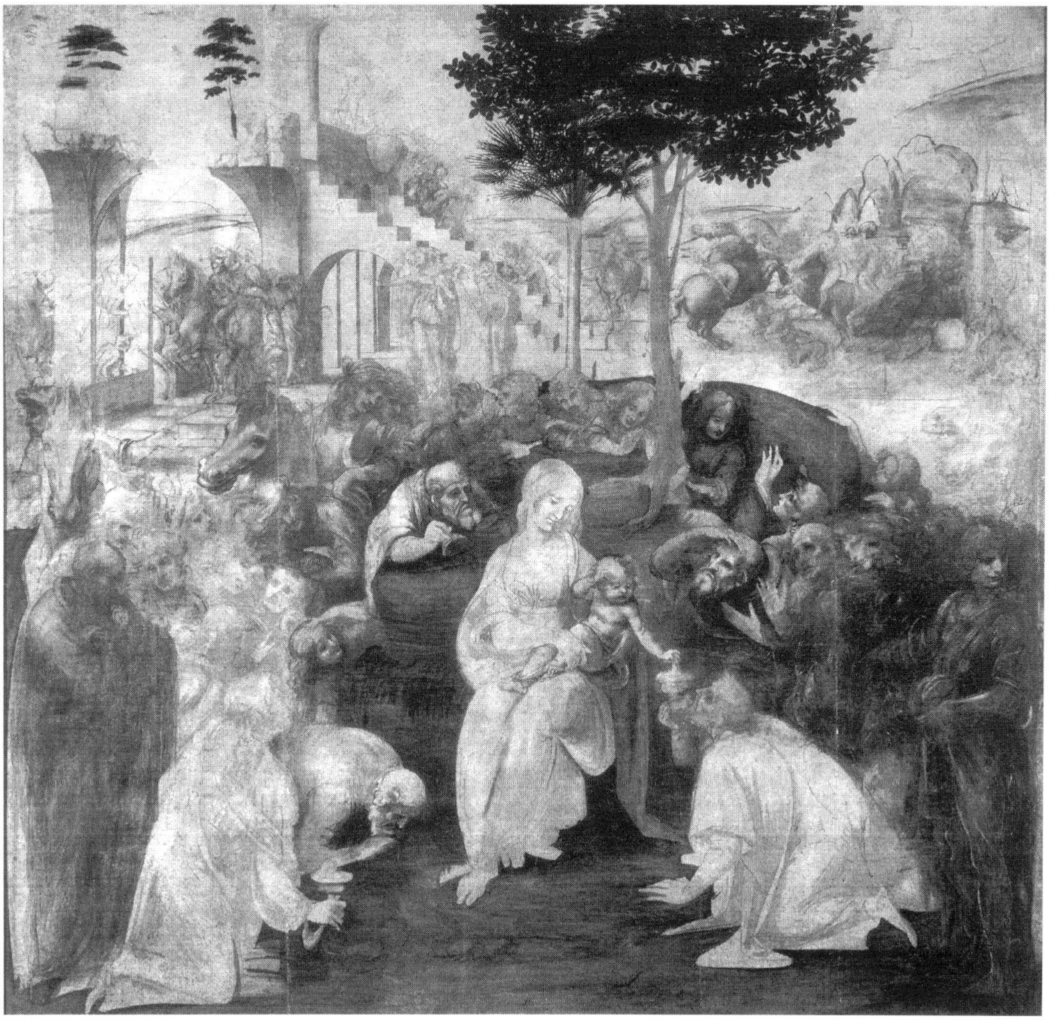

Figure 2.3: Leonardo da Vinci, *Adoration of the Magi* (ca. 1481), oil on wood, Uffizi Gallery (Alinari Archives, Florence, photograph by Nicola Lorusso, reproduced with the permission of Ministero per i beni e le Attivita Culturali [AGC-F-000514–0000]).

Other exemplary works include an interesting pair of early oil paintings on wood, *The Duke and Duchess of Urbino*, portraits of Battista Sforza and Federico da Montefeltro. Both double-sided portraits were executed sometime after the 1472 death of the duchess by Piero della Francesca (ca. 1410/20–92), an Umbrian artist who also worked in Florence. Several decades later, when the sixteenth century was dawning, paintings in oil became more prevalent. Fine examples of oil on wood on display in the Uffizi Gallery include (among many others) works by Luca Signorelli (*The Trinity, the Madonna and Two Saints*, ca. 1500–1510), Fra Bartolommeo (*Presentation in the Temple and Nativity*, ca. 1497), Raphael (*Portrait of Leo X with Two Cardinals*, ca. 1518), Pontormo (*Portrait of Cosimo the Elder*, ca. 1519–20), and Caravaggio (*Sacrifice of Isaac*, ca. 1592–1604, and *Medusa*, ca. 1595–98). By the beginning of the seventeenth century, canvas supplanted wood as the surface of choice, as exemplified by *Judith Beheading Holophernes* (ca. 1620), executed by Artemisia Gentileschi, one of the few Florentine women recognized as an artist in her own right.

2.3 Sculpture

Works of sculpture were created using various materials, either in the form of relief or free-standing. Some sculpture workshops were under the direction and ownership of a particular artist, while others were under the sponsorship of a board of works for a large building, the group that was responsible for construction and decorative art. Works of sculpture were created by carving, molding, and casting from a variety of materials such as stone, wood, terracotta, and bronze.

2.3.1 Marble

2.3.1.1 Formation, Composition, and the Sculpting Process

Marble is a metamorphosed form of a sedimentary carbonate rock, generally limestone or dolomite (i.e., the protolith). When acted upon by a combination of heat and pressure, the protolith's carbonate grains metamorphose, recrystallizing to form marble, which is characterized by interlocking carbonate crystals.[13] Relatively pure limestone, or dolostone, metamorphoses into white marble. Much of the marble used in Florence was mined in Carrara, a quarry that remains active. Particularly prized and often possessing a grayish blue hue, Carrara marble contains about 98 percent calcite (calcium carbonate, $CaCO_3$) with only traces of mica, dolomite, and quartz. This marble is characterized by isotropic grains of about 150 to 200 microns (micrometer, one millionth of a meter, μm) that in turn are composed of subgrains with a minimum size of about 3 microns.[14] Colored marble, also highly valued, arises from limestone or dolostone that, in its original form, contains iron, magnesium, and other metals and minerals.

The first step in the sculpting process involved the artist or one of the assistants carefully selecting stone at a quarry. The stone had to be of the correct size and free of apparent flaws, factors that could compromise the reliability and strength of the piece or the appearance of the finished surface. Once the stone was chosen, it had to be excavated from the mountain and then transported to the workshop in Florence. Moving large marble blocks from Carrara, for example, was not trivial. The distance from the quarry to Florence was only about sixty-five miles, but the travel routes required for transporting marble were considerably longer and involved travel by land, sea, and river, using oxcarts, boats, and barges. To address these difficulties, the Florentines developed innovative transport technology (see 6.2.3).

To give direction to their carving, artists worked from small wax or terracotta models and/or from sketches. Using these models, the artist would measure the distances between

13. Schmid, Paterson, and Boland, "High Temperature Flow."

14. Bruijn et al., "Mechanical and Microstructural Development of Carrara Marble."

points marked on the model and then scale them mathematically to the size of the desired piece. The artist might use calipers to mechanically mark the areas to be carved. Employing chisels and drills to remove the areas marked, artists worked slowly inward and used finer chisels as the work progressed. The entire process was precise and difficult and the stone was unforgiving. When the work with chisels was complete, the surface was covered with rough chisel marks that had to be smoothed with files and pumice, and finally polished to elicit the shine of the marble.

2.3.1.2 Examples of Florentine Marble Sculpture

A few of the numerous extraordinary Florentine works in marble are highlighted here, but as with those examples given in the painting section, they are by no means meant to be selective as to artistic superiority, but rather to highlight various styles and methods of execution.

a. The Church of Orsanmichele

The structure now known as the church of Orsanmichele was a loggia used for the storage of grain, but was reconfigured as a guild church with fourteen niches on the exterior walls sufficiently large to hold life-size statues. Guilds granted commissions for the statues—each depicting the guild's patron saint—to fill the niches, and the most accomplished Florentine artists were chosen to create the sculptures. Their works are magnificent. One such artist was Donato di Niccolò di Betto Bardi, known as Donatello (ca. 1386–1466). An acclaimed sculptor, Donatello's very early work included a marble statue of *David* done for the Duomo in about 1408/9. Shortly thereafter, he began work on his statue of *Saint George* (ca. 1410–15), which was commissioned by the Armor and Sword Merchants Guild (Arte dei Corazzai e Spadai) for Orsanmichele. His *Saint George and the Dragon*, the low-relief marble at the base of the statue, is significant for its use of perspective (see 3.1.2.3c). Donatello also created *Saint Mark* (1411–13), commissioned by the Linen Weavers Guild (Arte dei Linaioli e Rigattieri), for the Orsanmichele. His expertise and brilliance in carving marble are evident in this work, with the beautifully draped fabric clearly conforming to the contours of the body of the saint, the realistic veins in his hands, the strength of expression, and the classical *contrapposto* position (weight resting on one leg and foot) suggesting the potential of forward motion on an otherwise static form.

Nanni di Banco (ca. 1384–1421), another noted Florentine sculptor, fashioned four figures in the niche for the Master Stoneworkers and Woodworkers Guild (Arte dei Maestri de Pietra e Legname), marble statues depicting Castorius, Claudius, Symphorianus, and Nicostratus, the guild's patron saints (1415–17). The carving was a complex undertaking, with the four figures interconnected both physically and in their engagement with one another. Two of the saints each have a front foot extending over the edge of the niche, and the drape of the four togas with the many folds and contours shows a high level of expertise, documenting the technical ability of the artist.

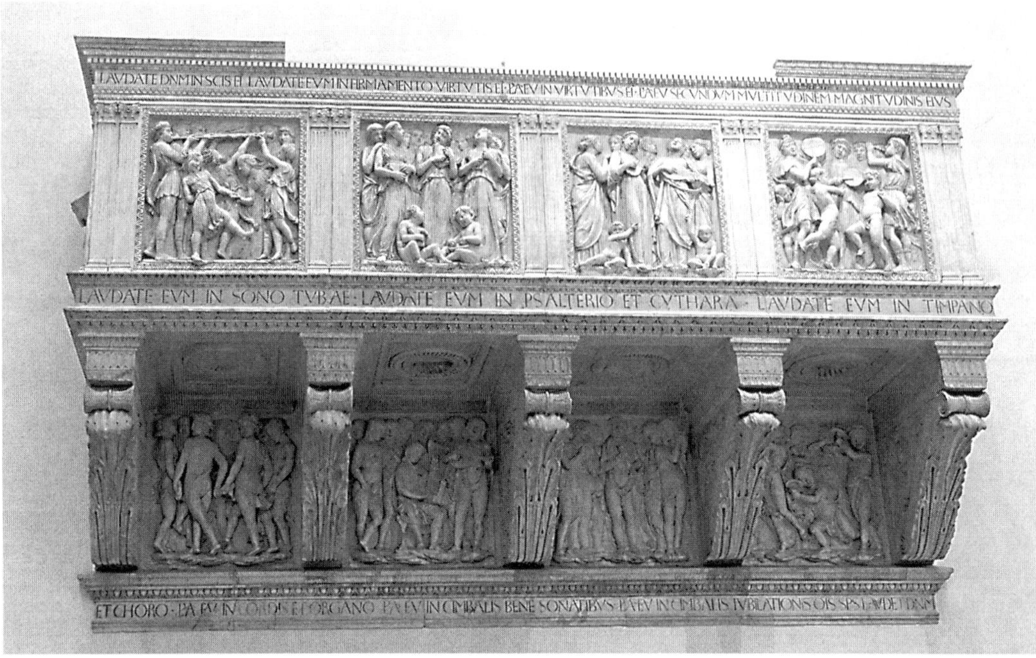

Figure 2.4: Luca della Robbia, *Cantoria* (1430–38), sculpture, Museo dell'Opera del Duomo (photograph by Jastrow, retrieved from Wikimedia Commons).

b. Duomo Museum

The Duomo Museum (Museo dell'Opera del Duomo) is the present site of two remarkable marble carvings commissioned by the Opera del Duomo to be placed above the doors of the two sacristies. Luca della Robbia (ca. 1400–1482), most famous for his innovative glazed terracotta sculptures discussed below (see 2.3.4.1), is also remembered for his marble masterpiece carved for one of the two sacristies. He worked on his *Cantoria* (10′9″ x 18′4″) the Duomo, which depicted a gallery of singing groups, from 1430 to 1438 (fig. 2.4). The singers, carved in relief, illustrate Psalm 150 in a classical framework wearing toga-like clothing, but having very human and unique expressions. The other *Cantoria* was carved by Donatello from 1433 to about 1440. It is a large frieze (11′5″ x 18′8 1/2″) representing *putti* and utilizing not only marble, but also bronze and mosaic.

c. Galleria dell'Accademia

Another of the many marble masterpieces that adorn the city of Florence can be viewed in the Accademia (Galleria dell'Accademia), the present home of Michelangelo's marble sculpture of *David* (fig. 2.5). Michelangelo Buonarroti (1475–1564) was born in Caprese to Lodovico di Leonardo Buonarroti and his wife, Francesca. Shortly after his birth the family returned to the village of Settignano, located a few miles from the center of Florence. As related by Vasari, the area around the family farm was abundant in greystone where local stonecutters and sculptors worked in neighboring quarries, and Michelangelo's wet nurse was the wife of one of those stonecutters. Michelangelo reportedly stated, "Just as with my mother's milk I sucked in the hammer and chisels I used for my statues."[15] Is it therefore surprising that of the five Buonarroti sons, Michelangelo was the one for whom sculpting was a calling?

Michelangelo apprenticed for two years in the workshop of Ghirlandaio before being invited by Lorenzo de' Medici in 1490 to live in his palazzo, where he was provided the financial freedom to perfect his talents. Fleeing the upheaval in Florence after the Medicis were driven from power, Michelangelo received commissions in Rome for his famous *Bacchus*, now in the Bargello Museum (Museo Nazionale del Bargello) in Florence, and the *Pieta* in Rome's Saint Peter's Basilica. When he returned to Florence in 1501, Michelangelo was given the opportunity to work on a sculpture depicting David to enhance the decoration of the Duomo. Later, he was recalled to Rome, where he executed the famous frescoes on the ceiling of the Sistine Chapel and other projects, including the tomb for Pope Julius II, which was never completed. It is, however, the carving of the *David* upon which we are focusing.

The marble Michelangelo used for the *David* was not a newly cut piece; Agostino di Duccio, an assistant of Donatello, had started the carving about forty years earlier. Ten years after that, another sculptor, Antonio Rossellino, was given a contract to complete the work. Both of these efforts had been abandoned with little actual progress. In 1501, the Wool Guild (Arte della Lana), the guild responsible for the decoration of the Duomo, and the Operai, commissioners of the cathedral works, gave the project to Michelangelo. In September 1501 he prepared his mallets and chisels, and began the process with an initial drawing on the block. He then gradually removed the marble from one side in much the same manner used to carve a relief. Michelangelo's technique involved using a claw chisel to crosshatch, as one would with a pen on paper, thus allowing for a fluidity of movement. Although he was somewhat restricted by the previous attempts and the depth of the marble piece, Michelangelo, working in solitude, completed the project in 1504. Because of its weight and size, it was no longer a viable candidate for display in the intended location, high on the Tribuna on the north side of the Duomo.

After a committee comprised of prominent artists of the time, including Leonardo da Vinci, Sandro Botticelli, and Giovanni Cellini, debated several locations, including the Loggia della Signoria, it was finally decided that the *David* should be placed at the entrance to the Palazzo della Signoria as a symbol of the Florentine Republic. Just transporting the *David* from the cathedral workshop to the Palazzo della Signoria proved to be a technological feat. Although the distance was only a few hundred meters, the size, weight, and relatively fragile nature of the sculpture required considerable attention to the logistical challenges. The *David* was encased in a wooden framework, held by ropes, and, using winches, was moved very slowly over greased planks. The entire process took four days. Because the *David* suffered some damage during the years in its vulnerable outdoor location, a copy now stands at the entrance to the Palazzo Vecchio and the original has been relocated to the Accademia.

15. Vasari, *Lives of the Artists*, trans. Bull, 1:326.

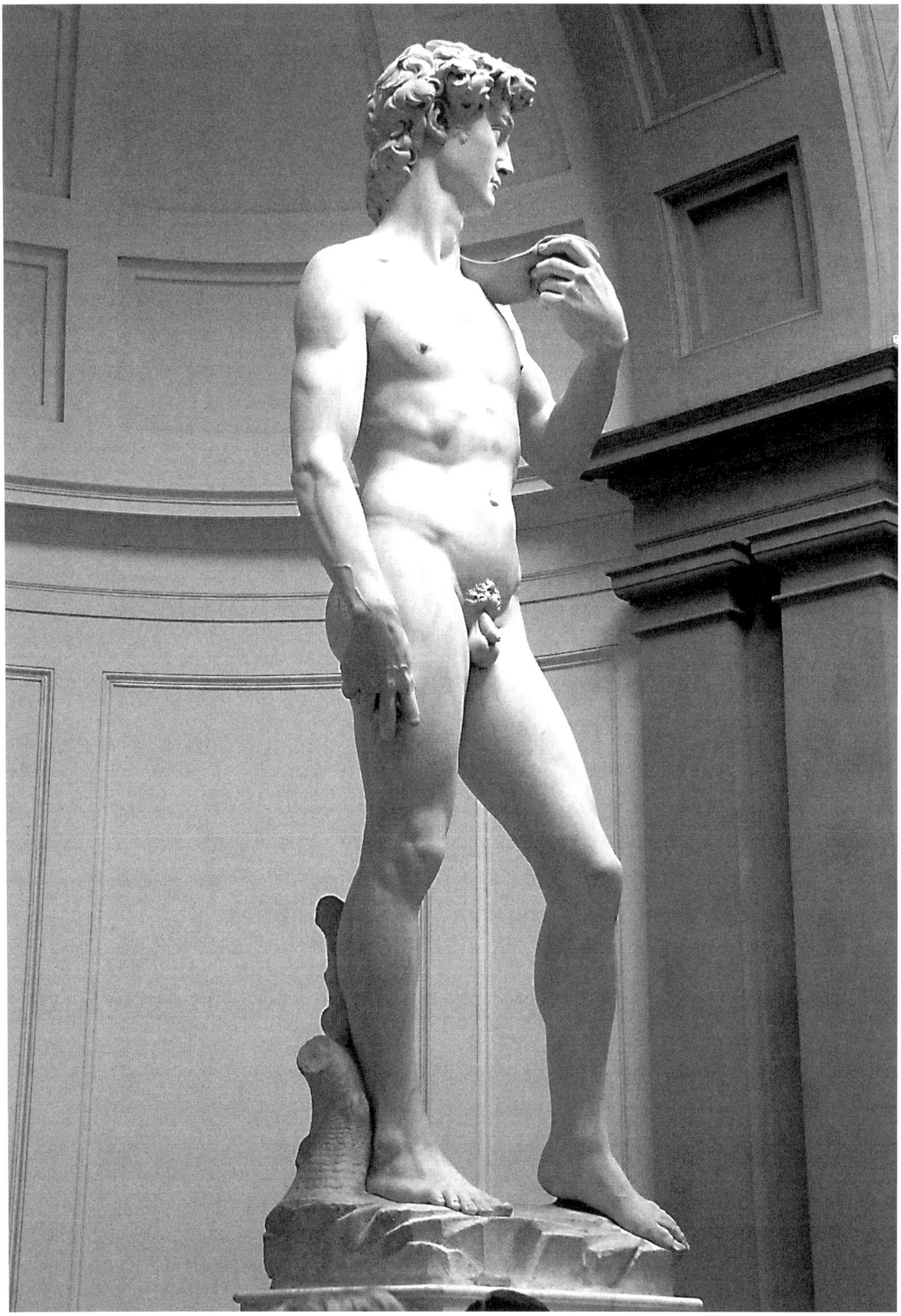

Figure 2.5: Michelangelo, *David* (1501–4), marble sculpture, Galleria dell'Accademia (photograph by Marcus Obal, retrieved from Wikimedia Commons, GNU Free Documentation License).

Nude, unadorned, and classical, Michelangelo's *David* remains an extraordinary example of what can be created from a solid block of stone.[16]

An interesting side note involves an unfinished sculpture of *Saint Matthew*, also displayed in the Accademia, which provides some insight into Michelangelo's technique in initiating the carving process. The statue was commissioned by the Opera del Duomo, and Michelangelo began work in 1503, moving from front to back as he carved the stone, using several types of single point and blunt edged tools. He completed only a portion, thus giving the impression of Saint Matthew still engaged in the act of emerging from the block of marble.

2.3.2 Porphyry

2.3.2.1 Formation, Composition, and Historical Perspective

While marble was the most common and desired stone used for sculpture, sculptors also worked in other types of stone, porphyry being an example. Porphyry is a variety of igneous rock with large grain crystals of at least two millimeters of quartz (SiO_2) or feldspar, a group of aluminum silicates, including potassium aluminum silicate ($KAlSi_3O_8$), sodium aluminum silicate ($NaAlSi_3O_8$), and calcium aluminum silicate ($CaAl_2Si_2O_8$). It is formed when rising magma is cooled in two stages: the magma first cools slowly in the crust of the earth, and then cools rapidly at a shallow depth after a volcanic eruption. The porphyry, first used by early Western civilizations, is said to have been quarried in Mons Porphyritis, a desert area east of the Nile and part of the Arabian-Nubian Shield that was formed when the Mozambique Ocean was closed off some 600 million years ago. Although the Egyptians made limited use of porphyry, Romans began to acquire the stone during expeditions that began around 19 AD. The road used to transport the stone from the quarry to a Roman site on the Nile appeared on a second-century map by Ptolemy, who mentioned the location of the porphyry quarry in his *Geographia*. After the fourth century, little mention was made of the quarry location. It was not until the early years of the fifteenth century, when Manuel Chrysoloras (ca. 1350–1417) translated Ptolemy's *Geographia* from Greek to Latin and the work was introduced to Florence, that the site was recognized. In the nineteenth century, the actual quarry discussed by Ptolemy was rediscovered.

Because of porphyry's purple color, relative rarity, and the durability of works carved from it, the stone was designated by the Romans for imperial use. Romans also used the stone for monumental columns and tombs, and Constantine I had columns of porphyry erected in Constantinople; he also had imperial rooms in his palace lined with the stone. This "royal only" label continued and was embraced by Christian art in the fourth century, where its color was seen as representing the blood of Christ. Gradually religious works were added to the list of porphyry sculptures. The ruling Medici family adopted the royal prerogative during the fifteenth century, reusing old Roman stones for Medici tombs.

16. Gill, *Il Gigante*.

2.3.2.2 Examples of Porphyry in Florence

a. San Lorenzo

One of the first important works commissioned by Lorenzo de' Medici was a tomb in the Old Sacristy for his father, Piero de' Medici, and his uncle, Giovanni de' Medici. The ancient tradition of using porphyry for a ruling family was followed. Andrea del Verrocchio (1435–88) fashioned the Medici tomb from 1470 to 1472 using various types of stone, including a porphyry sarcophagus with bronze leaf decoration and circular green inserts. The impressive sarcophagus rests on four bronze tortoises. Located in the center of the Old Sacristy, the tomb of the family patriarch, Giovanni di Bicci, and his wife was executed by Andrea di Lazzaro Cavalcanti (Buggiano) and topped by a porphyry disc. The tomb of Cosimo de' Medici, located in the crypt, is marked by a disc in the nave in front of the altar in San Lorenzo. Designed by Andrea del Verrocchio, the disc also uses the imperial stone. Although porphyry was incorporated to enhance the design of commissions in the early Renaissance, technical difficulties resulting from the hardness of the stone precluded any sophisticated sculpting.

b. Santa Maria Novella

It was not until the sixteenth century that more advanced technology for working with porphyry was developed and expanded in Florence. The Romans had been successful in carving porphyry, but their techniques were largely forgotten in the intervening years. During the Renaissance new tools were developed—including cutting wheels made of copper with diamond chips, hydraulic saws, drills made of steel, and hammers with steel points—that expanded the possibilities for carving an extremely solid stone like porphyry. A fine example of early carving on a slab can be seen on a flat porphyry surface set into the riser of a step in front of the main entrance to the church of Santa Maria Novella. The stone is inscribed with the Latin version of Bernardo Rucellai (Bernardo Oricellario) and marks the site of Rucellai's burial. The new tools allowed artisans to inscribe the stone, but did not allow for the option of more complex sculpting.

c. Palazzo Vecchio

When Cosimo I de' Medici wished to have a porphyry fountain created for the courtyard of the Palazzo Vecchio, he gave the commission to Francesco di Giovanni Ferucci (1497–1585), known as Tadda. The Florence native (he was born just outside of Fiesole) was instrumental in developing the stronger tools necessary for the more intricate sculpting in porphyry. He is often credited by Cellini and others with discovering a method for tempering steel tools to achieve the strength essential for the task. Vasari claimed that it was Cosimo who was actually behind the invention of tempering using liquor made from specific herbs into which the fiery hot tools were dipped, which raises some questions about Tadda's role in the process. According to art historian Suzanne Butters, "One cannot help but take seriously

Vasari's account of Cosimo's role in this innovation . . . [as he] would have witnessed the use of Cosimo's 'temper' while monitoring Tadda's progress. But Vasari also had a point of view, necessarily shaped by a wish to flatter his patron and to enjoy his continued protection." She further states, "Cellini's claim is as difficult to reject as Vasari's, for like Vasari, Cellini was in a good position to know."[17]

Regardless of with whom the tempering idea originated, it was most probably Tadda who developed the innovative technology for the actual design of forged tools used in Renaissance Florence. The size and shape of tools utilized were, because of the nature of porphyry, critical to the success of the tempering and the carving. Sculpting in porphyry was slow compared to sculpting in other media, but Tadda's accomplishments were impressive. In addition to the fountain in the courtyard of the Palazzo Vecchio, with its mid-sixteenth-century porphyry basin, his work includes a number of busts and bas relief portraits, many of which depict members of the Medici family.

d. Piazza Santa Trinita

The monumental figure of Justice (*La colonna della Giustizia*) in the Piazza Santa Trinita (fig. 2.6), which was begun in about 1569 and took some ten years to complete, is an imposing creation. The monolith, originally in the Baths of Caracalla in ancient Rome, was given to Cosimo I by Pope Pius IV to honor the victory of Florence over Siena. In 1565 it was placed in the Piazza Santa Trinita and about four years later, Tadda began to work on a figure of Justice to be placed atop the monolith. Carved from porphyry rather than marble, it was an artistic and scientific undertaking showcasing his expertise in both the design of tempered steel tools and the art of sculpting in such a difficult medium.

2.3.3 Wood

Wood sculptures also found favor during the Renaissance. Composed of tubular fibers, wood is comprised of three major biological polymers, cellulose (described above, see 2.2.2.1), hemicelluloses, and lignin, with cellulose imparting much of the strength and durability for artistic carving. A major problem presented by wood as a medium, however, is the fact that as the wood dries and ages, the likelihood of cracking increases. Artists found that they could reduce the possibility of cracking by hollowing the tree trunk from one side, making the wood less rigid and therefore less subject to problems posed by heat and humidity. Often several separate pieces of wood were attached together to allow for the arms of a crucifix or those of a human body. As was the practice in marble sculpting, artists made preparatory sketches and mock-ups from which they worked while carving. After the wood was shaped to the desired form, it was covered with a mixture of fine plaster and glue (*gesso*). In some instances, a linen fabric was spread on the wood before the *gesso* was applied. Paint was added after the wood had been covered, and it was not unusual

17. Butters, *Triumph of Vulcan*, 1:149–50.

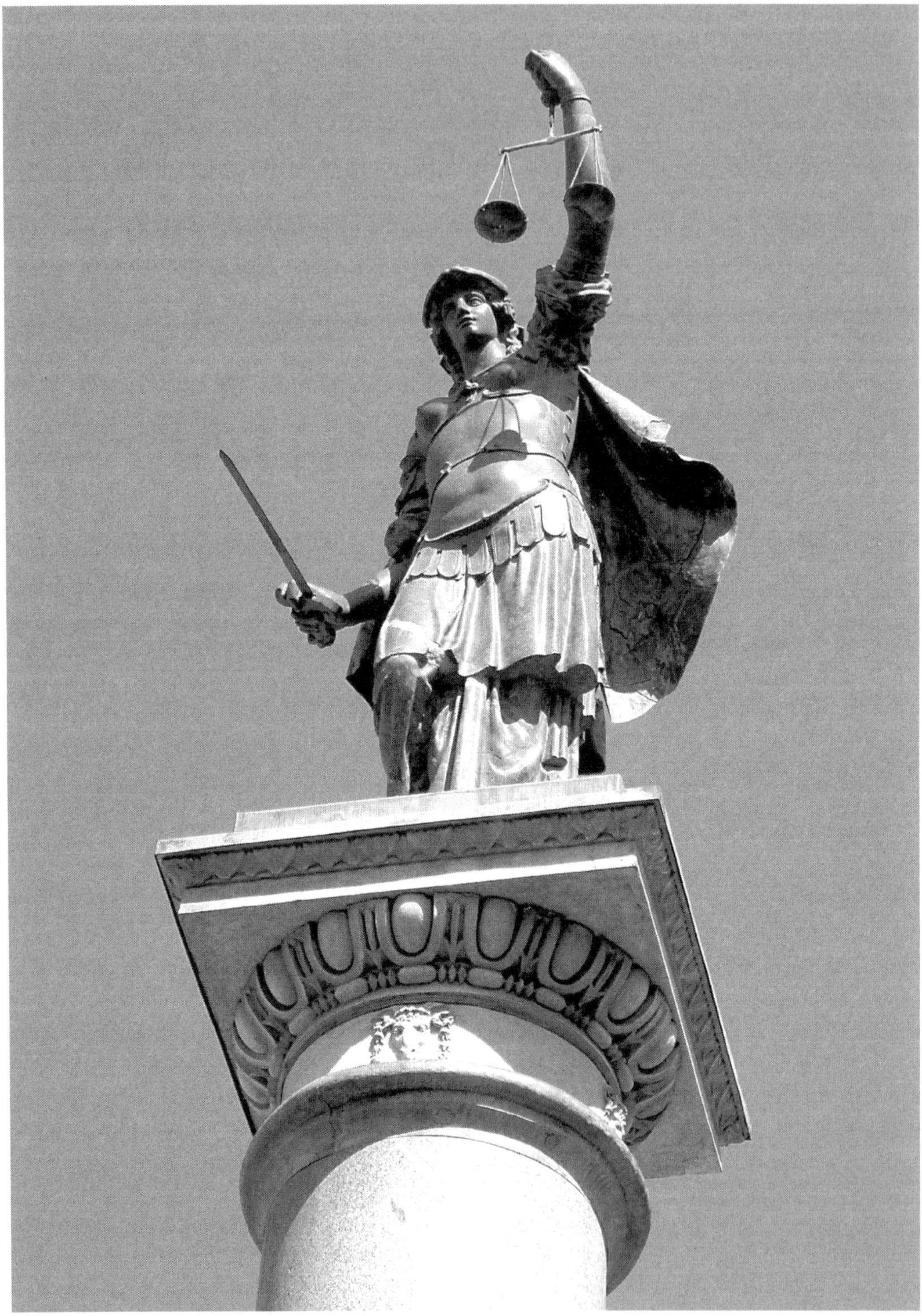

Figure 2.6: Tadda, *La colonna della Giustizia* (ca. 1569–79), porphyry sculpture, Piazza Santa Trinita (photograph by Marie-Lan Nguyen, retrieved from Wikimedia Commons, Creative Commons License).

to affix small pieces of metal (for example, to suggest a crown) or fabric to the completed wooden sculptures.

2.3.3.1 Representative Florentine Works

There are significant works in wood that can be seen today in Florence, a few of which are described below.

a. Santa Maria Novella and Santa Croce

In the Gondi Chapel in the church of Santa Maria Novella there is a remarkable *Crucifix* created by noted artist and architect Filippo Brunelleschi (1377–1446), who is discussed further in following chapters (see 3.1.2.1; 4.3.2; 6.1; 6.2.3; 6.3.1.2–6.3.1.7; 6.5.2). In *Lives of the Most Eminent Painters, Sculptors, and Architects* (ca. 1550), Vasari mentions that Brunelleschi carved the *Crucifix* in response to the *Crucifixion* executed by Donatello for the Bardi Chapel in the church of Santa Croce. According to Vasari's story, when Donatello asked for Brunelleschi's opinion of his completed work, Brunelleschi responded that Donatello "had placed a ploughman on the cross." Donatello challenged him to do better. After Brunelleschi completed his rendering of the crucifixion, Donatello visited Brunelleschi. In Vasari's narrative, Donatello, "having an apron full of eggs and other things for their common dinner, let it fall as he gazed at the work, beside himself with marvel . . . besides admitting himself beaten, proclaimed it a miracle."[18] Although this exchange is not documented in other sources, it speaks to the point that these two works, both created sometime in the 1420s, are remarkable examples of the possibilities for artistic expression using wood as a medium.

b. Duomo Museum

Another of Donatello's works in wood is *Saint Mary Magdalene* (date unknown, but possibly made before 1450), exhibited in the Duomo Museum (fig. 2.7). The sculpture, carved from white poplar, painted and with gilding in her hair, depicts Mary Magdalene's body covered only by her long matted tresses and emaciated from years of fasting and doing penance. By depicting Mary Magdalene's body in a realistic form, but eroded by deprivation, Donatello captured the essence of human suffering in this stark and powerful creation.

c. Bargello Museum

During the Renaissance in Italy, families often commissioned a wooden wedding chest (*cassone*) to hold dowry items or other possessions brought into the marriage. These *cassoni* were made by master woodworkers and often had intricate carving of the legs, top, and borders, with the panels on the sides and the front left smooth for the application of tempera painting. Three

18. Vasari, *Lives of the Most Eminent Painters*, transl. de Vere, 1:395.

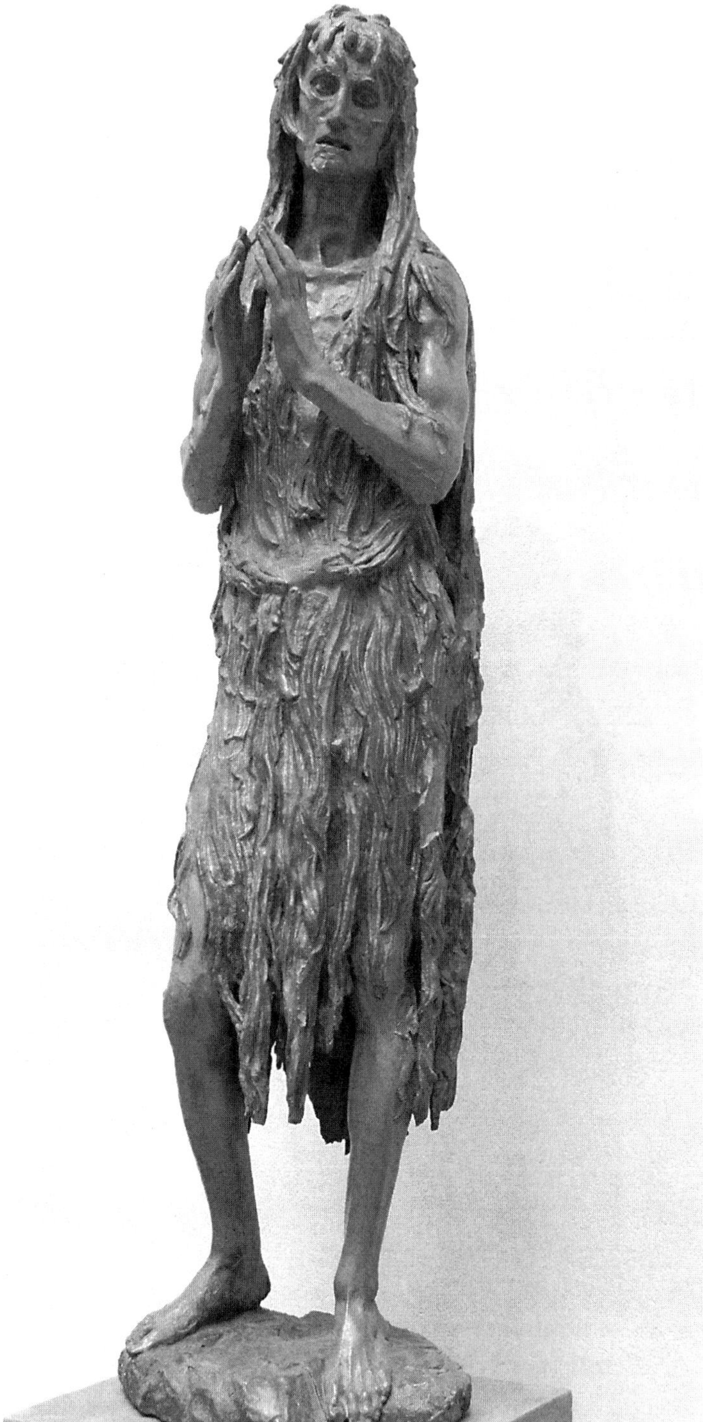

Figure 2.7: Donatello, *Saint Mary Magdalene* (ca. 1430s), wood sculpture, Museo dell'Opera del Duomo (photograph by Sailko, retrieved from Wikimedia Commons, GNU Free Documentation License).

interestingly carved examples can be found in the Bargello Museum, one from Santa Maria Nuova painted with plaster reliefs by Giovanni Boccaccio (fifteenth century), a second *cassone* also from Santa Maria Novella by an unknown artist (fifteenth century), and a third, probably from the Palazzo Davansati, that is undated.

2.3.4 Florentine Terracotta

2.3.4.1 The della Robbia Workshop

Historically, artwork made from terracotta (from the Italian, literally meaning "cooked earth") was very fragile and easily cracked, but the innovative skills of Florentine sculptor Luca della Robbia (ca. 1400–1482) led to a method of transfiguring cooked earth into durable terracotta pieces capable of being displayed both indoors and out. Born in Florence in about 1400 to a wool merchant, Luca is believed to have trained in the workshop of the Florence Cathedral. His family workshop continued to use his formulation to create sculptures for nearly a century; after Luca himself, his nephew Andrea (1434–1525) and then Andrea's sons, Giovanni (1469–1529) and Girolamo (1488–1566), produced works in terracotta. Although there were some limits to the number of colors that could be used to paint on terracotta, the glazed pieces were in demand for a variety of venues including altarpieces and building façades, and as freestanding figures or reliefs.

a. Duomo Museum

The beautiful *Cantoria* carved in marble in about 1430 to 1438 and currently on view in the Duomo Museum is believed to be the first important work of Luca della Robbia, but his thoughts were occupied by his desire to solve the problem of producing his distinctive terracotta sculptures in a form resistant to dampness and sufficiently strong and durable to be of lasting value. Luca's *Resurrection of Christ*, created from 1442 to 1445, was his first known major sculpture of glazed terracotta. Commissioned by Brunelleschi as a lunette to be placed above the door of the North Sacristy in the Duomo, it is a remarkable demonstration of how Luca's technique could be utilized to create large and impressive terracotta works. For the *Resurrection of Christ*, Luca used a blue glaze for the background and white for his figures; his formulation for the glaze remained a workshop secret for many years. The huge success of the technology he developed is illustrated by the work's blue and white coloring, which makes a striking visual statement, and the fact that his terracotta is much less susceptible to collecting dust and soot than other examples. Another feature that made terracotta an attractive choice of material was that it could be produced more quickly and at a lower cost than a marble sculpture.

In 1446 Luca was commissioned to create a second lunette for the area above the door of the South Sacristy and for this undertook the production of the *Ascension*. This work has a benign quality, and the figures are dressed less like Romans, but more in the manner of contemporary Florentines. For this lunette, Luca again used the blue background with white figures, but experimented with additional coloration, adding a shade of green for the plants and another for the earth.

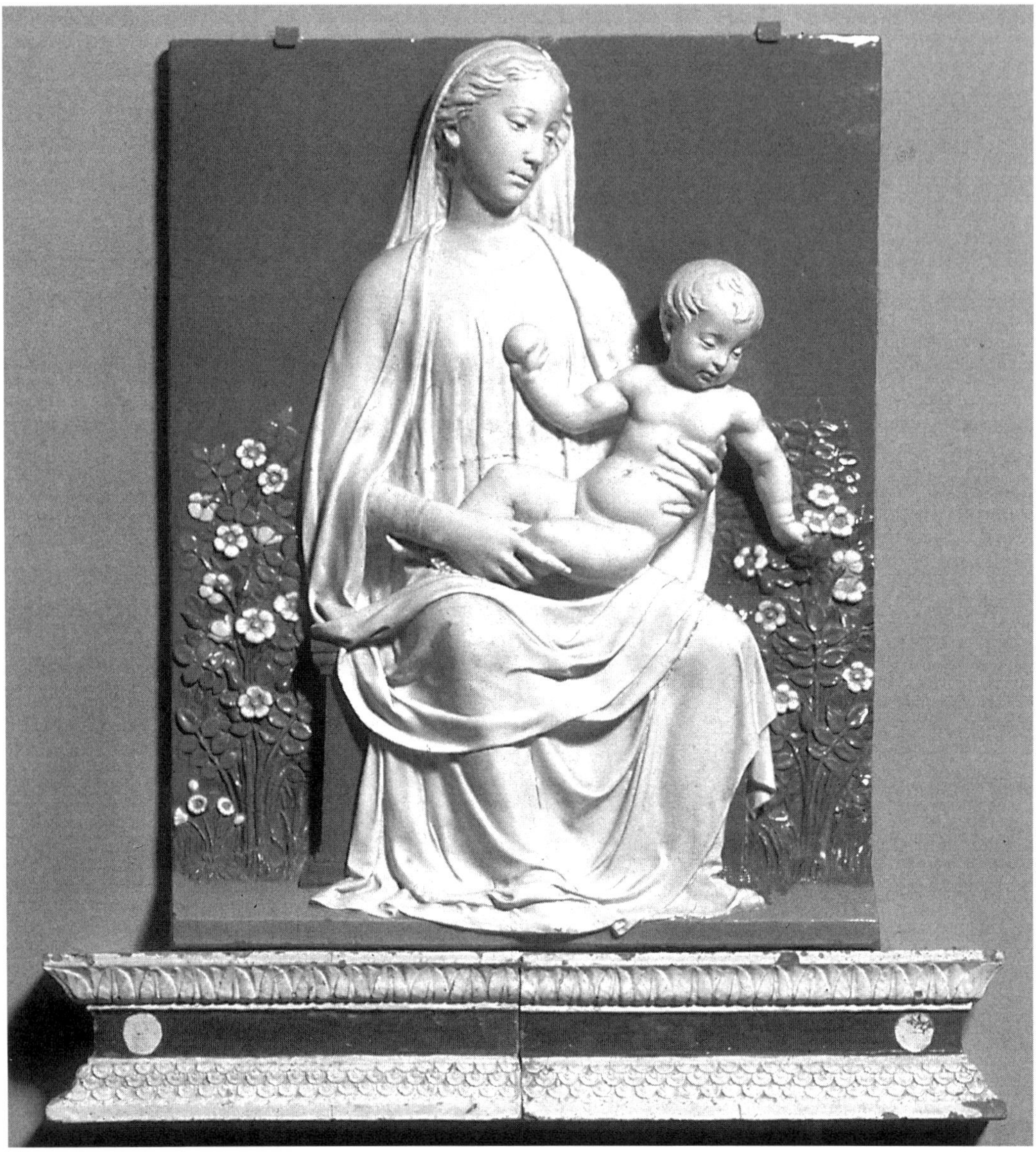

Figure 2.8: Luca della Robbia, *Madonna and Child in the Rose Garden* (late 1450s), terracotta sculpture, Bargello Museum (Alinari Archives, Florence, photograph by Daniela Cammilli, reproduced with the permission of Ministero per i beni e le Attivita Culturali [CAL-F-003744–0000]).

b. Bargello Museum

Working with his new innovation, Luca continued to produce numerous reliefs of the Virgin Mary. One of his most memorable is the *Madonna of the Apple*, which he executed in about 1460. Using only blue and white glazes, he created figures that are simple, beautiful, and realistic, with the Christ child appearing as a baby rather than the often-used "little man" figuration. Luca's *Madonna and Child in the Rose Garden* (fig. 2.8) exhibits somewhat less simplicity. In addition to the blue background and white figures, that work incorporates small amounts of green, yellow, and violet. Both of these Madonnas, as well as a third *Madonna and Child*, are on display in the Bargello Museum. The della Robbia workshop was extremely productive, and the resulting output can be seen in museums, churches, and buildings throughout Florence.

Luca's studio was eventually passed on to his nephew Andrea, who had worked alongside his uncle perfecting the family workshop's technique. Under Andrea's tutelage, the workshop's terracotta sculptures consisted of the same basic blue and white glazes used by Luca, but often were highlighted by additional colors. A particularly beautiful example, *Madonna of the Stonemasons* (1475–80) with its green plants and yellow flower centers, is exhibited in the Bargello Museum. Andrea also created the magnificent roundels for the porticos of several hospitals, including the Ospedale di San Paolo in Piazza Santa Maria Novella and the Ospedale degli Innocenti. For the Innocenti, each roundel portrays an infant to represent the orphaned or abandoned babies taken in and cared for by the facility (for more on that commission, see 4.3.2.2). After his death Andrea's oldest son, Giovanni, continued to operate the workshop in Florence, and many of his creations can also be seen in the Bargello Museum. Giovanni's works include elaborate compositions in terms of the numbers of figures, background decoration, and variety of colors. Giovanni's younger son, Girolamo, then took the technique to Paris where he was commissioned to decorate numerous buildings.

2.3.4.2 Contemporary Analyses

The formulation of the della Robbia glazes has long been of interest, and recent studies have used a variety of modern analytical techniques. A study by Zucchiatti et al. based on proton induced X-ray emission (PIXE) technology was employed to make a non-destructive analysis of inorganic elements, including the nature of the cobalt minerals responsible for the brilliant blues. This research has resulted in new information on the composition of the glazes and on the dating of terracotta sculptures. The researchers found that the della Robbia workshop changed the composition of the blue glazes in about 1520. Thus, the work by Luca and Andrea della Robbia can be distinguished from that of Andrea's sons by discerning the relative amounts of arsenic, bismuth, iron, and nickel, probably arising from the difference in treatment of the cobalt ores. The researchers were also able to define the basic glaze composition: "The Renaissance glazes are heterogeneous vitreous covers, coloured and opaque, prepared from a mixture of ingredients: a 'frit' made of river sand (the glass matrix) and potassium-rich wine dregs (a flux), the calcina (lead and tin burned powder, used both as a flux and as an opacifier) and 'pigments' containing metallic oxides."[19]

In another study, Zucchiatti et al. used energy-dispersive spectroscopy (EDS) via scanning electron microscopy (SEM) to determine the many compounds in the glaze oxides.[20] Surprisingly, in the blue glazes, the percentages of cobalt oxide (CoO) and ferric oxide (Fe_2O_3) vary widely, for example 17 to 53 percent for cobalt oxide and 1 to 58 percent for

19. Bouquillon et al., "PIXE Analysis of Artistic Blue Glazes."

20. Zucchiatti et al., "Elemental Analyses of a Group of Glazed Terracotta Angels."

ferric oxide, with cobalt being absent in the white glazes. Some of the other compounds, mostly oxides, identified were sodium oxide, magnesium oxide, aluminum oxide, silicon dioxide, sulfur trioxide, chloride, potassium oxide, calcium oxide, titanium dioxide, manganese oxide, nickel oxide, copper oxide, tin dioxide, and lead oxide.[21]

Using a different technique, micro-Raman spectroscopy, Sendova et al. analyzed the pigments used to create the different colors in order to determine the chemical components that made the della Robbia glazes unique. The researchers found the blue for which della Robbia glazes became known appears to be the result of a combination of cobalt and a sulfur ion (S_3^-) from a lead silicate matrix rather than an aluminosilicate matrix only, as expected from lapis lazuli. They attributed the yellow glaze to Naples yellow ($Pb_2Sb_2O_7$) and hematite (Fe_2O_3) micro-particles. Green was from a mixture of blue and yellow pigments; brown was derived from hematite and other components; and the white glaze contained tin oxide (cassiterite, SnO_2) as its primary opacifier.[22]

In other studies using EDS-SEM on della Robbia glazes, Tite found cobalt oxide only in the blues and greens; copper oxide was present in the blues, greens, and golden brown, but not the whites; manganese oxide was in trace amounts in golden browns only; the whites, blues, greens, and golden browns all contained significant percentages of silicon dioxide, lead oxide, and tin oxide; and other metal oxides were present in smaller amounts in all colors. A major conclusion of Tite's work was the identification of the della Robbia glaze components that distinguished their formulation from the *maiolica* glazes in pottery. In particular, Tite reported that the lead and tin contents of the della Robbia glazes were higher, while the potassium oxide/sodium oxide ratios were significantly lower than those of the *maiolica* glazes.[23]

2.3.5 Bronze

The use of bronze, an alloy of copper (ca. 90 percent) and tin (ca. 10 percent), has a long history dating back to 2500 to 3000 BCE—during the Bronze Age—in western and eastern Europe and in southern Asia. Evidence of bronze castings dating from about 1700 BCE has been found in China. Most of the early works in bronze were vases, small statues, and vessels used for worship, bells, tools, and, later, helmets and other armaments. During the Middle Ages in Europe it became increasingly common to create bronze pieces such as utensils, candlesticks, and vessels for household and religious uses, as well as castings for armaments, but it was in Renaissance Florence that bronze sculpture reached a level of artistic prominence. Most Florentine sculptures were executed using the lost-wax (from the French *cire-perdue*) method, a technology that had been in existence for centuries.

21. Zucchiatti et al., "'Della Robbia Blue.'"

22. Sendova et al., "Micro-Raman Spectroscopy," 663.

23. Tite, "Production Technology of Italian Maiolica," 2073–78.

2.3.5.1 The Lost-Wax Method of Bronze Casting

Exact descriptions of the lost-wax processes are somewhat difficult because explanations vary from source to source. Stone explains one reason for the apparent discrepancies: "The technical study of bronze casting in the Renaissance is as yet in its infancy, and considerable confusion is generated by what at best can be described as inconsistent terminology."[24] However, several variations of the lost-wax technique were used in Florence and can be described in general terms.

Florentines used the **direct method** of the lost-wax process well into the sixteenth century. In this technique, a model of the piece to be made was created in wax and rods were inserted into the model at various points. The whole work, with the exception of the tips of the rods, was then covered in clay or plaster and left to dry. When the resulting clay or plaster work, with its inner core of wax, was fired, the wax melted and drained out through the holes left by the melted rods (called spues). The result was a hollow clay mold in the desired shape with several holes or vents. After the bronze was liquefied (depending upon the copper/tin ratio, a temperature of at least 950°C [1,750°F] is required to melt bronze), the molten metal could be poured through one of the holes and allowed to fill the area previously occupied by the wax model. The heated air inside the mold was vented out through one of the other holes. Once the molten bronze cooled and hardened, the clay or plaster mold was broken away, revealing a bronze sculpture ready for refinement by the artist. The lost-wax method was a technically exacting procedure, and workshops often used the services of professional "casters" to aid in the process prior to the final steps. The finishing steps in the bronze process were closely analogous to those followed in sculptural carving. Artists spent considerable time finishing the bronze surface by chiseling, filing, incising, chasing or embossing, and polishing their work. In some instances, parts of the surface areas were gilded, covered with gold leaf or a combination of gold and mercury.

Artists began to use a second method, known as **indirect casting**, toward the end of the fifteenth century, first in northern Europe and somewhat later in Florence. The direct method resulted in the loss of the original wax model, never to be recovered if something should fail in the process or if future castings were desired. The indirect casting method eliminated that problem. Although the Greeks used a form of indirect casting in the seventh century BCE, it is not certain whether the Renaissance method was a revival of the Greek bronze work or a reinvention of a previous technology.

This process began with using clay or wax to create a model that was then carefully, precisely, and fully covered with sections of clay or plaster of Paris. After the sections were dry, they were disassembled and removed from the original model, thus preserving the original. The sections of clay or plaster of Paris were accurately reassembled and joined together to create a hollow clay/plaster of Paris mold that was filled with molten wax. After the wax hardened, the clay/plaster of Paris sections were removed, leaving a second, identical wax model to which rods could be attached. This process allowed the artist to continue with the

24. Stone, "Antico and the Development of Bronze Casting," 89.

technique utilized in the direct method while retaining the original model.

Another method, called **hollow casting**, allowed artists to make large works that were hollow shells. Small items such as medals, vessels, and statues could be made of solid bronze, but it was not practical for large pieces. The hollow casting technique was somewhat more exacting, but resulted in a creation that was lighter and required less bronze. There were several variations of the hollow casting technique, but basically it was accomplished by pouring wax into a plaster or clay mold, and allowing it to begin hardening. Some of the wax was poured out after the outer layer had hardened, thus leaving a thin layer inside the clay/plaster of Paris mold. More plaster of Paris was then poured into the hollow space. When the plaster of Paris had dried, the mold was heated and the melted wax flowed out of the mold through the holes left by the rods. This process created a mold with an outer and an inner portion, with an area between where the wax shell had been. Molten bronze was poured into that area, and once the bronze had hardened, the plaster of Paris was removed, leaving a bronze sculpture that was hollow rather than solid.

2.3.5.2 Illustrative Florentine Bronzes

Three outstanding Florentine artists who perfected the lost-wax method of bronze sculpture were Ghiberti (1378–1455), Donatello (ca. 1386–1466), and Cellini (1500–1571). Having been taught in his father's goldsmith workshop, Lorenzo Ghiberti's training provided excellent schooling in the techniques required for working with metals. He then apprenticed in the workshop of Bartoluccio de Michele, but the bubonic plague forced him to immigrate to Rimini for a short period where he was engaged in fresco painting. Returning to Florence a year later, he entered the Baptistery door competition.

a. Baptistery

In 1401, there was a competition to select an artist to create panels for the Baptistery doors. The two finalists were Ghiberti and rival Florentine artist Filippo Brunelleschi (1377–1446, see 6.3.1.2a for biographical details), chosen based on their submitted bronze competition panels, each depicting the *Sacrifice of Isaac* (figs. 2.9 and 2.10). Both panels, exhibited in the Bargello Museum, show Abraham, ordered by God to sacrifice his son, being stopped in the act by an intervening angel. After the judges from the Arte del Calimala, the leading merchants' guild, which was responsible for maintaining and adorning the Baptistery, studied the submissions, Ghiberti was awarded the commission in 1403.

Although both finalists' works were worthy, and the decision was by all accounts difficult, Ghiberti's entry, with its aura of movement and realism, was evidently deemed to be the more deserving work. The manner of execution needed for the project clearly favored Ghiberti's panel, which was made in a single piece, unlike Brunelleschi's. Ghiberti's panel was lighter, less time-consuming to produce, and less costly. Asked to craft the doors with panels depicting scenes from the New Testament, Ghiberti spent twenty-one years on the project, completing the lengthy molding and finishing process in 1424 when the doors were installed on the Baptistery's east side. One account claims that the judges had decided

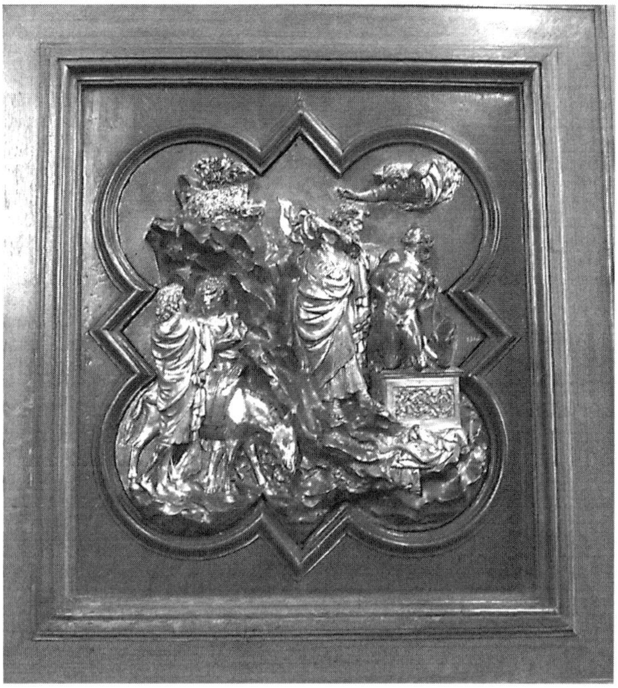

Figure 2.9: Lorenzo Ghiberti, *Sacrifice of Isaac* (1401–ca. 1403), cast bronze panel for the competition, Bargello Museum (Web Gallery of Art, retrieved from Wikimedia Commons).

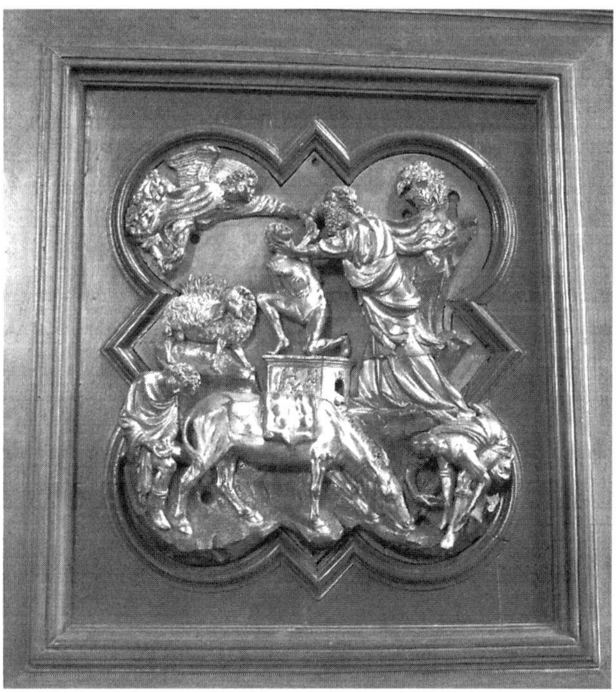

Figure 2.10: Filippo Brunelleschi, *Sacrifice of Isaac* (1401–ca. 1403), cast bronze panel for the competition, Bargello Museum (photograph by Richard Fabi, retrieved from Wikimedia Commons, GNU Free Documentation License).

Ghiberti's and Brunelleschi's panels were equally artistic and granted the commission to both, expecting them to work together on the project. Refusing to share the work with Ghiberti, Brunelleschi left Florence with his friend Donatello to study the ruins in Rome.[25]

In 1425, after realizing the impact of the finished project, the Arte del Calimala commissioned Ghiberti to cast a second set based on Old Testament narratives. Those doors, made from about 1429 to 1452, were so stunning that, according to Vasari, Michelangelo declared them a paragon of perfection appropriate for the *Gates of Paradise*, a designation that over the centuries has become synonymous with the work. Equally impressed, the Arte del Calimala felt the extraordinary design and artistry dictated that their placement should rightfully be on the east side of the Baptistery in order to face the main entrance to the Duomo. Thus, Ghiberti's New Testament doors were painstakingly moved to the north portal, and the Old Testament depictions were placed on the east side for all to see as they departed the cathedral.

The exact lost-wax process followed by Ghiberti is not known, but most probably he started with sketches and translated them to wax models. There is some disagreement as to whether he modeled his panels directly or used an indirect wax casting. What is known about his execution of the east doors is that each panel was made as a single piece, as had been the one he submitted for the competition. Ghiberti's method was clearly more economical and less labor-intensive than Brunelleschi's. Following completion of the wax model, rods were placed on the back before covering the entire relief with a fire resistant material (clay or plaster) that was allowed to dry. As described above, heating the relief melted the wax and left a space into which molten bronze was poured through the rod-created holes. Interestingly, the channels (sprues) left by the rods can still be seen on the back of the panels. After the bronze hardened, the clay was cut away, leaving the difficult and time-consuming tasks of chasing, chiseling, filing, incising, and polishing the newly cast bronze. At the end of the finishing process, Ghiberti gilded each panel with a mixture of mercury and gold dust, then burned off the mercury, thus leaving the gold attached to the panel. The ten panels were surrounded by a frame with gilded bronze figures and high-relief busts, one of which is claimed to be the head of Ghiberti. The resulting bronze doors have a truly magnificent aesthetic and rank as one of Renaissance Florence's finest examples of melding art and technology. Copies of the doors are now installed in the Baptistery portal and the original panels have been placed in the Duomo Museum. For a description of the panels in terms of content and artistry, see chapter 3 (3.1.2.3b).

The Baptistry doors were studied and analyzed during many years of restoration at the Metal Conservation Department of the Opificio delle Pietre Dure after the 1966 flood in Florence. In order to determine Ghiberti's process for creating the doors, better assess the cleaning techniques, and establish the best strategies for exhibiting the doors, researchers initiated a non-invasive study of the structure and the composition of the subsurface. The investigation used prompt gamma-ray activation imaging combined with neutron radiography (PGAI/NR). In the process of the study on the east doors, the neutron beam was

25. For a nice description of the competition, see Walker, *Feud That Sparked the Renaissance*, 17–25.

angled to a depth of about one milimeter in order to ascertain the composition of an east high-relief bust and a north bust. For the east bust, there seemed to have been an extended remelting from the flat base to the front of the relief. This finding suggests that the punctilious Ghiberti was not satisfied with the original melting and therefore repeated the procedure. The process of pouring the remelted bronze inside the original space resulted in more metal accumulating in the front section of the head and much less at the nape. The elements identified in the east bust proved to be copper, gold, mercury, and chloride. A comparable study of the north bust indicated the presence of the same elements, as well as potassium. The finding of chloride, in particular, is an indication of corrosion from the environmental exposure.[26] The copper used in the production of bronze, as well as the gold and mercury used for the gilding, are as expected. The information concerning the remelting is an interesting indication that Ghiberti experienced many of the same problems that plagued other workshops and had to make several attempts to melt the bronze satisfactorily. It was not a straightforward process, but rather one that was technically intricate and often beset with unforeseen complications.

b. San Lorenzo

The church of San Lorenzo contains two bronze pulpits with elaborate relief panels created by Donatello. The two pulpits are known as the Resurrection Pulpit and the Passion Pulpit. While Ghiberti was at work on the first set of doors for the Baptistery, the Florentine Donatello may have been an assistant in his workshop, probably from about 1404 until 1407, most likely exposing him to bronze casting techniques. Donatello had begun his career as an artist sculpting in marble, but soon extended his expertise to include works in wood, clay, and bronze. He accepted commissions in Prato, Pisa, Siena, and Padua, as well as Florence. It was while working on a commission for Cosimo de' Medici to create these bronze reliefs depicting scenes from the life of Christ that Donatello died in 1466. Donatello's creations distinguished him as one of the most significant sculptors of his time. Praised for his innovative reliefs as well as his large free-standing works, Donatello became a master at creating sculptures using pioneering techniques. He was given the distinct honor of being buried in San Lorenzo, next to his patron and dear friend Cosimo de' Medici.

c. Orsanmichele

Saint Louis of Toulouse, created for a niche on the façade of Orsanmichele (the original now in the Museum at Santa Croce), was probably the first large bronze attempted by Donatello, and as such was a major undertaking and accomplishment. Technically, the statue was a challenge. It was commissioned in about 1422 by the Guelf party, which in the early 1400s was a wealthy organization with more prestige than power. Nonetheless, the Guelfs had been granted a niche to adorn in the Orsanmichele, along with the Florentine guilds. Saint

26. Festa et al., "Nondestructive Stratigraphic and Radiographic Neutron Study."

Louis was the patron saint of the Guelfs, and they wished to have a bronze statue more grandiose than those commissioned by the guilds.

Donatello chose to cast *Saint Louis of Toulouse* in separate pieces rather than a single work. He adopted the sectional method because he wished to gild the statue, and the procedure had never been used for a large bronze sculpture. The gilding process was both dangerous and taxing. Ghiberti had used this method on the door panels for the Baptistery, but his panels were flat, making it somewhat less complicated than Donatello's curved sections that had to be bolted together to form a unified sculpture. Like Ghiberti, Donatello combined gold dust with mercury and applied the mixture to the bronze, heating it over coals to make a permanent bond with the bronze sections. The necessary intense heat created fumes from the mercury that were hazardous to breathe, and the shape of the bronze sections could be somewhat altered by the roasting. It was an intricate technological balance that Donatello executed extremely well, resulting in a splendid work with fine details, notably the way the vestments were expertly given the appearance of having been draped.[27]

d. Bargello Museum

Possibly cast in the mid-1440s, Donatello's *David* (fig. 2.11) is believed to be the first freestanding nude to be sculpted since antiquity, a remarkable achievement. As no documents exist regarding the commissioning of the work, the date, or the location for which it was originally intended, the history of the sculpture is as enigmatic as the statue itself. It is known that at some point the statue was placed in the courtyard of the Palazzo Medici before it was moved to the Bargello Museum. With its sensuous, youthful body, Donatello's *David* is nude except for a hat and leather leggings, and stands with his left leg on the head of Goliath. There are no extant records of the procedures for casting; however, when *David* was undergoing restoration (2007 to 2008) interesting facts about the original finishing of the statue came to light. In the eighteenth century, a thick layer of wax and oils had been applied to the *David* to darken the bronze so that its shade more closely matched other statues being displayed in the Uffizi Gallery. Restorers using a medical-grade laser to clean the bronze verified that not only had *David* originally been lighter in color, but that Donatello had finished the newly hardened bronze by adding gold leaf to decorate parts of the sculpture.

e. Loggia dei Lanzia (Loggia della Piazza Signoria)

A third master of the bronze technique was Benvenuto Cellini (1500–1571). His father was a mason who taught Cellini to play the flute at a young age with the hope that he would become a part-time musician and part-time mason. Cellini, however, preferred drawing, and in 1515 his father finally agreed that he could pursue a career in art and apprenticed him to a goldsmith, Antonio di Sandro. Between 1519 and 1545, Cellini received commissions in

27. Andres, Hunisak, and Turner, *Art of Florence*, 1:398–99.

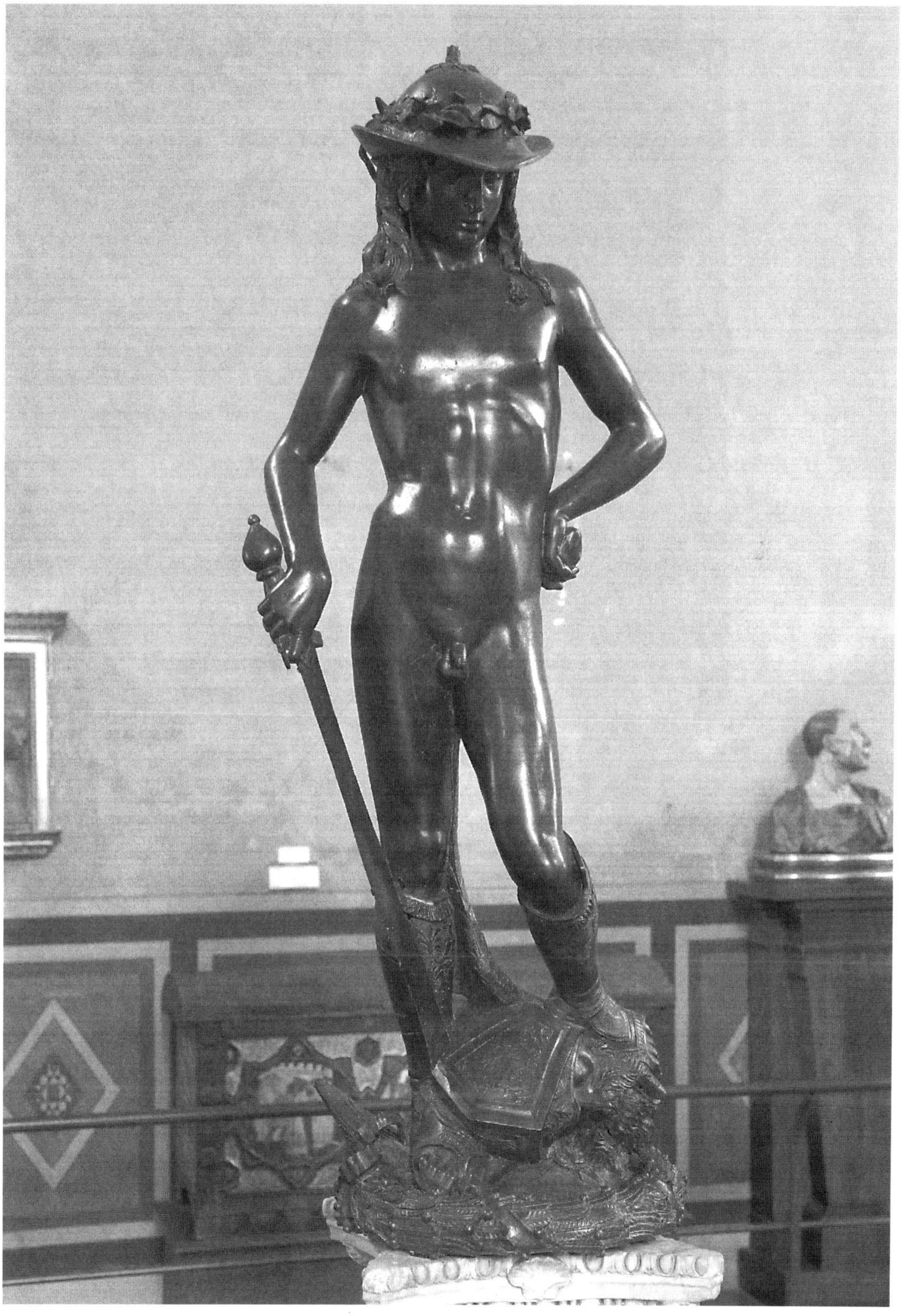

Figure 2.11: Donatello, *David* (ca. mid-1440s), bronze sculpture, Bargello Museum (photograph by Rufus46, retrieved from Wikimedia Commons, Creative Commons License).

Rome, Florence, Venice, and France. When Cosimo I offered him the opportunity to create the bronze *Perseus* in 1545, Cellini returned permanently to Florence.

From his autobiography, written between 1558 and 1566, we know much about the casting of the *Perseus with the Head of Medusa* (fig. 2.12), which spanned the years from 1545 to 1554. The sculpture was created using the lost-wax method. The process was complicated to execute because Cellini's design proposed to depict Perseus holding the head of Medusa aloft in his outstretched left hand. Cellini first cast the body of Medusa, which was to be writhing under Perseus's feet. His autobiography shows that he was committed to using the best clay possible and to taking great care in the casting process: "I was determined to make use of all the skill and experience I had acquired in order to avoid any error. As a result the first cast I made in my little furnace came out superlatively well."[28]

Cellini wrote that he hoped the casting of the figure of Perseus would go as smoothly as it had with the figure of Medusa. He was pleased with the figure's wax model, but the firing of the covering clay figure was fraught with difficulties. The technology of lost wax casting more often than not presented unforeseen challenges. The remainder of the Cellini narrative regarding the creation of *Perseus* vividly attests to the problems he encountered.

> When all the wax was gone and the mold well baked, I at once began to dig the pit in which to bury it, observing all the rules that my art demands . . . . I had it [the mold] filled with a great many blocks of copper and other bronze scraps . . . so piled up that the flames would be able to play through them, heat the metal more quickly, and melt it down. Then, very excitedly, I ordered the furnace to be set alight . . . what with the greasy resin from the wood and the excellence of my furnace, everything went so merrily that I was soon rushing from one side to another, exerting myself so much that I became worn out . . . the workshop caught fire and we were terrified that the roof might fall in on us, and at the same time the furnace began to cool off because of the rain and wind that swept in at me from the garden. I struggled with these infuriating accidents for several hours, but the strain was more than even my strong constitution could bear, and I was suddenly attacked by a bout of fever—the fiercest you can possibly imagine.[29]

Cellini reported that he took to his bed after instructing his assistants in the procedures they should follow. After several hours, the melting process was not going well and one of his assistants came to tell Cellini that the work was ruined, insisting, "Look here, Benvenuto, what you want done is beyond the powers of art. It's simply impossible." Enraged, the ardent Cellini rose from his bed and found that the metal had indeed "curdled" and "caked." Cellini then decided to restart the furnace and burn oak to reinitiate the process. Before long, the metal began to melt and Cellini reported, "By this means, and by piling on the fuel and stirring with pokers and iron bars, the metal soon became

28. Cellini, *Autobiography*, trans. Bull, 329.

29. Ibid.

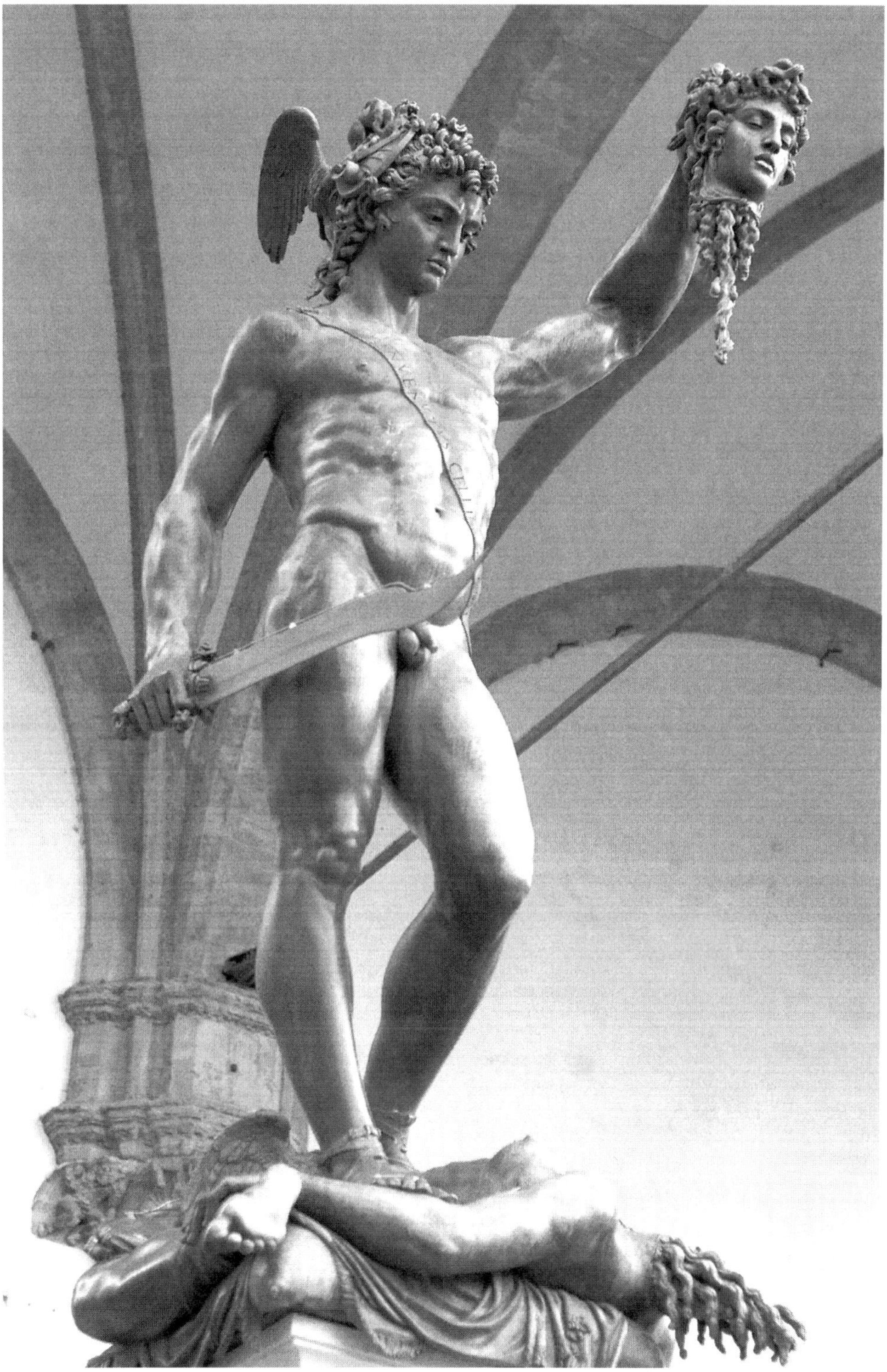

Figure 2.12: Giovanni Cellini, *Perseus with the Head of Medusa* (1545–54), bronze sculpture, Loggia della Signoria (photograph by Marie-Lan Nguyen, retrieved from Wikimedia Commons, Creative Commons License).

molten." But the work did not progress as hoped: "At this point there was a sudden explosion and a tremendous flash of fire, as if a thunderbolt had been hurled in our midst . . . [I] then realized that the cover of the furnace had cracked open and that the bronze was pouring out." After Cellini addressed that problem and stabilized the situation, the bronze eventually melted sufficiently and Cellini was able to fill the mold. After leaving the cast to cool for two days, Cellini very carefully uncovered the casting and declared, "The first thing that I found was the head of Medusa which had come out beautifully because of the air vents, just as I had said to the duke that the nature of fire was to ascend. Then I began uncovering the rest, and came to the other head—that is the head of Perseus—which had also succeeded beautifully."[30]

The stunning bronze, *Perseus with the Head of Medusa*, commissioned for the Loggia della Signoria in Florence, can be viewed at that site. The original carved base with small bronze statues is presently located in the Bargello Museum.

2.4 Glass

The process of glassmaking is known to have existed in Mesopotamia and Egypt as early as the fourth millennium BCE and was later undertaken by the Greeks and Romans. During the time of the Roman Empire, the early steps in the glassmaking process were done in Egypt and the Levant, where the raw materials were combined and formed into blocks that were transported to secondary furnaces to be melted and finished into the desired objects. Later, the process for glassmaking entered Italy, probably from Alexandria, and expertise in glass artistry reached outstanding levels in Venice. In order to prevent the spread of furnace fires to neighboring buildings in Venice proper, the city mandated in 1271 that all glassmaking be moved to the nearby island of Murano. This move also enabled the Venetians to maintain close control over this vital and lucrative industry of glassblowers and craftsmen who excelled in creating utilitarian objects as well as magnificent works of art. Beginning in the thirteenth century, Florence also developed its own glass industry, although theirs was less innovative than that practiced in Venice.

2.4.1 Composition

The major constituent of glass is silicon dioxide (SiO_2), which is obtained from quartz or sand that commonly contains a variety of impurities. Of those impurities, iron, in the form of the oxide Fe_2O_3, is particularly undesirable since it imparts a greenish hue to the product. The early glassmakers learned that adding manganese, often in the form of the mineral oxide pyrolusite (MnO_2), would decolorize the final product, yielding a clear glass or glass with a slightly yellowish hue. High temperatures, approaching or exceeding 2,000°C, are required to effectively melt quartz and high-quality sand. This process of vitrification is followed by cooling the molten form of silicon dioxide to make a non-crystalline solid.

30. Ibid.. Bull, 348–51.

In view of the high temperatures needed for vitrification, other compounds were generally mixed with these starting materials to achieve a lower temperature at which melting (vitrification) occurs, specialized properties of the glass, and/or various colors. A common additive was potash (potassium carbonate, K_2CO_3), which was obtained by leaching plant ashes and then evaporating the liquid to yield a white residue. Two types of ash were frequently used: Levantine ash, imported from the eastern Mediterranean in an area that is now Lebanon and Syria, and Barilla ash, imported from the western Mediterranean including regions of Alicante, Spain, Sicily, and Sardinia. Levantine ash was obtained by burning the plants from the Amaranthaceae family, including, to name but a few, *Salsola jordanicola*, *Salsora soda*, *Salsola vermiculata*, *Arthrocnemum strobilaceum*, *Anabasis syriaca*, and the *Halopeplis*, *Chenopodium*, and *Salicornia* genera, which contain 1.5 to 4 percent potassium oxide (K_2O) and 6 to 11 percent calcium oxide (CaO). Possessing a higher potassium content than Levantine ash, the regional and thus less expensive Barilla ash was derived from several of the same plant species but found growing in different soil conditions than Levantine ash, including those from the *Salsola*, *Chenopodium*, and *Salicornia* genera.[31] These chemical differences make it possible for modern researchers to analyze Florentine glass and determine the origin of the ash that was used.

2.4.2 Glassmaking in Florence

In the Florentine territory, most of the glassmaking was concentrated in the Valdese, including the towns of Colle di Esta, San Gimignano, Gambassi, and others. This region of Italy was blessed with abundant oaks used to fuel the furnaces, while sand was transported to the sites from the Arno and the Pisan beaches. With the import of Venetian glassmakers into the Valdese, the industry developed rapidly and was able to supply glass in various forms for artistic needs, windows, eyeglasses, and scientific uses, such as magnifying lenses. Toward the end of the fifteenth century, glassmaking was taking place in Florence proper. A painting by G. M. Butteri in the Studiolo of Francesco I in the Palazzo Vecchio depicts a glasshouse that opened in 1569, and in 1612, the first treatise on glassmaking was published. The manual, written by Antonio Neri, a Florentine abbot and an accomplished glassmaker, alchemist, and herbalist, explained and discussed all phases of working with glass.

In the early 2000s, researchers used a variety of analytical, physicochemical, and microscopic techniques to analyze raw materials, frits, crucibles, working wastes, and glass artifacts, to investigate seven kilns in three medieval and Renaissance glassmaking sites in the Valdesa. Their comprehensive study provided considerable information concerning the nature of the materials used for glass production. As expected, silicon dioxide was the major constituent of the starting materials and the resulting glasses contained about 50 to 60 percent silicon dioxide, while many different oxides of aluminum, lead, magnesium, iron, calcium, sodium, potassium, titanium, and others were also present. The researchers also determined the nature of the chromophores used to produce various colors such as green, blue, and yellow.[32]

31. Barkoudah and Henderson, "Plant Ashes from Syria and the Manufacture of Ancient Glass."

32. Casellato et al., "Medieval and Renaissance Glass Technology"; Bianchin et al., "Medieval and Renaissance Glass Tech-

Other researchers used modern analytical technology, including scanning electron microscopy with an energy dispersive X-ray detector (SEM-EDX) and inductive coupled plasma-mass spectrometry with sampling accomplished by a laser ablation system (LA-ICP-MS), to determine the materials used in glassmaking. This was done by measuring the amounts of silicon dioxide, aluminum oxide (Al_2O_3), phosphorus pentoxide (P_2O_5), iron III oxide (Fe_2O_3), manganese oxide (MnO), calcium oxide (CaO), sodium oxide (Na_2O), potassium oxide (K_2O), chloride (Cl), and trace elements in Tuscan vessel glasses. Compositional analysis identified three groups of glasses, labeled groups I, II, and III. All groups were found to contain 60 to 65 percent silicon dioxide and about 1 percent manganese oxide; the latter was probably added in the form of the mineral pyrolusite for decolorization.

The researchers postulated that glasses from group I had been made with fluxing material from a potassic ash obtained from the western Mediterranean Barilla ash and those from group II with fluxing material from a sodic Levantine ash from the eastern Mediterranean, similar to that of certain Venetian glasses. The glasses in group III were made from a sodic ash that may have been purified before use or arose from using remelted glass. This last set of glasses was similar in some respects to Venetian glass. Moreover, a distinction could be made between a local Tuscan source of silica and silica obtained from river pebbles from other areas. The relative proportions of trace elements, e.g., arsenic (As), zirconium (Zr), zinc (Zn), and others, provide valuable evidence for the use of different ashes, reflecting distinct purification treatments and local soil geochemistry. The respective potassium oxide and sodium oxide contents were about 6 percent and 14 percent for group I, 3 percent and 14 percent for group II, and 3 percent and 18 percent for group III. In addition to silicon dioxide, all the glasses had the following approximate contents of other components: 3 to 4 percent aluminum oxide, 0.5 percent phosphorus pentoxide, 0.5 to 0.9 percent chloride, and 0.1 to 0.2 percent sulfur trioxide. Calcium oxide was present at about 9 percent in groups I and II, and at 5 percent in group III. Magnesium oxide was present at 3 to 4 percent in groups I and II and at 2 percent in group III; iron III oxide varied from 0.7 to 1.4 percent in all three groups.[33] These chemical signatures, obtained centuries after the initial glassmaking, provide definitive information on the types and origins of starting materials in the manufacture of glass in Tuscany.

2.4.3 Mosaics

Mosaic art, prevalent in the Byzantine and Hellenistic empires, was brought to Italy by the Greeks in the first century. The Romans used mosaics to enhance the beauty of their villas, and later Christians adopted mosaics to decorate their churches, spreading the mosaic tradition northward. The Venetian artists working with mosaics were influenced by the Byzantine artistic styles, as can be seen in the mosaics of Saint Mark's Basilica.

nology, Pt. 2"; Bianchin et al., "Medieval and Renaissance Glass Technology, Pt. 3"; Brianese et al., "Medieval and Renaissance Glass Technology, Pt. 4."

33. Cagno, Janssens, and Mendera, "Compositional Analysis of Tuscan Glass Samples"; Cagno et al., "Raw Materials for Medieval to Post-Medieval Tuscan Glassmaking."

The art of converting a drawing to a mosaic was an exacting skill requiring considerable expertise. After the artist drew a cartoon or preliminary sketch, the mosaicist worked from that sketch to assemble small pieces of colored glass, called *tesserae*, and began the painstaking process of placing them in a base of plaster following the color palette specified by the artist. The mosaicist could achieve a glistening effect by inserting a layer of silver or gold leaf between layers of the glass and also by angling the *tesserae*.

2.4.3.1 An Illustrative Florentine Work

Perhaps the most noteworthy example in Florence is the arresting interior of the dome of the Baptistery, which is magnificently covered in mosaics (fig. 2.13). The project was started in about 1225 under the leadership of the Franciscan Jacopo da Torrita and took more than a century to complete. Although the identities of the mosaicists are not known, the work was probably, judging from the Byzantine style, executed in part by craftsmen from Venice. The artists Cimabue, Coppo di Marcovaldo, and Giotto are also said to have been involved. With a diameter of about 26 meters, the dome has a series of eight concentric registers in which mosaics are set in a gold background. A heavenly host of angels adorns the top, and the area beneath includes depictions of the Last Judgment, a 7.9-meter-high figure of Christ in majesty and the Virgin Mary with apostles and saints, episodes from Genesis, and scenes from the lives of Mary, Joseph, and Saint John the Baptist. The intent of the mosaics is to convey the Christian doctrine beginning with the Creation and continuing through Last Judgment, with the patron saint of Florence, the venerated John the Baptist, highlighted throughout. The iridescent effect makes the mosaics resplendent in their presentation.[34]

2.4.3.2 Contemporary Scientific Studies

A recent study analyzed glass *tesserae* from the Nehemiah panel in the Baptistery, which were obtained during restoration work by the Opificio della Pietre Dure. Researchers used a variety of techniques, including electron microprobe analysis, X-ray diffraction, scanning electron microscopy, and fiber optic reflectance spectrometry in the visible-near infrared region to study twenty-three samples of glass *tesserae*. Most of the samples had a silicon dioxide content of about 60 percent and could be divided into sodium-rich and potassium-rich with sodium oxide contents of 9 to 14 percent and potassium oxide contents of 13 to 25 percent. These differences were attributed to the use of plant ashes: the sodium-rich *tesserae* used ashes from halophylic (salt tolerant) plants of the genus *Salicornia* or *Salsola* (or perhaps ferns), while the potassium-rich *tesserae* may reflect a somewhat later trend of using wood ash as a source of flux. The different colors in the *tesserae* were associated with varying levels of specific metallic oxides such as iron, manganese, copper,

34. Wirtz, *Art and Architecture, Florence*, 48.

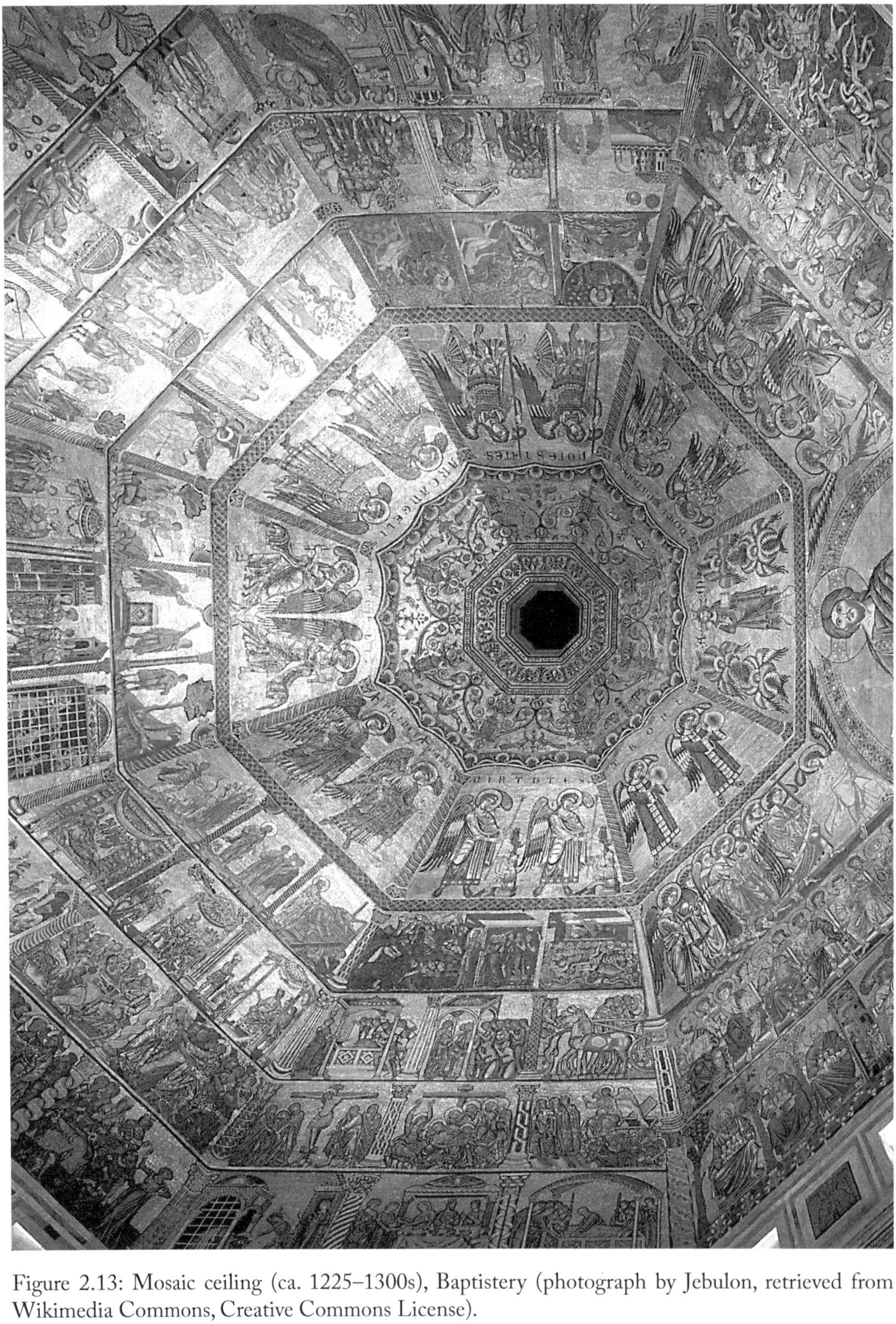

Figure 2.13: Mosaic ceiling (ca. 1225–1300s), Baptistery (photograph by Jebulon, retrieved from Wikimedia Commons, Creative Commons License).

cobalt, and others.[35] Investigations such as this and others continue to provide information on the chemical nature of the glass, the raw ingredients used, and the sites and techniques of production.

In another study, using glass mosaic *tesserae* from the Grotta del Buontalenti of the Boboli Gardens (behind the Pitti Palace), researchers were able to identify the composition of the glass and, in addition, provide information on periods of restoration and the materials used in the restoration process. Employing different analytical techniques, including various forms of spectroscopy and scanning electron microscopy, researchers found that, as expected, silicon dioxide was the major constituent averaging about 60 percent, although its content ranged from approximately 40 percent to one sample having 98 percent. Of the glass *tesserae* attributed to the Medicean period, the mean silicon dioxide content was 53 percent, and sodium oxide and lead IV oxide (PbO) were each present at 10.5 percent. Other components included calcium oxide, potassium oxide, tin II oxide (SnO_2), aluminum oxide, copper II oxide (CuO), manganese oxide (MnO), and iron III oxide, each present at 1 to 4 percent, and antimony III oxide (Sb_2O_3) and sulfur trioxide (SO_3), each present at less than 1 percent. The *tesserae* were probably fluxed with plant ash, as in the Venetian tradition, and the content of the manganese and iron suggest they were intentionally added to the glass melt. The presence of manganese is consistent with the fact that the early glassmakers added manganese (e.g., as manganese III oxide, Mn_2O_3) to decolorize glass, particularly to remove colors resulting from iron contaminants. On the other hand, manganese and iron could also be used to obtain intense dark colors.[36] From early glassmakers, the craftsmen of Florence had inherited considerable knowledge on which to build. It is intriguing to consider how modern glass chemistry and analytical techniques can be used to explain the materials and production techniques that were empirically discovered centuries or even millennia ago.

2.4.4 Stained Glass

By the tenth century, artisans were using stained glass in churches in England, France, and Germany. Like mosaics, stained glass was not a preeminent form of artistic expression in Renaissance Florence, but it did gain some popularity in chapel design, especially in the high altar or choir chapel, private chapels, and special purpose chapels. The methodology for creating stained glass windows has changed little over time, although the content of the windows themselves has tended to become more oriented toward pictorial composition and less concentrated on pure design. Interestingly, during the Renaissance, glass windows were sometimes painted white rather than featuring colored glass.

35. Arletti et al., "Florence Baptistery."

36. Costagliola et al., "Mineralogical and Chemical Characterisation."

2.4.4.1 The Creative Process

As in the process for creating a mosaic, an artist started a stained glass design by drawing a cartoon that was used as the blueprint for the window decoration. A glazer used the full-size cartoon to size and cut the pieces from sheets of colored blown glass; then the glazer joined the pieces of glass together with lead according to the color pattern of the cartoon. After the window was completed, the artists often used a monochrome dark gray pigment (*grisaille*) to paint any desired shadows or lines. The *grisaille* was then fused onto the glass with high heat. Paoletti and Radke described the impact of stained glass windows when placed in a church: "the heavenly light reached the viewer through the saintly figures depicted on the stained glass windows, figures who were literally transmitters of the divine to humankind through their own bodies."[37]

2.4.4.2 Examples of Florentine Stained Glass

a. Duomo

The stained glass windows in the Duomo were designed by prominent Florentine artists, including Donatello, Ghiberti, Uccello, and del Castagno, and executed by master glassmakers. Windows line the nave and eight glass roundels adorn the drum of the dome (for details, see 6.3.1.7). Entering the cathedral, the first window on the left is *Santo Stefano and Two Angels* (1405–15), created by stained glassmaker Niccolò di Pietro from a cartoon by Ghiberti. Ghiberti and Niccolò di Pietro also decorated the window beneath the vault of the central nave with *Assumption of the Virgin* (fig. 2.14) and created the stained glass window on the right side of the nave, *San Lorenzo and Angels*. All three were done between 1405 and 1415. The window in the fourth bay on the right side, *Saints* (1395), was designed by Agnolo Gaddi and made by Antonio de Pisa. The cathedral transept has three tribunes, each with five chapels illuminated by stained glass windows. The cartoons for these windows are mainly credited to Ghiberti, and the windows themselves were made by several different master glassmakers between 1439 and 1442. The double windows in the tribunes depict the saints and the prophets. When stained glass windows were commissioned, the glassmakers—not the artists designing the cartoons—were hired first, demonstrating the importance placed on their skills.[38]

b. Santa Maria Novella

The church of Santa Maria Novella has some interesting stained glass, for example the large window in the façade designed by Andrea di Bonaiuto da Firenze and made in 1365. He may also have prepared some of the glass in the center of the window, a section known as the *Coronation of the Virgin*. This window is discussed further in chapter 5 (see 5.2.3).

37. Paoletti and Radke, *Art in Renaissance Italy*, 28.

38. Di Cagno, *The Cathedral*, 74–80.

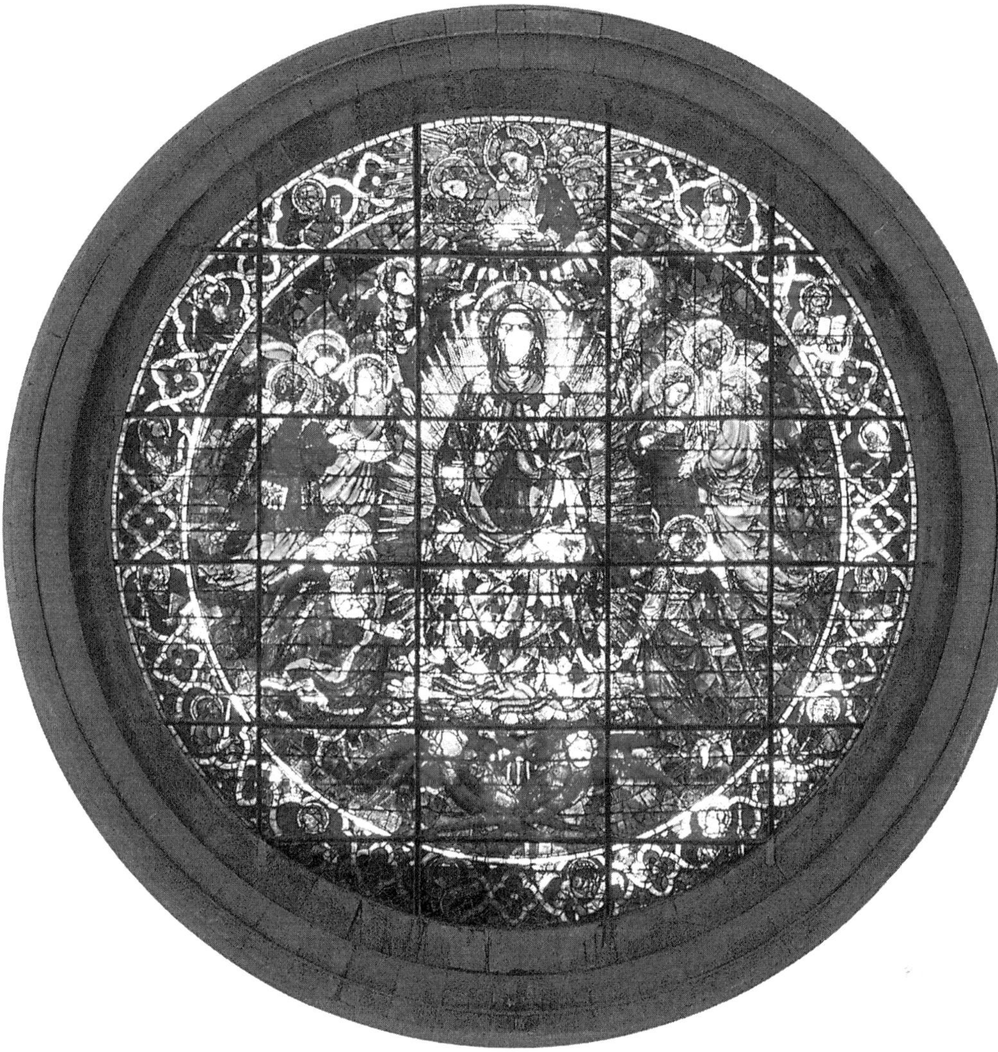

Figure 2.14: Lorenzo Ghiberti and Niccolò di Pietro, *Assumption of the Virgin* (1405–15), stained glass window, Duomo (Alinari Archives, Florence, photograph by Nicolò Orsi Battaglini, reproduced with the permission of Ministero per i beni e le Attivita Culturali [OBN-F-000349-0000]).

The Cappella Maggiore in Santa Maria Novella contains a three-part stained glass window designed by Domenico Ghirlandaio and completed in 1491 by Alessandro Agolanti. The left portion of the window includes sections titled the *Coat of Arms of the Tornabuoni Family*, *Saint Dominic*, *Saint John the Baptist*, and *Saint Peter the Apostle*. The *Foundation of the Basilica of Santa Maria Maggiore in Rome*, *Circumcision of Christ*, and *Our Lady of Assumption Handing the Girdle to Saint Thomas the Apostle* are shown in the center. To the right, there is a second set of sections called *Coat of Arms of the Tornabuoni Family*, *Saint Thomas Aquinas*, *Saint Lawrence*, and *Saint Paul the Apostle*.

The Strozzi family commissioned a chapel in Santa Maria Novella dedicated to Saint Thomas Aquinas that was begun in about 1337 and completed in 1350. The walls of the chapel were frescoed by Nardo di Cione (1320–65/66) and his brother Andrea (ca. 1308–68). Nardo di Cione also designed the stained glass window located at the end of the chapel illustrating the Madonna and Child and Saint Thomas Aquinas.

2.5 Semi-Precious Hardstone

Art created using semi-precious hardstone (*pietre dure*) developed very early in Europe and Asia, and was refined during the Renaissance. By the fourth century BCE, the Greeks were creating mosaic art using small colored stones. Ancient Romans were adept at executing beautiful marble wall embellishments using colored stones, and their technique spread as the Roman Empire grew. Byzantine and Islamic artists perfected the art of inlaying stone on walls and ceilings and used stone as well as glass and ceramics in their mosaics. The art form was revived in Renaissance Florence where artisans of *pietre dure* were able to achieve a greater degree of finesse and precision than their predecessors.

The Medici family had a history of being enamored with *pietre dure*. Lorenzo the Magnificent (1449–92), continuing an interest of his father Piero (1416–69), amassed a collection of objects containing semi-precious stones, and Cosimo I, Grand Duke of Tuscany (1519–74), had a particular fondness for Egyptian red porphyry and Roman marble. Cosimo I and his son Francesco I (1541–87) were responsible for bringing artisans to Florence adept at stoneworking, including carvers, engravers, and stonecutters. In 1588 Francesco's brother Ferdinando I de' Medici (1549–1609), organized and brought leadership to the artistic workshops in the Uffizi started by Francesco I.

The Opificio delle Pietre Dure became a reality, and under Ferdinando's direction the workshop began to specialize in crafting semi-precious stone into beautiful works of art, including vases, cameos, furniture inlays, and most superbly, landscapes and portraits. Anything that could be painted could be produced in *pietre dure* with all of the varying shades of color, shadowing, and light. The Opificio (workshop) developed an extraordinary level of technical expertise that was soon recognized throughout Europe and continued for three centuries. It was not until the Kingdom of Italy was founded in 1861, when the workshop could no longer find the clientele and patrons willing or able to support this costly form of art, that the direction of the Opificio delle Pietre Dure changed. Today it is a recognized center for art preservation and restoration.

2.5.1 The Technology

2.5.1.1 Opificio delle Pietre Dure

It was the Opificio delle Pietre Dure's mosaic technique that gave the workshop its reputation for craftsmanship and technical skill. Artists first produced a basic design, which may have been in the form of either a water-based or oil painting on canvas, as with a portrait of Cosimo I commissioned by Ferdinando I. The original design was then traced and transposed by the workshop craftsmen into the chromatic color scheme to be reproduced in stone. The Medici workshop had collected stones from many parts of the world, and so a wide variety was available. Choosing the stones to be used from the hundreds at his disposal, a stone specialist/artist looked for color tones and gradations suitable for the particular picture. The paper pattern was glued to each stone slice, which was then cut into the exact shape using a saw and an abrasive to work a piece held steady in a vise. The workbenches were designed specifically for the Opificio, allowing specialists to cut the stones with such precision that, when they were laid side by side to create the semi-precious stone mosaic (*commesso di pietre dure*), it was difficult to discern the lines between each individual piece. The sections were then glued to a flat surface, most often slate, and finally, the surface of the stone sections was polished to give the resulting picture its luster. So subtle were the color gradations, shading, and opacity, and so exact were the cuts, that the finished work could be described, as Ferdinando I called it, a "painting in stone."[39] In 1598, Francesco Ferrucci of Fiesole, one of the early stone mosaic specialists working in the Opificio, created a portrait of Cosimo I that is on display in the museum of the Opificio delle Pietre Dure. Also on display in the museum are an impressive number of *pietre dure* artworks, including vases, cameos, furniture pieces, other portraits, and landscapes completed during the Renaissance and beyond.

Another type of *pietre dure* work made in Florence was known as a *glyptic*. The term is derived from the Greek and refers to the art of carving cameos and incising jewels using semi-precious stones. During the sixteenth century in Florence, however, the term most commonly referred to the creation of vases using semi-precious stones and later to denote *pietre dure* sculptures in the form of reliefs or small statues. The technology for carving a vase involved using a vertical stand (*castelletto*) to which metal discs were attached. The rotation of the discs, together with abrasives, carved the form of the vase, and incised sections were created with very fine pointed instruments. The Opificio delle Pietre Dure was one of the few workshops in which the technology for small sculptures in *pietre dure* was developed. The *castelletto* was also used for these works, although somewhat differently. The *pietre dure* were placed on the *castelletto* into which drill points had been fixed to a shaft. The craftsman or his assistant rotated the drill points by using pulleys hooked to a wheel that could be turned directly or with a pedal. Multiple sections of the *pietre dure* were assembled and attached with glue.

Still later, in the seventeenth century, another form of stone art became fashionable: painting with oils on a stone slab, incorporating the natural colors and markings in the slab for the background of the painting. Artisans used local Florentine stone for this purpose.

One of the early workbenches used in the Opificio delle Pietre Dure was discovered in Florence in the nineteenth century. The bench was probably constructed in the latter part of the sixteenth century and may have been commissioned by Francesco I. Made of solid walnut, the workbench is of a simple design and measures 114 cm high x 85 cm deep x 92.5 cm wide. It was built to be used by the carver in a standing position, with the edge on the carver's side recessed. In the center of the bench is a lapidary lathe for cutting the selected stones to the shapes indicated by the craftsmen.[40] The workbench is currently exhibited at

39. Giusti, *Museum of the Opificio delle Pietre Dure*, 7.

40. Bedini, "Renaissance Lapidary Lathe."

the museum of the Opificio delle Pietre Dure, along with tools and other workbenches, most post-Renaissance, but similar in design to those utilized by the early stone carvers. The collection includes a workbench used for *glyptics*, a bench designed for mosaics with a vice to hold the stone steady and an attached saw, and seven other types of benches.

2.5.1.2 San Lorenzo (Chapel of the Princes)

After establishing the Opificio delle Pietre Dure, one of Ferdinando I de' Medici's major projects was the embellishment of the Chapel of the Princes (Cappella dei Principi) adjoining the church of San Lorenzo. The chapel, begun during the lifetime of Cosimo I, was designed to immortalize the Medici grand dukes and is grandiose both in proportion and decoration. The octagonal chapel walls and floor are ornately adorned with marble and *pietre dure*. The tombs are made of granite from Asia and green Corsican jasper with the altar inlaid with *pietre dure*. The coats of arms have inlays of lapis lazuli, mother-of-pearl, and coral. Fernando initiated work on the chapel in 1604, and it was continued for two centuries with changes and redirected plans, thus accounting for the lack of a coordinated aesthetic. An adjacent room holds Medici tombs, including those of Lorenzo the Magnificent and his brother Giuliano. In marked contrast to the Chapel of the Princes, that area, with its classical architecture and adorned with marble statues by Michelangelo, is a space of symmetry and cohesive style. Today the Chapel of the Princes remains an extravagant mausoleum open to the public.

2.5.2 Stones Utilized

With collections from various parts of the world, hundreds of different hard stones were available to the craftsmen. A look at some of the most common ones gives insight into the varieties and depth of choices.

Rock crystal, a type of quartz, has been used since ancient times. Found primarily in the Alps, it is usually colorless, but can be tinged in purple (amethystine quartz). **Chalcedony** is a form of quartz, several varieties of which were accessible to the Florentines. Agate, a striated semi-transparent type of chalcedony, is found in different areas of the world, including Sicily. A light-colored and fairly uniform type of chalcedony was often used for flesh tones; Sienese agate with multiple layers provided the brown and ochre shades; and a translucent chalcedony yielded many colors including red and brown.

Basalt is a volcanic rock that exhibits black or very dark tones. **Jasper**, which is opaque and has the hardness of chalcedony, is found in many varieties and colors, including a deep red jasper, called blood jasper, that came from parts of Italy and Germany, greenish blue Corsican jasper, and a kind with white and pink markings that was available locally. The blue of **lapis lazuli** contains small particles of pyrite that create a golden glistening effect. The lapis lazuli stones most valued during the Renaissance, because of the consistency of color and expense in purchasing, were from Afghanistan. Artisans also used **malachite**, which ranged from mottled green to a more consistent dark green, extensively. Some of the other stones comprising the most common selections were **granite**, containing quartz, easily polished to a high luster and used for sculpture; **alabaster**, a softer stone with a play of iridescence (opalescence); **slate**, formed by a metamorphism of shale; **porphyry**, red volcanic rock from Egypt and much prized; **marble**, a metamorphic rock with impurities resulting in a range of colors; **petrified wood**, having its origins in tree trunks that have hardened and then crystallized; and **limestone**, which is mainly calcium carbonate (most often white, but also having an array of markings in the yellow, brown, and gray-brown range) and easily found in Florence along the Arno River. Limestone was often used for inlay and background work.[41]

2.6 Ceramics

Ceramics have served both a utilitarian function and an artistic one for thousands of years. The earliest type of ceramics is earthenware—clay that has been fired or baked to a degree of hardness. Many of the resulting vessels were also incised, impressed with patterns, or decorated with colored pigments and, in later periods, glazed. Early cultures throughout the world developed techniques for producing earthenware. Recent excavations from the Xianrendong Cave in the Jiangxi Province of China uncovered examples of pottery that were dated, using radiocarbon technology, to 19,000 to 20,000 years ago, making it one of the oldest-known pottery finds in the world.[42]

A second type of ceramic vessel production was developed by the Chinese, with the technique becoming increasingly refined around 200 BCE. In this technique, the Chinese used decayed feldspar, primarily kaolin (a white clay) and petuntse (micaceous or feldspathic rocks), both of which are found in China, to make what became known as Chinese porcelain. The proportions of kaolin to petuntse determined the appearance of the porcelain, with tones ranging from a mellow to a stark white. Also known as hard-paste porcelain, Chinese porcelain had a smooth, glassy, white finish achieved by firing the body with applied glaze at a high temperature. Over time these works became increasingly more sophisticated in terms of shape and the manner in which they were decorated.

With both earthenware pottery and porcelain, the successful outcome was determined by the use of the correct amounts of clay components, the skill in shaping the vessels, and the empirical determination of the optimum firing temperature.

2.6.1 Pottery in Italy

The history of pottery in Italy spans many centuries, which is particularly evident from the excavations bringing to light Etruscan works from the period of the tenth to the fourth century BCE. Roman pieces dated from about the eighth century BCE have also been identified. Excavations from the Umbrian town of Orvieto have yielded archaic examples of a style of painted earthenware now known as *maiolica*, dated from the end of the thirteenth century, and probably named later for traders who imported pottery from the Spanish island

41. Giusti, *Museum of the Opificio delle Pietre Dure*, 88–89.

42. Wu et al., "Early Pottery at 20,000 Years Ago."

of Majorca beginning in the early fourteenth century. In 1301, Abu'l Qasim from Kashan, a Persian city known for its ceramics, wrote an essay on the technique for producing what was later known as *maiolica*. Qasim's work was intended as a manual for creating the type of pottery that was made in Mesopotamia as early as the ninth century and was probably influenced by Tang Dynasty pieces. The manual is brief and not particularly clear in all aspects, but it does speak to a technique that made its way through the Middle East to the Moors in North Africa, to Spain, and then to Italy. That technique is characterized by the low firing temperature of the clay and the application of a design over a white background of a tin oxide glaze. Qasim's work discusses two types of glaze: one is a single opaque glaze layer with tin oxide particles less than 1 μm in width or collecting in groups up to 10 μm across; the other is composed of an inner layer similar to the single opaque glaze layer but also with quartz and feldspar particles, and an outer transparent layer devoid of particulate crystalline units. In this second type of glaze, lead antimonite particles can often be found between the two layers. The tin oxide particles are responsible for the reflectance and light scattering associated with the glaze.

The decoration of the vessels, including incised designs, paints, and glazes, was based on motifs and applications common during the classical and medieval periods. In the Renaissance, improvements in technology enabled advances in the art of pottery. While artisans retained the same basic technical protocols, Renaissance creators of vessels experimented with formulas in terms of slips, colors, and glazes, and made improvements in the design and function of kilns; by the sixteenth century, craftsmen were able to control kiln temperatures more effectively, which meant that the outcome of their firing was more predictable and they were able to experiment with more difficult designs.

Ceramic artists in Florence, as well as other locations, used a technique known as *zaffera a rilievo* during the Renaissance. The word *zaffera*, meaning sapphire, reflects the use of a cobalt blue glaze to produce the signature vessels with blue designs on a white background, often with small brown or green detailing. Pieces from the Florentine workshops of Giunta di Tuglio and Piero di Mazeo, created in the early to mid-1400s using the *zaffera a rilievo* technique, can be found in museums including the Louvre in Paris.

Pottery decoration that featured narratives often mirrored paintings and prints being created during the same period. The most prominent Renaissance decorator of pottery was Francesco Xanto Avelli of Rovigo (ca. 1486–1542). Working primarily in Urbino in the early 1500s, Xanto decorated entire surfaces of his vessels with historical, biblical, and literary themes.

In Renaissance Florence, workshops and kilns were located mainly in Montelupo Fiorentino. Situated outside central Florence, Montelupo Fiorentino was an ideal location because it both was a trading crossroad and boasted plentiful high-quality clay. Pottery workshops concentrated on producing utilitarian pieces during the fourteenth and fifteenth centuries, and by the dawn of the sixteenth century they were also creating vessels and plates intended solely for artistic display. The desire of wealthy families for the *maiolica* ware transformed pottery workshops into flourishing businesses. Since the fourteenth century, the town has been known for its innovative works sold throughout Europe.

The art of *maiolica* was an exacting science as the colors were very unstable. In this process, the clay pieces were fired twice. After the first firing, the piece was decorated and then dipped in a glaze of lead or tin oxide containing a silicate of potash, which permanently fixed the glaze and pigments to the works. A second firing followed. After centuries of producing *maiolica*, Italian craftsmen had refined this style of pottery and raised it to a sophisticated level of artistic expertise. In 1556, a potter named Cipriano Piccolpasso wrote *The Three Books of the Potter's Art* (*Li tre libri dell'arte del vasaio*). In his discourse, Piccolpasso takes the reader through the *maiolica* process, from the basics of clay preparation to the making of the vessel, including pigments to be used, specific preparation of the white glaze used as a base, and methods of firing. Piccolpasso's discussions and illustrations, such as ones showing the design and arrangement of a typical potter's studio, provide a better understanding of the process.[43]

An interesting study was conducted on a variety of Italian *maiolica* ceramics spanning the thirteenth to the sixteenth centuries using analytical scanning electron microscopy and energy dispersive spectroscopy. The compositions of the glaze, body, and pigments, as well as microstructure, were found to be consistent with those methodologies given by Piccolpasso. For example, analysis of a white, uniformly opaque glaze and the clay used for the body of an early fifteenth-century Florentine Renaissance blue bowl showed the content to be as expected for the period.[44] Another investigation used Raman spectroscopy to analyze nine pottery fragments from the Castle of Cafaggiolo in Tuscany, which was owned by the Medicis and converted to a villa by Michelozzo in 1443. In 1495 a workshop was added for the manufacture of *maiolica*. From an analysis of the engobe, glaze, and clay body of each fragment found in the workshop, the authors concluded that the artisans were experimenting with new techniques. Decorative colors were achieved primarily by using gypsum ($CaSO_4 2H_2O$) and calcite ($CaCO_3$) for white, carbon for black, massicot (PbO) and Naples yellow ($Pb_2Sb_2O_7$) for yellow, and ochre (Fe_2O_3) for red.[45]

2.6.1.1 Bargello Museum

The Bargello Museum has on display interesting pottery from Montelupo made between the fourteenth and seventeenth centuries, including small pharmacy jars and jugs (*boccali*) with emblems from the Santa Maria Nuova pharmacy. Pieces made in Florence include *boccali*, several spouted pitchers (*orcioli*), and both small and large plates (*piatti*). Many of the Florentine pieces are decorated with a cobalt blue pigment (*a zaffera*).

2.6.2 Porcelain

Chinese porcelain, traded in the West beginning primarily during the Ming Dynasty (1368–1644), has been found in what is now Iran and Iraq dating from the ninth and tenth cen-

43. Piccolpasso, *Three Books of the Potter's Art*.

44. Tite, "Production Technology of Italian Maiolica." The chemical compositions of the fifteenth-century Florentine *maiolica* glaze were determined to be SiO_2 (45.1 percent), PbO (34.4 percent), SnO_2 (5.5 percent), and CaO (4.2 percent), with lower amounts of MgO, AlO_3, FeO, CuO, and MnO. The total alkali content was 5.4 percent (K_2O, 4.6; Na_2O, 0.8). The body of the bowl examined exhibited chemical compositions similar to others of the period.

45. De Santis et al., "Micro-Raman and Internal Microstratigraphic Study."

turies. This porcelain was greatly admired by Europeans, and attempts to replicate it were begun in several Italian cities early in the sixteenth century. It was not until Grand Duke Francesco I de' Medici (1541–87) brought his interest and resources to the challenge that a concerted effort was initiated to discover the Chinese formulation and create vessels of equal quality. To that end, an experimental workshop, financed by Francesco I and housed in the Casino of San Marco in Florence, was started in 1575 and remained in operation until 1587. Through trial and error, this workshop produced the first Florentine porcelain, which differed from the Chinese hard-paste formulation that was a mixture of kaolin and petuntse. The Florentine porcelain, known as Medici porcelain, had a slightly glass-like quality. For this porcelain, Florentine artists borrowed shapes from the *maiolica* pottery and adopted the blue and white motifs used by the Chinese. The cost of producing those porcelain plates, cruets, vases, bowls, flasks, jars, and jugs was extremely high, however, and manufacturing was halted upon Francesco's death in 1587.

A collection of Medici porcelain created from 1575 to 1587 is currently held in the Musée National de Céramique in Sèvres, France. In a recent effort to determine the composition of the Medici porcelain and discern whether the designation of soft-paste as opposed to hard-paste was accurate, the collection was made available for scientific analysis. Raman spectroscopy revealed that the Medici porcelains were characterized by a feldspar signature. Feldspar is a group of minerals with slightly varying chemical compositions and structures, but all containing aluminum, silicon, and oxygen, with either sodium, potassium, and/or calcium. The feldspar signature discovered in the Medici porcelains was indicative of hard-paste and soft-paste β-wollastonite ($CaSiO_3$) and α-wollastonite, signifying that the Medici porcelains were made by an intermediary technology between hard- and soft-paste. The use of calcined bone was indicated by the presence of calcium phosphate that was probably used in Venetian glass and known to be present in Islamic enameled glass. The study also revealed the presence of α-quartz grains.[46]

The Medici porcelain workshop, active for twelve years in Florence, resulted in an abundant array of vessels. Those pieces held by the grand dukes and their families were spread among the various grand ducal residences and also stored in the Guardaroba generale, storage rooms located in the Palazzo Vecchio. In about 1723, after the death of Cosimo III, an inventory of items held in the Medici Guardaroba included a listing of porcelain made in Florence ("porcellane fatte in Firenze"). From that inventory, it is known that 256 pieces of Medici porcelain were stored in the Guardaroba. Other items located in private residences were, of course, not listed, making it impossible to know the full extent of the porcelain holdings. While some inventories of Medici villas exist, they are neither definitive nor complete.

By the eighteenth century, ceramicists in cities across Europe, including Germany, Austria, France, and England, were beginning to produce fine porcelain and interest in Medici porcelain waned, not to be revived to any great extent until the latter half of the nineteenth century. In 1772, a large auction was held in the Palazzo Vecchio to dispose of pieces no longer considered to be of great value. The items offered for sale included both Tuscan

46. Colomban, Milande, and Lucas, "On-Site Raman Analysis."

maiolica and Medici porcelain. Other auctions were held in the intervening years with the result that few Medici porcelain pieces in the family collection remain in Florence.[47]

2.6.2.1 Bargello Museum

Displayed in the Bargello Museum is a large porcelain basin with Saint Mark depicted in the center. This basin, designed by G. Penez and engraved by Aldegrever, was made in the Florentine workshop, Casino di San Marco, in circa 1580. In addition, a small fragment of porcelain showing the top of the Duomo (ca. 1580) is also on view. Other examples are currently displayed in major museums and continue to be held in private collections.

2.6.3 Decorative Italian Technique of *Sgraffito*

The technique of *sgraffito*, which involves incising, was probably adapted from early Chinese porcelain artisans working in the Tang Dynasty (618–906). Their pieces were exported to the Islamic world along the Silk Road and by sea routes. From the Near East, the works were eventually traded in Europe. Considerable skill was required of those who incised the vessels. In the fifteenth century, Italian potters began utilizing a type of incising called *sgraffito* in earthenware vessels. The clay pieces, known as *a punta*, were of a buff or red color and were dipped into a clay slip. After the slip had dried, incising was done so that the color (red or buff) of the vessels was exposed where lines were incised. The vessel was then fired for a first time, followed by the addition of a transparent glaze, sometimes after decorating with colored pigments. The pieces were then fired a second time.

From the second half of the fifteenth century through the seventeenth century, Florentine towns, including Montelupo, employed the *sgraffito* technique. Interestingly, wasters (damaged pieces of pottery) from the workshop on Via Bertini, Castelfiorentino (Florence), were recently discovered with thousands of pieces of dishes, bowls, and cups, providing an opportunity to study this technique. Scientific examinations were performed to determine the types of *sgraffito* employed, the mineralogical and chemical compositions, and the sources of the raw materials. Information from those studies showed that the *sgraffito* slipware was made of a fine-grained clay mixture with the clay probably coming from Pliocene clay deposits near Castelfiorentino that had been quarried since the time of the Etruscans. The slip mixture was made from two different materials derived from various markets outside of the Florence area, but matching those used in and around Florence from the late fifteenth century until the start of the seventeenth century. The raw materials used for the vessels, the slip, and the glaze were amazingly homogeneous, testifying to the availability of a consistent supply of raw materials and little variation in recipes utilized during the Florentine Renaissance.

According to the study, the *sgraffito* procedure for the Via Bertini workshops seems to have been similar to the protocol described above. The potter first shaped the piece on a

47. Spallanzani, "Medici Porcelain."

potter's wheel, then dried the clay, and then covered the piece with a slip solution either by immersion or pouring the slip over the item. After the slip dried, it was scratched or incised, then fired. Following that first firing, colors were applied and covered by a lead glaze mixture. The piece was fired for a second time. The workshops also produced a more complicated and time consuming type of *sgraffito*, called *a fondo ribassato*, which denoted the incising and scraping of large portions of a vessel to create a bas-relief effect. By the end of the sixteenth century, however, that procedure was less common and the more simplified process used by the Via Bertini workshop became the norm.[48]

2.7 Textiles

As mentioned in chapter 1 (see 1.3), the wool industry in Florence became less profitable as artisans from other areas of Europe developed the skills in which Florentines had excelled. With the increased cost of importing raw wool from England and Spain and the decimation of much of the Florentine population by the Black Death, the silk industry, less labor intensive than wool, began to flourish. Florentine silk workers, part of the Arte della Seta guild, added greatly to their ranks when Pisa sacked Lucca, the preeminent center of Italian silk production, and the weavers from Lucca were welcomed by the Florentines. Silk production in Florence increased throughout the fourteenth century and by the fifteenth century it had outstripped the wool industry in Florence. Adding to the industry's profitability, Florence imported silkworms that were fed by the mulberry trees planted throughout the region, thus becoming less dependent on the East for raw materials. Additional materials still needed to augment Florentine supplies were more easily accessible due to the increase and ease of sea trade.[49] Florentine silks woven into brocades, velvet, and taffeta for clothing and purses were in great demand. According to historian Carole Frick, "Clothing as a metaphor for the dream (or nightmare) of transformation was central to the society of Renaissance Florence from Boccaccio to Machiavelli."[50]

2.7.1 Florentine Tapestries

The Medici family had an early interest in tapestries, as evidenced by the commissions and acquisitions of Giovanni, Piero, and Lorenzo during the mid-fifteenth century. Records show the Medicis purchasing tapestries from Netherlandish weavers, and those tapestries often show up in the cartoons of works commissioned from Florentine artists.[51] The Florentine wool and silk industries contributed to artistic innovation with the advent of tapestry production in Florence in 1545. In the process of renovating the Palazzo Vecchio as the family residence, Cosimo I consulted his agent in The Netherlands, Giovan Battista Ricasoli,

48. Brianese et al., "Incised Slipware Production."

49. Schevill, *History of Florence*, 400–401.

50. Frick, *Dressing Renaissance Florence*, 1.

51. Campbell, *Tapestry in the Renaissance*, 88–89.

bishop of Cortona, about the availability of appropriate tapestries. At about that same period, Cosimo was contacted by a Florentine silk merchant in Ferrara who had been approached by a group of about fifty weavers from Brussels. The weavers expressed an interest in relocating with the specific goal of introducing this art form in Florence. This idea appealed to Cosimo, who wished to broaden his economic base, and he decided to hire the Netherlandish weavers and inaugurate tapestry production. Workshops were opened in 1545, led by principal weavers Janni Rost and Nicholas Karcher, both of whom were working in Ferrara at the time.

In the opening year, Janni Rost and fifteen weavers produced a tapestry based on a design by Angolo Bronzino (1503–73) entitled *Abundance* (now in the Pitti Palace). The next tapestries from Rost were *Spring* and *Justice Liberating Innocence*, also Bronzino designs. Later in the year, Nicholas Karcher established a second workshop a short distance from that of Rost. In 1546, he delivered to Cosimo *Lamentation*, the cartoon for which had been prepared by Francesco Salviati (1510–63). With the workshops operational, many more tapestries followed in rapid succession.[52] Cosimo requested twenty tapestries depicting the *Life of Joseph* to adorn the Palazzo Vecchio. Artists Bronzino, Salviati, and Jacopo Carucci Pontormo (1494–1556) designed sixteen of the tapestries, and the weaving was done by Rost and Karcher between 1546 and 1553. The tapestry cycle relates the story of Joseph as cited in Genesis.[53] There is plausible speculation that Cosimo considered the *Joseph* tapestries to be an allegory for his life and achievements.

Florentine tapestries were created with a combination of wool and silk and have deteriorated badly over time. Ten of the *Life of Joseph* tapestries were displayed in the Palazzo Vecchio until 1983, and the remainder had been relocated to Quirinale Palace in Rome.[54] In 1983 the tapestries were deemed to be in such poor condition that they were removed to the Opificio della Pietre Dure for one of the longest restoration projects undertaken by the Opificio. The project, twenty-seven years in length, was completed in January of 2013.[55]

One study has analyzed the dyes used in two of the tapestries from the *Joseph* series that had been in the Quirinale Palace before removal for restoration by the Opificio delle Pietre Dure. The dyes were extracted from the threads with dimethylsulfoxide, then hydrolyzed in an acidic methanolic solution (containing hydrochloric acid), and analyzed by high-pressure liquid chromatography equipped with a diode array detector to determine the absorption spectra of the compounds. A number of dyes were identified that produced a variety of colors, including red, pink, violet, blue, green, yellow, and others. The dyes were expensive and derived from a number of sources: coccid insects from the superfamily Coccoidea; a variety of plants, *Reseda luteola* (weld), *Rubia tinctorum* (madder), *Indigofera* (indigoid), *Cotinus tinctoria* (young fustic), bark from a number of trees for tannins, and *Carthamus tinctorius* (safflower for a pinkish-orange color). The detection of safflower, used for millennia in Egypt, Indonesia, and China, represented the first identification of this material in a European

52. Ibid., 494–96.

53. Smith, "Cosimo I and the Joseph Tapestries."

54. Campbell, *Tapestry in the Renaissance*, 498.

55. Donati, "Precious Renaissance Tapestries."

fabric. The far-reaching sources of the dyes testify to the widespread commerce of Florence during this period.[56]

2.8 Printmaking and Engraving

Artistic printmaking and engraving arrived in Florence in the mid-fifteenth century, with Florentines developing innovative styles, somewhat different from that of the Germanic community.[57] Woodcut refers to the technique in which the artist draws a picture on a smooth wooden surface, then removes areas of wood around the lines, leaving the picture in high relief. The raised areas are inked and pressed onto paper. By repeatedly inking, the artist could produce multiple prints of the same image. This technique was used to produce identical prints for collectors and consumers, but more importantly for the burgeoning book trade.

For engraving, the artist worked on a metal plate, typically of copper. The artist sketched an image on the metal and then incised those lines into the plate. After forcing a thickened ink into the engraved lines and then removing any excess ink from the remainder of the plate, the plate was placed on a press and stamped onto paper, which absorbed the inked drawing. As with woodcuts, the metal plates could be reused, making it possible to produce multiple identical prints.[58]

For woodcuts, artists used tools called gouges and burins (or gravers) in a variety of widths and shapes. They created shading by making some lines thicker than others and varying the distance between lines. In a workshop, after the artist drew the image, assistants could do the carving, inking, and pressing, thereby making multiple editions in less time. The woodcuts were particularly useful for illustrating books.[59]

2.8.1 Illustrative Florentine Works

The Uffizi Gallery has a valuable collection of engravings, including the two highlighted here. Antonio del Pollaiuolo (1429/33–98) is the Florentine artist who best exemplified the technological innovations made in the field of metal engraving. A painter, sculptor, and goldsmith, he used the burin to hatch lines and create a shading that made for convincing contours and a depth that was unique. His relatively large (approximately 42 x 60 cm) work, *The Battle of the Ten Nudes*, took the art form of engraving to a new level, adding specialized technique to single line-point drawing, including his innovative zigzag strokes. Created in about 1470, the engraving depicting "the human nude in action . . . contains the germ of important stylistic developments"[60] that influenced Renaissance artists in other genres.

56. Degano, Lucejko, and Colombini, "Unprecedented Identification of Safflower Dyestuff."

57. For an in-depth discussion of printing as it relates to the production and circulation of texts, see Richardson, *Print Culture in Renaissance Italy.*

58. Paoletti and Radke, *Art in Renaissance Italy*, 37.

59. Stokstad, *Art History*, 615.

60. Andres, Hunisak, and Turner, *Art of Florence*, 1:703.

Florentine goldsmith and engraver Baccio Baldini (1436–ca. 1487) perfected the engraving art of incising complex images, often collaborating with Sandro Botticelli for design ideas. He executed a series of highly intricate engravings depicting the personifications of the Ptolomaic planets: the sun, moon, Mars, Mercury, Venus, Saturn, and Jupiter. Planet Venus is on display in the Uffizi Gallery, and the other six are located in the British Museum in London.

Chapter 3

Applying Mathematics to Art and Cartography

> "It would please me if the painter were as learned as possible in all the liberal arts, but first of all I desire that he know geometry."
>
> —Leon Battista Alberti, *On Painting*

> "Like printed maps, the Florentine . . . maps were conceived as marvelous images on the boundaries between art and science that conveyed the knowledge of the world spatially and graphically."
>
> —Francesca Fiorani, *The Marvel of Maps*

Prior to the Florentine Renaissance, artists produced exquisite paintings that, however, lacked three-dimensionality. For example, works by Byzantine artists, although vibrant in color choices, had the appearance of a level plane without perceptible depth. Masterpieces from the Florentine school remain that clearly demonstrate the progression from the Byzantine genre to an empirical perspective in which artists explored various means of suggesting depth in their paintings to a more scientific and more uniformly applied method of depicting space and distance. With the reexamination of classical works focused on geometry, Renaissance artists began experimenting with these principles as they applied to their artistic endeavors. This integration of art and mathematics enabled the production of magnificent perspectival paintings and cartographic images with an accuracy never before portrayed. An understanding of the geometric and mathematical approaches the Florentine artists used to create these masterpieces gives us a deeper appreciation of those transformative works.

3.1 Use of Geometric Perspective in Art

3.1.1 The Origins of Modern Geometry

Beginning in the fourth millennium BCE, the early cultures of Egypt and Mesopotamia were using basic mathematical principles, and these, along with considerable astronomical observations, were surely known to the early Greek mathematicians. The story of geometric perspective as it relates to the study of art begins in about 295 BCE, the year Euclid founded the school of mathematics in Alexandria. Although it is thought that Euclid taught there, few personal details about his life are available. His legacy was his conceptual approach to mathematical formulations, and his *Elements of Geometry,* arguably one of the most influential books written, remains in use today. In presenting his theory of numbers and geometry, Euclid brought a logical progression and organization to a discipline that had been previously more philosophically oriented.

In addition to being a mathematician, physicist, and music theorist (although the two harmonic treatises attributed to Euclid may have actually been written by someone else), he was also well known for introducing geometric theories to optics, the study of light production and behavior. In his *Optica* (ca. 300 BCE), Euclid applied laws of geometry to explain, among other things, how people perceive their surroundings.[1] Euclid was not writing a scientific guide for painters, for example in explaining the use of a vanishing point, but rather presenting a series of systematic geometric proofs demonstrating his theories about light. According to Euclid, the eye emits straight-line rays of light that lead to the object being viewed. Those rays diverge as they extend from the eye to create a virtual cone whose tip is the eye and whose base is the entity under observation. Euclid's resulting theory was that close objects, whose base of the cone is close to the eye, appear larger because of the wider angle, whereas distant objects, whose base of the cone is farther from the eye, appear to diminish in size because of a smaller visual angle. Although the assumption that rays of light project from the eye is invalid, Euclid's insightful hypothesis (stated in Theorem 8) gave optics an important mathematical structure.

Some four centuries later, a Greek/Egyptian scholar expanded on Euclid's theories and thus made valuable contributions to both art and cartography. Claudius Ptolemaeus (ca. 100–170), known to us as Ptolemy, was an astronomer, mathematician, geographer, and optical theorist. In addition to his works on astronomy (see 5.1) and geography (discussed later in this chapter; see 3.2.1), Ptolemy wrote a treatise on optics, *Optica*, in the mid-second century. In that work he attempted to explain the way in which images are shaped by reflection and refraction, as well as the perception of color and dimensions. Ptolemy also provided a further explanation of Euclid's theory of the centric visual ray principle and noted that the shorter visual rays are near the center of the cone and lead to clearer imprints of the objects being viewed, which explains why objects observed from a greater distance or a greater angle may appear to be less distinct. Ptolemy was aware of the relevance of his theories to artistic

1. Tobin, "Ancient Perspective and Euclid's Optics."

endeavors; he suggested that artists use brighter colors for objects in the foreground and less vivid hues for those in the distance.

Euclid and Ptolemy's work on optics became important influences on Muslim scholars of the Middle Ages, who enlarged upon optical principles based on geometrical considerations. One of the most important of these was Ibn al-Haitham, known as Al-Hazen. Born in Basra in about 965, he wrote extensively on astronomy, physics, mathematics, and medicine. His treatise, *Perspectiva,* which appeared in about 1000, integrated the knowledge of Euclid, Ptolemy, and others; his experiments with lenses, refraction, and reflection added insights on the properties of light and the manner in which objects are perceived.

3.1.2 Mathematical Perspective in Renaissance Art

The principles of optics and geometry found a rebirth during the Renaissance in Florence, where artists began to use the ideas of Euclid, Ptolemy, and other scholars of the past to develop painting styles that not only appeared to have proportional depth, but were mathematically and geometrically correct in their presentation. This method of creating an illusion of depth on a flat surface using a combination of horizontal lines and mathematically determined converging lines that appear to meet in the distance became known as linear perspective. The pioneers of linear perspective dramatically altered the artistic landscape, and the story of the emergence of that new method is extraordinary, especially when one considers its relatively sudden development and rapid acceptance. A possible contributing factor is presented by art historian Samuel Edgerton, who framed it in terms of the period in which it appeared, when the general public's interest in the church was diminishing. "Linear perspective could therefore retool the visual arts," Edgerton wrote, "refurbishing them to present the Christian message more convincingly and help shore up the sagging beliefs of an increasingly cynical population."[2]

3.1.2.1 Establishment of the Principle

Paolo dal Pozzo Toscanelli (1397–1482) was a mathematician, geographer, and astronomer, as well as a medical doctor. After completing his medical studies in Padua, Toscanelli returned to his native Florence in about 1424 and became friends with Filippo Brunelleschi (1377–1446), already known as the brilliant architect of some of Florence's most impressive buildings, including the famous dome of the Florence Cathedral (Duomo), Santa Maria del Fiore, the construction of which is discussed in chapter 6 (see 6.3.1.2). It is believed that Toscanelli influenced and mentored Brunelleschi in mathematical principles. In the first quarter of the fifteenth century, Brunelleschi created several painted panel illustrations demonstrating the principles of perspective. Being an artist/architect interested in structures and their design, Brunelleschi naturally used buildings as subjects for his work, selecting the Florentine Baptistery in front of the Duomo and the Palazzo Vecchio in the Piazza

2. Edgerton, *Mirror, the Window, and the Telescope*, 7.

della Signoria.[3] Sadly, his panel illustrations have been lost over time; however, from the writings of others it is possible to recreate them with some degree of confidence. Brunelleschi's biographer, Antonio Manetti (1423–97), was only a few years old when Brunelleschi demonstrated his experiment; nonetheless, he is a source of much information on the subject. Manetti claims to have had firsthand knowledge of the mirror device Brunelleschi used: "I have had it in my hands and seen it many times in my days and can testify to it."[4]

There are multiple descriptions of Brunelleschi's perspectival study at the Baptistery, but most seem to agree with the one put forth by Manetti. He wrote that Brunelleschi painted a small panel picture of the Baptistery while standing outside the central portal of the Florence Cathedral. After establishing an ideal place from which one should view the Baptistery, Brunelleschi drilled a small hole through his painting at the centric vanishing point. According to Manetti, "The hole was as tiny as a lentil bean on the painted side and it widened conically like a woman's straw hat to about the circumference of a ducat, or a bit more, on the reverse side."[5] Brunelleschi set the painting at the same location where he stood while creating it. A viewer standing behind the painting and holding a mirror at arm's length could look at the backside of the picture and, through the hole, see an image reflected in the mirror that was essentially identical to the actual scene.

Brunelleschi used the mirror image to develop his rules of perspective, geometrically showing that the distances between horizontal parallel lines lessen proportionally as they recede into the distance toward the horizon and that distant objects appear proportionally smaller than objects in the foreground. He also calculated the converging points of the two sides of the Baptistery, thereby adding vanishing points in those locations. Brunelleschi applied a similar system to his drawing of the Palazzo Vecchio viewed from the point at which Via dei Calzaiuoli enters the Piazza della Signoria. Thus, he showed that an artist could create paintings that, to the observer, appeared to be replicas of an actual building, scene, or object.

Although it is unlikely that Brunelleschi used complicated calculations to demonstrate that lines perpendicular to the plane would meet at a centric point,[6] his basic rules for perspectival painting worked as a geometric exercise. Since Brunelleschi was not scientifically trained, some art historians have speculated that it was probably Toscanelli who imparted the knowledge that enabled Brunelleschi to add mathematical rigor to his approach to perspectival drawing. Unfortunately, no writings remain from Toscanelli that would document his suspected contribution to this important topic.

While Brunelleschi demonstrated methods that applied mathematics to linear perspective, he was not the first artist to employ perspective to simulate depth.[7] In the Uffizi

3. Trachtenberg, "What Brunelleschi Saw."

4. Manetti, *Life of Brunelleschi*, ed. Saalman, 42–44.

5. Ibid., 44.

6. Field, "Piero della Francesca's Mathematics," 264n10. "Centric point" was a term used by Alberti in his treatise *De Pictura*. Later, the term "vanishing point" was used synonymously.

7. Edgerton, *Mirror, the Window, and the Telescope*, 109.

Gallery, displayed in close proximity, are three similar paintings portraying the Madonna and Child. Viewing these works by Cimabue (1240–ca. 1302), completed about 1280, Duccio (ca. 1255/60–ca. 1318/19), created in about 1285, and Giotto (1266/67–1337), painted in about 1310, gives one insight into the evolution of a more visually correct portrayal of depth, particularly as demonstrated by Giotto. His frescoes in Padua's Arena Chapel (ca. 1305–10) have a realistic depth that, although lacking mathematical precision, may be based on Euclidian visual theory.[8] As the fourteenth century progressed, works by various Sienese artists, including Duccio (*Maesta,* 1307–11), Simone Martini (*Maesta,* 1315), and Lorenzetti (*Allegories of Good and Bad Government*, 1338–39), manifest a marked three-dimensionality. Lorenzetti's 1342 *Presentation of Christ in the Temple* (now in the Uffizi Gallery) achieved an empirical perspective in the depiction of the tiled floor. It was, however, in Florence that Brunelleschi, for the first time, initiated a process by which artists could employ consistent mathematical principles to their art.

3.1.2.2 Seminal Treatises on Perspective

After Brunelleschi's demonstration of the use of precise linear perspective to achieve realism in painting, others began to examine and quantify means by which to consistently reproduce the desired results. A treatise titled *On Painting*, by **Leon Battista Alberti** (first printed in 1435 in Latin as *De pictura* and in 1436 in Italian, entitled *Della pittura*), is the earliest surviving written explanation of the methodology of linear perspective as it related to artistic works, establishing him as the preeminent scholar in the field of artistic expression. Alberti (1404–72), a brilliant architect, mathematician, and scholar, was the illegitimate child of a wealthy merchant and a noblewoman from Genoa. Because his family was exiled from Florence during his youth, he was educated first in Padua and then in Bologna, where he studied law. The ban that had led to the exile of his family was lifted in 1428, and he relocated to Florence, where he lived for several years before accepting a position as a secretary in the papal court. It was during his time in Rome that Alberti began his quest to understand the forms of ancient Roman architecture and in the process developed his expertise in that field. Probably influenced greatly by Vitruvius, the famous Roman architect who lived some 1,500 years earlier, Alberti began to work on his own *Ten Books of Architecture* (published in 1452), which would single him out as an authority on architectural style. After returning to Florence, Alberti's first major commission as an architect was for the façade of the Palazzo Ruccellai (1446–51), followed several years later with a commission for the façade of Santa Maria Novella (1448–70).

In addition to his work in architectural design, Alberti was interested in exploring ways in which a painter could depict three-dimensional objects on a two-dimensional surface in correct proportions to match those perceived when viewing the objects themselves. He wrote, "I do not believe that infinite pains should be demanded of the painter, but paintings which

8. Hoffmann, "Giotto and Renaissance Perspective," 5–32.

appear in good relief and a good likeness of the subject should be expected."[9] To accomplish this goal using the principles and language of optics and geometry, Alberti suggested that the artist should imagine that the eye is the tip of a pyramid to which linear visual rays extend from the object he wishes to paint. A viewing point must be determined, taking into account the average height of persons observing the painting and the distance from the painting from which the artist has the viewer optimally stand. While dismissing the bipolar nature of human vision, Alberti nonetheless developed a method for systematically projecting a three-dimensional object viewed from any angle onto a flat plane.

To create volume and form in a painting, Alberti urged artists to use shading in order to emulate the gradations of dark and light that appear naturally. To him, the artist's role was to present a narrative (*istoria*) in which the emotions of the subjects depicted would be revealed as accurately as possible in a chosen scene. Painters should, according to Alberti, depict a scene in a manner that is both aesthetically pleasing and a mathematically correct representation of three dimensions. In both painting and architecture, Alberti promulgated the concept of *concinnitas*—achieving an overall harmony such that nothing could be added or removed without jeopardizing the quality of the work. An artist, in Alberti's view, needed to be well versed in both mathematics and humanism. His emphasis on a well-rounded education must have seemed a revelation to many for whom the practice of art had been considered a craft rather than a profession.

Although Florentine artists had begun to use geometric principles in their paintings, it was Alberti who first delineated a clear methodology for using those guidelines. Unfortunately Alberti rarely included figures and mathematical equations in his writings, but his verbal descriptions were clearly stated and have, over the intervening centuries, permitted artists and historians to recreate his approaches, although not without some controversy.

In giving an explicit method for representing any three-dimensional object onto a flat surface, Alberti offered two practical approaches an artist could take to achieve an accurate likeness of the object. The first of these involved drawing a grid of squares on the painting surface and also having an upright clear grid situated between the artist and his subject. These two grids, used together, could aid the artist in mapping a coordinate system for the object being drawn. The second and much more versatile approach did not require that the artist be looking at the actual object to be portrayed. Rather, the artist could use his or her imagination to draw any desired scene with mathematical precision. Alberti based this second approach on the creation of a preparatory drawing that included a perspectival checkerboard grid, often referred to as the *construzione legittima*. The resulting preparatory drawing gave a mathematically and geometrically correct representation in perspective with the tiles being used as a projected grid to serve as a guide for adding any type of object, either real or imagined.

In order to achieve correct proportional relationships, Alberti placed the observer (artist or viewer) in front of an open window intended to represent the frame of the painting surface. The observer should imagine that the floor (or base) of the image to be viewed is a

9. Alberti, *On Painting*, trans. Spencer, 69.

continuation of the floor on which he is standing, the floor having a checkerboard pattern. This is illustrated in figure 3.1, in which there is an observer of height (*h*). The picture frame is designated *ABCD*, and the observer (denoted by the eye) is standing at a specified distance (*d*) from the frame. The artist positions a centric vanishing point (CVP) on the painting surface at a location above the baseline at the same height of the central object to be painted and that corresponds to the eye level of the observer. If the observer (eye), standing in front of the frame, moves 90° either to the right or left (maintaining the same distance from the CVP), so that he is standing to the side of the window/frame (represented by the two figures labeled "O"), the line of vision is still at the level of the centric vanishing point. Focusing on the figures positioned on the sides in the diagram, the height and distance are apparent. On the diagram the two O designations are located at the eye level of each of the observers and are designated as the right and left vanishing points (RVP and LVP, respectively).

Verticals are lines drawn from the line *CD* (bottom of the window/frame) to the CVP; the central vertical line is perpendicular with the line *CD*, as of course are the *AC* and *BD* lines. Orthogonals are lines beginning at the same point on the *CD* line as the vertical and drawn to RVP and LVP. Transversals (horizontals) are lines parallel to the ground or floor (*CD* in the diagram). The distance between the transversals is determined by the intersection of the verticals and the orthogonals. The resulting pattern clearly shows the receding appearance of a checkerboard floor. For purposes of illustration, we will consider projecting a tiled floor, say six tiles across and six tiles in depth, to proper perspective on the surface of the frame. We are showing equal-sized square tiles, but the procedure could be done with rect-

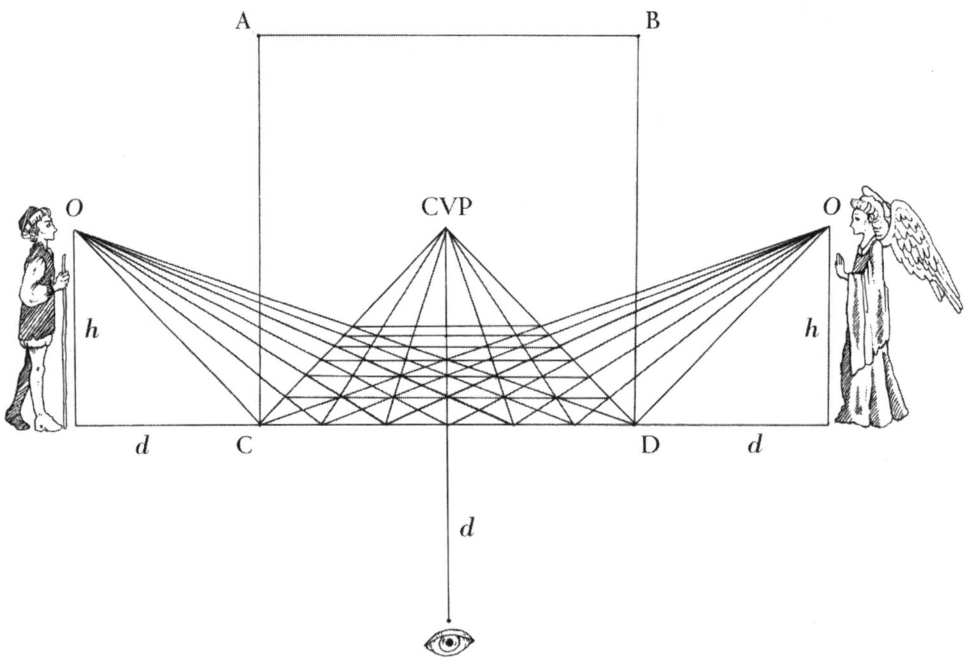

Figure 3.1: Depiction of Alberti's perspective method (drawing by Justine E. Stevens, commissioned by the authors).

angular tiles or tiles of any uniform shape. The edges of the first row of tiles on the ground floor line of the frame, *CD*, are connected to the CVP. While figure 3.1 shows both the RVP and the LVP for clarity, in practice either one of these would suffice.

Among the many presentations of the Albertian method for construction of a perspective image, a particularly lucid description for the projection of a four-by-four square tile pattern has been given by Field, Lunardi, and Settle.[10] Salgado also presented another very clear interpretation of Alberti's technique of perspective applied to receding flooring consisting of squares involving eight steps. It is useful to itemize these steps as they illuminate the process of developing Albertian perspective and pictorial space in an ordered manner.

> Step 1: "The visual field of the observer is demarcated as if it were a window."
>
> Step 2: "The projection of the central vanishing point on the projective plane—in other words, the point where the visual symmetry intersects the projective plane—is defined."
>
> Step 3: "From the modulation of the lower border of the projective plane, the visual rays leading to the central vanishing point are drawn one by one—in other words, the pavement plane vanishes at infinity."
>
> Step 4: "The observer's eye is located at the same height as the central vanishing point, and the depth visuals are drawn from the observer to the floor base line modulation."
>
> Step 5: "The distance between the observer and the projective plane is defined in a lateral view. Where the depth visual rays intersect the projective plane, they define the succession of the transversal lines on the plane."
>
> Step 6: "The depth of the transversal lines is drawn. In order to do this, lines are drawn parallel to the floor base line from the intersections of the depth visuals with the projective plane. The floor becomes modulated when this design is superimposed on Step Three."
>
> Step 7: "Given that the floor modulation is the same and proportional in both directions, it must be legitimized in perspective by the diagonal line of the reticulate."
>
> Step 8: "The visual horizon is drawn on the projective plane, running necessarily through the central vanishing point. This line is parallel to the floor line and runs along the entire width of the visual field."[11]

It has been pointed out that analyses of Alberti's method applied to a checkerboard pavement can lead to somewhat different results.[12] This caveat notwithstanding, the impact of Alberti's

10. Field, Lunardi, and Settle, "Perspective Scheme of Masaccio's *Trinity* Fresco," 83–85 (app. 1).

11. Salgado, "Geometric Interpretation of the Albertian Model."

12. Green and Green, "Alberti's Perspective," 641–45. This paper is based on a mathematical analysis of reports by Cecil

work on Renaissance art and architecture cannot be overstated, particularly as he was the first to offer a comprehensive treatise on the application of optics and geometry to art.

Following the appearance of Alberti's *On Painting*, other artists began to take an interest in writing about the principles being applied to the science of perspective. In about 1450, **Lorenzo Ghiberti** (1378–1455), Brunelleschi's nemesis and a sculptor of extraordinary talent, wrote *I Commentarii*. A history of art in Italy, it included information on perspective gleaned from the works of early scholars such as Al-Hazen and others, much of which Ghiberti copied verbatim and then integrated into a cohesive treatise. According to Martin Kemp, Ghiberti succeeded in his anthology in "bringing the mediaeval science and the artists' system into harmony."[13] Ghiberti's own perspectival masterpieces are discussed in depth in chapter 2 (see 2.3.5.2) and later in this chapter (see 3.1.2.3b).

Although he was not a Florentine, **Piero della Francesca** (ca. 1410/20–92) arrived in Florence at age twenty-four, and he periodically resided and worked there throughout his career. While there, he studied the artistic traditions and evolving techniques. Written several decades after Alberti's *De pictura*, Piero's significant treatise on perspective in painting, *De prospectiva pingendi*, formulated the mathematical and geometrical structure underlying the principles of proportion and foreshortening. Piero used the term "proportion" to refer to the contours and outlines of objects placed in correct relationship to their locations on the painting surface. Foreshortening was used to denote the proper depiction of an object that, on the painted surface, is receding away from the observer. Piero was well versed in Euclidian geometry, and one of his major contributions was to show how, with the use of accurately determined coordinates, the artist could apply the rules of mathematical precision to objects of any shape one wished to paint in proper proportion. In two other books, Piero focused more specifically on pure mathematics and geometry, demonstrating his proficiency in both disciplines.

De prospectiva pingendi is a practical text in which Piero provided basic instructions for diagraming a proposed painting. He discussed painting as being composed of three components: drawing, color, and proportion, primarily focusing on the methods of producing proportionately accurate works of art. The diagram in figure 3.2, commissioned by the authors solely for the purpose of demonstrating the principle, shows only one object to be painted. In this diagram, the tree shown on the right side of the figure is to be projected onto a painted surface in the correct mathematical proportion to its actual size. The observer (denoted by the eye) is viewing a tree, the height of which is shown by the line *def*, or L_1, which is to be projected accurately in proportion onto the canvas, shown from the side and denoted by the line *AB*. The figure is drawn such that the reader is looking parallel to the plane of the picture frame, and thus it appears as a straight line. The observer is facing the picture frame and views the line *def*, which forms the base of a triangle whose tip (point *o*) is at the observer's eye level. A line projecting straight from the observer's eye to the tree is represented by line

Grayson ("L. B. Alberti's 'Construzione legittima,'" *Italian Studies* 19 [1964]: 14–28) and Alessandro Parronchi ("La 'Construzione legittima' e uguale alla 'construzione con punti distanza,'" *Rinascimento* 4 [1964]: 35–40).

13. Kemp, *Science of Art*, 26.

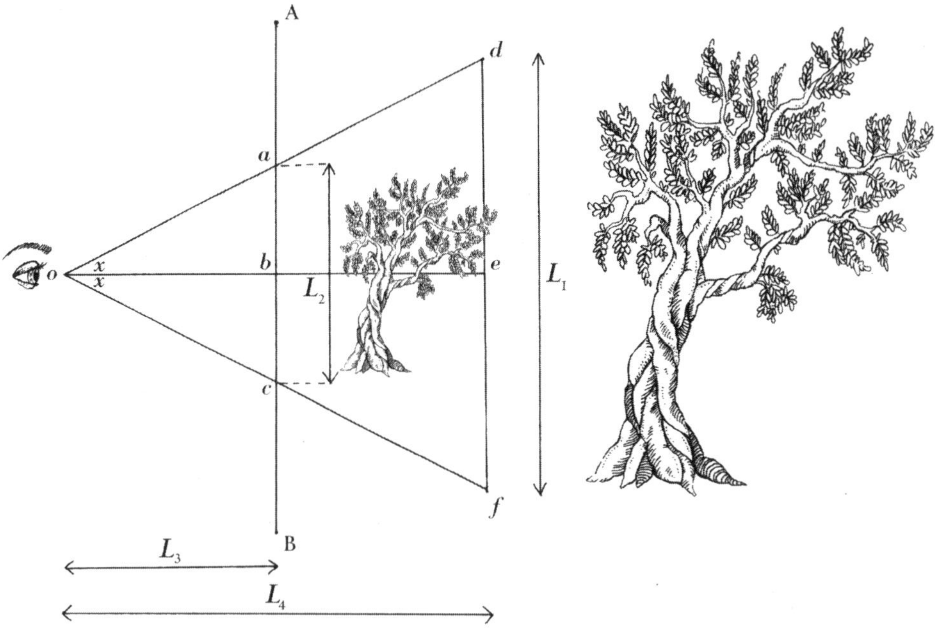

Figure 3.2: Depiction of foreshortening (drawing by Justine E. Stevens commissioned by the authors).

obe, which intersects line *def* at its center at point *e*. This creates two identical, albeit inverted, triangles each with the angle *x*. With line *AB*, which intersects the triangles at points *a* and *c* and line *obe* at point *b*, four right triangles are created: *oab* and the identical albeit inverted *obc*, and *ode* and the identical but inverted *oef*. Lines *oa* = *oc*, *ad* = *cf*, *de* = *ef*, and *oad* = *ocf*. For convenience, line *def*, the height of the tree, is defined as L_1. Line *abc*, the proportional height of the tree on the surface to be painted, is denoted as L_2. Next, define the line *ob*, the distance from the observer's eye to the surface, as $L_{3.}$ The line (*obe*), the distance from the observer's eye to the tree, is represented as L_4. Each of the smaller right triangles is similar to the larger right triangles. Since the two right triangles defined by *oab* and *ode* share a similar angle *x*, one can write the tangent of *x* (ratio of the opposite side to the adjacent side) as follows: tan $x = (ab)/(ob) = (de)/[(ob) + (be)]$, or upon rearranging and substituting the defined Ls (1, 2, 3, and 4), the equation reduces to the following: $L_2 = (L_3/L_4)L_{1.}$

This simple relationship shows that the painted tree should appear foreshortened by the ratio of the distance the painter is from the frame (L_3) to that of the painter to the original object (L_4). For illustrative purposes, this diagram represents only one object to be painted, whereas in reality an artist using this technique could paint a more complex scene containing multiple objects, each of which would be placed in proportion to the others depending on their location. A similar approach as outlined holds for any number of objects regardless of their orientation and position.

Art historian Kirsti Andersen has presented a rigorous discussion and mathematical analysis of perspective and foreshortening, as well as a thorough treatment of Alberti and

Piero della Francesca's methods for achieving accurate perspective.[14] Two other eminent art historians, Samuel Y. Edgerton Jr.[15] and Martin Kemp,[16] have also provided lucid explanations of this subject with helpful diagrams and insightful discussion delineating the progression of the field from basic geometric optics through the use of perspective in art.

3.1.2.3 Applying the Principles of Perspective

a. Masaccio

Tommaso di ser Giovanni Cassai (1401–28), known as Masaccio, was the first Florentine painter known to use geometric linear perspective incorporating a centric point and a viewing distance in his work. The young artist was a friend of Brunelleschi and a member of the Florentine Guild of Doctors and Apothecaries (which originally included artists), which he joined in 1422. Between about 1425 and 1427, Masaccio painted his now-famous frescoes in Santa Maria Novella and Santa Maria del Carmine. He traveled to Rome in 1427 and died shortly thereafter in 1428, his life tragically cut short at only twenty-seven years of age.

Although Masaccio's Florentine frescoes revolutionized the art world, he enjoyed neither fame nor financial stability in his all-too-brief career. Contemporary documents show that he lacked the resources to pay the full enrollment fee required for joining the guild. He paid only a small portion of the fee and remained indebted to the organization for the balance. The struggling young artist, beset with financial problems, was also attempting to support his widowed mother and brother; he owed money to various merchants; and he was in arrears on the rent for his workshop.[17] It was not until about nine years after his death that Alberti, in his treatise *On Painting*, gave Masaccio and his Florentine masterpieces the recognition they so well deserved.

Masaccio created the *Trinity* in the church of Santa Maria Novella (fig. 3.3) soon after Brunelleschi completed his perspective drawings and optical devices using the Baptistery and Palazzo Vecchio. Masaccio's likely use of Brunelleschi's rules sent traditional artistic forms in a new direction by incorporating both the existence of a point of convergence and the illusion of depth. According to historian Thomas Settle, Masaccio's fresco in Santa Maria Novella could arguably be called "the first 'modern' science" in that it represented "a physical world using another 'language' and in another medium."[18] The *Trinity* is a depiction of God the Father, Christ the Son on the cross, a dove representing the Holy Spirit, and Mary and Saint John the Evangelist set in the interior of an alcove. The two figures pictured outside the alcove are probably the donor, Domenico Lenzi, and his wife. Interestingly,

14. Andersen, *Geometry of an Art*, 17–40.

15. Edgerton, *Renaissance Rediscovery of Linear Perspective*, 32–90; Edgerton, *Mirror, the Window, and the Telescope*, 117–50.

16. Kemp, *Science of Art*, 21–35.

17. Ahl, *Cambridge Companion to Masaccio*, 5.

18. Settle, "Traces of Science and Engineering" (http://www.imss.fi.it/~tsettle/traces.html).

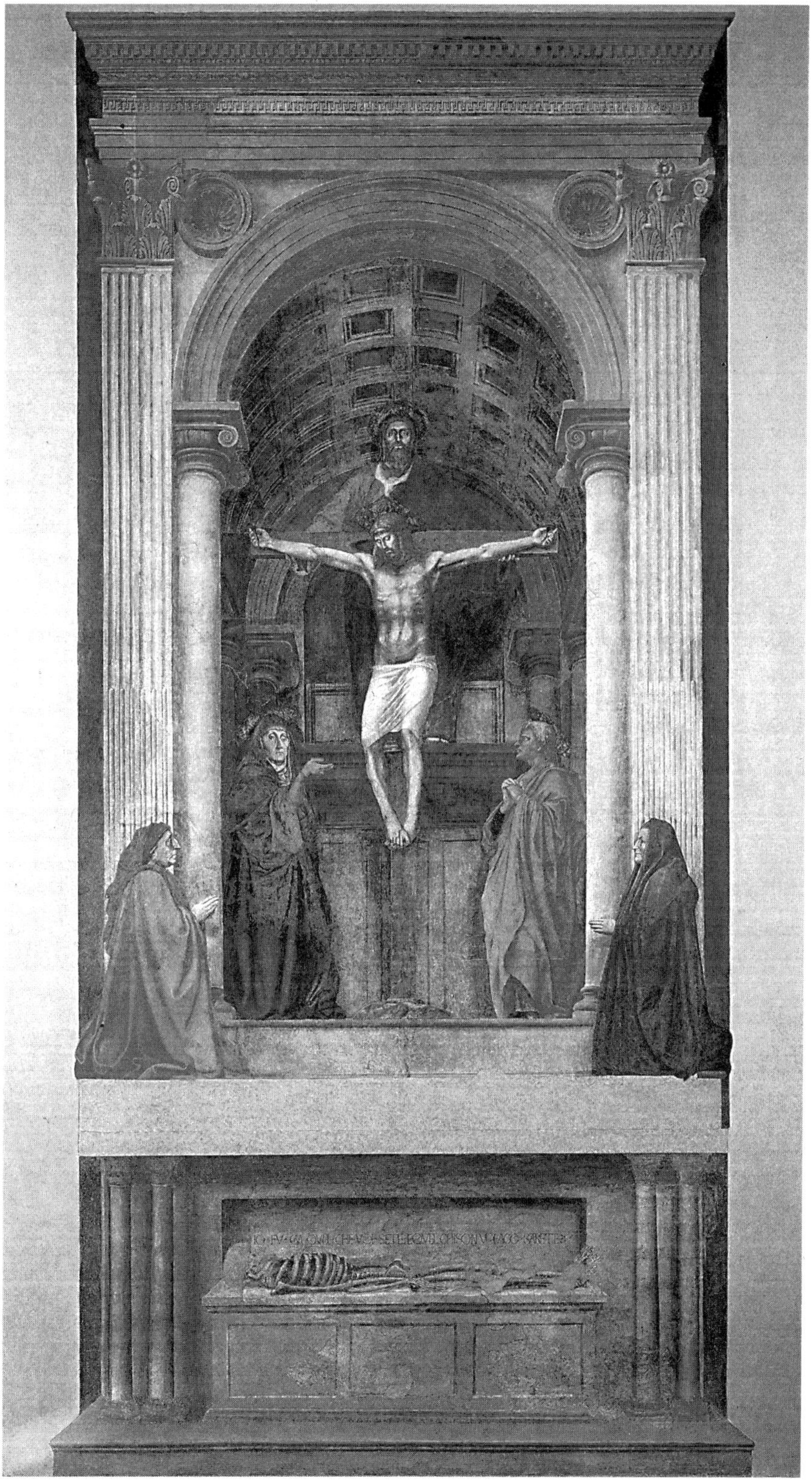

Figure 3.3: Masaccio, *Trinity* (ca. 1425–27), fresco, Santa Maria Novella (Alinari Archives, Florence, photograph by George Tatge [CAL-F-009620–0000]).

although the figures of Mary and Saint John follow the rules of perspective with their foreshortened forms, the portrayal of God the Father does not, perhaps reflecting Masaccio's desire to depict God as outside human boundaries. An altar with a sarcophagus bearing a skeleton supports the fresco and above the skeleton is an inscription generally translated as, "I once was what you are now, and you will be what I am."

Art historians have attempted to establish as definitively as possible Masaccio's use of Brunelleschi's principles regarding linear perspective. In order to analyze the mathematical techniques Masaccio employed in the *Trinity*, one should focus on the architectural elements in the fresco. Given that both Euclid's *Optica* and the astrolabe were available to Masaccio and his contemporaries, what set Brunelleschi's work apart was his establishing a mathematically correct viewing distance and a centric point, a method adopted by Masaccio. A particularly thorough and extensive study was conducted by Field, Lunardi, and Settle to determine a viewing distance and centric point, using calculations one could reasonably deduce were used by Masaccio. From direct measurements made on the painting, Field, Lunardi, and Settle located the centric point, from which converged the perpendicular lines (orthogonals) at the edge of objects such as the architraves on the left and right sides and the lines that passed through the edges (longitudinal) of the columnar abaci.

Because Masaccio, when painting the fresco, did not strictly adhere to all of the geometric parameters, it is unlikely that scholars will ever precisely know how Masaccio constructed his plan for the perspectival layout. Most accept the following scheme presented by Field, Lunardi, and Settle. The centric point found by them, and in agreement with others, is level with the lower step on which the donors are kneeling.[19]

In addition, Field, Lunardi, and Settle determined the ideal viewing distance: "When one stands in front of the original fresco, the minimum viewing distance for the *Trinity* seems to be about the width of the aisle of the church (at this point it is 686.25 cm, about 11.75 braccia)." They further stated they were "confident that this distance must have been incorporated into the perspective construction as the distance of the ideal observer from the picture."[20] When standing at that distance the eye of the observer is level with the centric point. The investigators calculated the centric point and optimal viewing distance using their meticulous measurements, first utilizing the abaci on the columns and secondly those of the barrel vault rib system shown in the fresco, including the locations of the semicircular transverse ribs and arcs. From the positioning of the longitudinal ribs, they were able to "estimate the 'true' transverse dimension of the coffers," which led them to agree with previous assumptions that the coffers in the vaulted ceiling were intended to be square. Field, Lunardi, and Settle concluded that the viewing distance discerned by observation was consistent with the distance determined by making mathematical calculations.[21] Results from this investigation confirm the prevailing view that Brunelleschi provided perspectival input into the painting of the *Trinity*.

19. Field, Lunardi, and Settle, "Perspective Scheme of Masaccio's *Trinity* Fresco," 37.

20. Ibid., 39.

21. Ibid., 40.

When the viewer stands in the correct position, he becomes part of what appears to be a three-dimensional scene due to Masaccio's innovative and effective presentation of depth in his fresco painted on a flat wall. The placement of the figures in the fresco, including those of the donor and his wife, further add to the impression of three-dimensionality, as does the sarcophagus that appears to be recessed in a niche. Although attempts to exactly measure and recreate Masaccio's perspective scheme have left some questions unanswered, these analyses provide indisputable evidence that a detailed mathematical component leading to the establishment of a centric point was an integral part of the *Trinity* fresco. That there is not unanimity relating to the fresco's asymmetry is of little consequence when one considers that the *Trinity* was not intended as an exercise in geometry. Rather, it is a revolutionary work of art in which mathematical principles were used to present a perspectivally correct scene.

The other remarkable work by Masaccio is the frescoed Brancacci Chapel in Santa Maria del Carmine. Masaccio created and directed these frescoes in collaboration with Masolino (ca. 1383–ca. 1447), and they were completed by Filippino Lippi (1457–1504) in the 1480s after Masaccio's death. The Brancacci Chapel walls are frescoed with scenes portraying the life of Saint Peter from his calling to his martyrdom. Two additional frescoes adorn the room: the temptation of Adam and Eve and their expulsion from the Garden of Eden by Masolino and Masaccio, respectively.

There are two rows of frescoes, one above the other on three walls, with the altar in the center of the back wall. The top row is framed by the two frescoes of Adam and Eve. The remainder of the top row contains four panels, and the bottom row has six panels, all depicting the life of Saint Peter. Although the ceiling originally contained frescoes by Masolino, in the mid-eighteenth century it was repainted by Vincenzo Meucci with an unremarkable Baroque-like scene containing oversized figures.

The story of the life of Saint Peter, arranged in rows on the walls, begins at the top row on the left and is situated to the right of the *Expulsion from Paradise* by Masaccio. His first Saint Peter panel is the *Payment of the Tribute Money*, a narrative presented in three stages. With the centric point at the head of Christ and the orthogonal lines emphasized by the outstretched arm of Peter and the direction of the light, the figure of Christ clearly commands the scene (fig. 3.4). After the next fresco, Masolino's *Sermon of Saint Peter*, Masaccio created the *Baptism of the Neophytes* (to the right of the altar) in which he used light, perspective, and human expression to achieve a realism generally not present in the work of Masolino. Particularly effective in this fresco is Masaccio's depiction of the naked young man who appears to be shivering from the cold. That top row of frescoes finishes with a cycle scene depicting *The Healing of the Cripple and the Raising of Tabitha*, followed by the *Temptation of Adam and Eve*, both by Masolino.

The bottom row on the left wall of the chapel continues the St. Peter cycle directly under the *Expulsion* fresco and begins with *Saint Peter in Prison* by Lippi, followed by the *Raising the Son of Theophilus and Saint Peter Enthroned.* Started by Masaccio and later completed by Lippi, this panel contains portraits of contemporaries of Masaccio as well as one that is believed to be a self-portrait. The next two panels also show the work of Masaccio. *Saint Peter Healing with His Shadow* and *Saint Peter Distributing Alms and the Death of Ananias* are

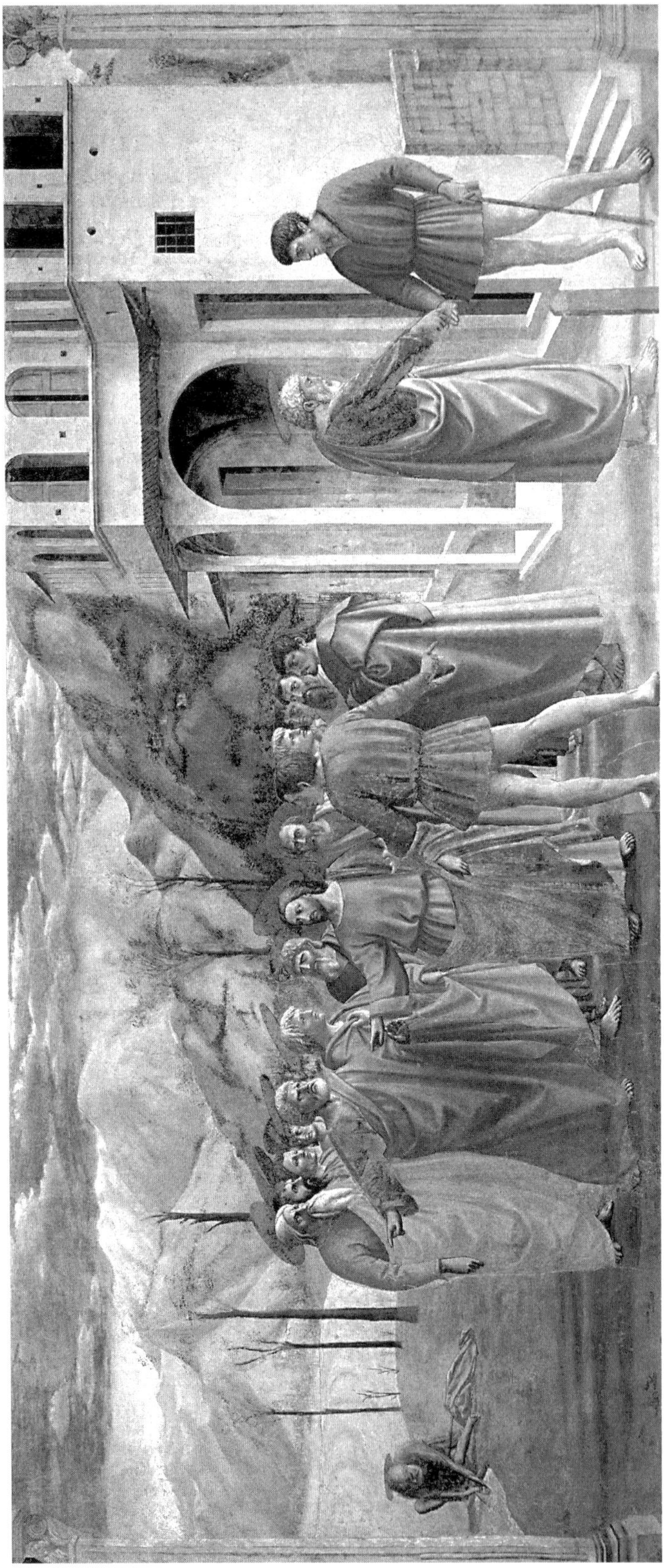

Figure 3.4: Masaccio, *Payment of the Tribute Money* (ca. 1425–27), fresco, Brancacci Chapel, Santa Maria del Carmine (Alinari Archives, Florence, photograph by Serge Domingie [AGC-F-000550-0000]).

located on either side of the altar, with the vanishing point for these two panels appearing to be in the altarpiece itself. In *Saint Peter Healing with His Shadow*, Masaccio used the gradual increased elevation of the positions of the cripple from crouching to kneeling to standing to aid in the sense of depth. Additionally, the eye is drawn upward by the architectural elements depicted in the painting. By contrasting the forward movement of Peter and John with the receding architecture and the cripple in gradually rising stances, Masaccio achieved a dramatic effect. It is not only his use of perspective, however, but his interplay of light and shadow that makes the scene memorable. The fresco to the right of the altar and the last in the cycle attributed to Masaccio is *Saint Peter Distributing Alms and the Death of Ananias*. In this panel, Masaccio draws the viewer's eye upward and toward the background of the fresco, from the body of Ananias on the ground to the standing figures and receding buildings and mountains. The final two panels, *Peter Debates Simon/Crucifixion of Peter* and *Freeing Peter from Prison*, are the works of Lippi.

In the Brancacci Chapel fresco cycle, Masaccio adroitly employed the principles set forth by Brunelleschi. He created orthogonal lines that clearly converge in the scenes, reminiscent of his work on the *Trinity*. That Masaccio's calculations may not have been exact in all instances does not detract from the overall three-dimensionality of the frescoes. Masaccio and his fellow artists also used perspective techniques to unify the chapel walls. The artists applied the rule of "horizon line isocephaly," so termed by Samuel Edgerton to describe "the phenomenon whereby, if we see other persons standing on the same plane as ourselves, the apparent diminution in the size of more distant figures begins with the feet; the heads of all figures standing on the same level as the viewer are always seen aligned with his own head on the common horizon."[22] Thus, the viewer is able to follow the scene from one panel to the next with ease. The artist enhanced this unity by extending the landscape across three of the top-row panels. Masaccio's use of light helped give the illusion of three-dimensionality to human forms, and the drape of their clothing and the employment of less distinct colors in the background certainly enhanced the realism he was able to so effectively achieve.

Masaccio's *Expulsion from Paradise* is both innovative and moving with its image of raw human suffering. The trauma experienced by Adam and Eve is apparent, and the interplay between the light and shadows adds to the drama of the fresco. This fresco by Masaccio, on the left wall directly opposite Masolino's traditional and rather emotionless treatment in the *Temptation of Adam and Eve,* creates a striking artistic contrast.

Giorgio Vasari's comments concerning Masaccio's frescoes highlight the significance his art had on the painters who were to follow. Vasari (1511–74; see 2.2.1.3) came to Florence from Arezzo at about thirteen years of age to embark on his broad humanist education. After 1531, Vasari lived in both Rome and Florence where he excelled as a painter, architect, and writer. While in Florence, Vasari was appointed to renovate the Palazzo Vecchio for Cosimo de' Medici, creating magnificent frescoes throughout; he was also the architect of the Uffizi. Vasari may be best remembered, however, for his book entitled *The Lives of the Most Eminent Painters, Sculptors, and Architects*; the first edition appeared in about 1550, followed by a later

22. Edgerton, *Renaissance Rediscovery of Linear Perspective*, 26.

expanded edition in about 1568. In *Lives*, Vasari gives the reader interesting personal information about the artists and their contributions. Although some doubt has been cast on the accuracy of his historical details, his analysis of the works of those talented people and the manner in which they were perceived is enlightening and has served as a valuable resource for art historians. Writing about Masaccio, Vasari commented, "Masaccio, introducing many new methods, made foreshortenings from every point of view much better than any other who had lived up to that time."[23] When discussing his work in the Brancacci Chapel, Vasari summarized the importance of Masaccio's contribution to the use of perspective: "In short, all those who have sought to learn that art [use of perspective] have ever gone to this chapel to learn and grasp the precepts and the rules for good work from the figures of Masaccio."[24]

Remarkably, Masaccio's two seminal works were almost destroyed. During a renovation of Santa Maria Novella in the mid-sixteenth century, a portion of the *Trinity* was hidden when a new altarpiece was installed. It was not until about 1860 that the top part of the fresco was rediscovered, detached, and moved to a location alongside the rear door. About a century later, when the bottom section was found in its original position behind the altar table, the two sections of the fresco were reunited. The Brancacci Chapel frescoes were also threatened when, in about 1690, a proposal was made to replace the frescoes that were thought to be outdated. Fortunately for the art world, those opposing the suggestion prevailed. Then, in 1771, a fire spread through the main section of the church. Parts of the Brancacci Chapel were damaged by the flames, and soot darkened the work of Masaccio and the other artists. Interest in the chapel was lost, and the frescoes remained unrestored for two centuries. In retrospect, this neglect served the art world well, since restoration techniques improved dramatically during the intervening years. The successful five-year restoration project begun in 1983 returned the frescoes to their original state with their vibrant colors and minute details.[25]

b. Lorenzo Ghiberti

The Florentine Lorenzo Ghiberti was apprenticed in the goldsmith shop of his stepfather and also received training as a painter. It was his expertise in the use of bronze, however, that made him one of the major artistic figures of the quattrocento. When the Arte del Calimala held a competition to choose an artist to create doors for the Baptistery, Ghiberti won the commission in 1403. The technical methods used for creating the doors are detailed in chapter 2 in the section on bronze sculpture (see 2.3.5.2a). Here the focus is on a different aspect of Ghiberti's contribution to art and science: his use of perspective.

Ghiberti completed the doors now on the north portal, primarily devoted to the New Testament, in about 1424, and he executed the magnificent east doors from about 1429 (or earlier) to 1452. (The doors now installed at the Baptistery are copies; the original panels

23. Vasari, *Lives of the Most Eminent Painters*, trans. de Vere, 1:382.

24. Ibid., 1:388.

25. Ahl, *Cambridge Companion to Masaccio*, 6–7.

have undergone extensive restoration since the flood of 1966 and are preserved in the Museo dell'Opera del Duomo.) Illustrating scenes from the Old Testament, the east doors are moving testimony to Ghiberti's commitment to the challenging goal of melding the principles of perspective with the depth that a relief naturally provides. In that endeavor he succeeded magnificently. Using mathematical calculations, he created the background landscapes and architectural depictions in correct relationship to the figures in the foreground, diminishing the size of figures as they recede toward the background.

The east doors—which Michelangelo referred to as the *Porta del Paradiso* (*Gates of Paradise*)—have ten remarkable and unique square panels. The grandeur of the *Gates of Paradise* makes the viewing of each individual panel an extraordinary experience. Adding to the impact of the doors is Ghiberti's framing of each scene with relief figurines and heads in roundels. The panels comprising the left side of the east doors portray (from top to bottom) *Adam and Eve*, *Noah*, *Jacob and Esau*, *Moses*, and *David*. The right side (from top to bottom) shows *Cain and Abel*, *Abraham*, *Joseph*, *Joshua*, and *Solomon and the Queen*. According to Martin Kemp, the panel that contains the "most harmoniously perspective relief" is *Jacob and Esau* (fig. 3.5). Kemp states that the scene has a "proportional grace" and "visual measure" that stand the test of careful scrutiny with orthogonals consistently converging at the center of the panel.[26] Furthermore, when *Jacob and Esau* is viewed together with the *Joseph* panel directly to its right, the perspectives of both panels work in unison to bring the two sides of the doors in harmony with one another. It is of little wonder that visitors to Florence crowd the area between the Cathedral and the Baptistery to see the doors (despite their being copies), a true testament to the skills required and the lasting impact of Ghiberti's accomplishment.

c. Donatello

Another prominent Florentine artist to employ perspective techniques was Donato di Niccolò di Betto Bardi, known as Donatello (ca. 1386–1466). One of the greatest sculptors of the quattrocento, his early works were primarily created from marble, but after about 1420 he began to concentrate on the use of bronze. In Siena, around the same time that Masaccio was completing the *Trinity*, Donatello finished the *Feast of Herod* for the Baptistery font; it can still be seen in that location. There, Donatello applied perspective to produce a distinctive effect, *rilievo,* making an important contribution to the technique of integrating perspective and relief. The art of relief refers to the use of protruding parts on an otherwise flat sculpture or painting. That technique, in practice since classical times, gives the appearance of depth, but is not to be confused with *rilievo*, a more visually accurate relief. Donatello used a flattened or shallow relief (*rilievo schiacciato*) expertly to create optical perspective within the plane of the sculpture, a departure decidedly distinct from that of traditional relief.

Donatello's magnificent freestanding sculpture and art can be seen throughout Florence, but two of his works provide excellent examples of his use of perspective in shallow relief. Donatello's *Saint George and the Dragon* (1417) on the base of the statue of *Saint George* was

26. Kemp, *Science of Art*, 24.

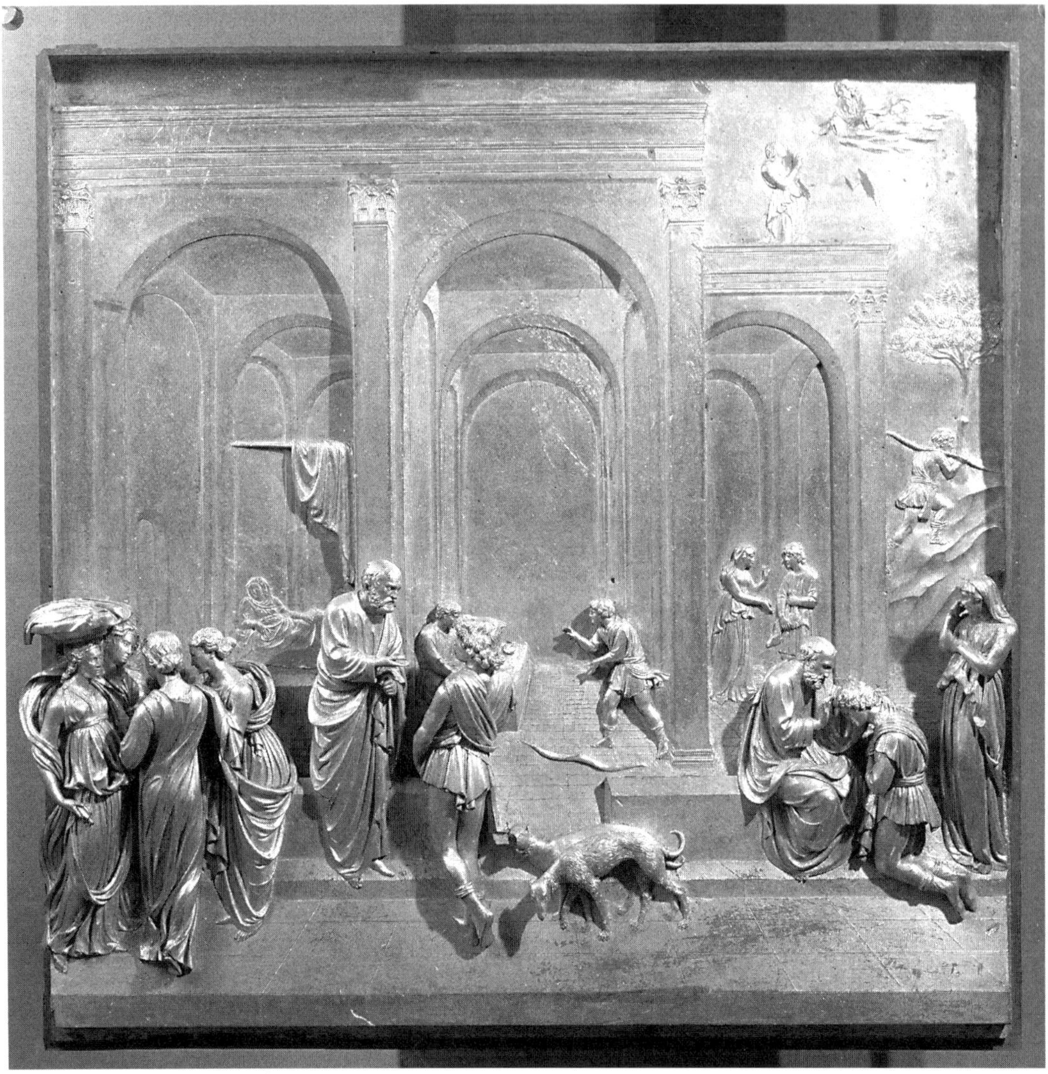

Figure 3.5: Lorenzo Ghiberti, *Story of Jacob and Esau* (ca. 1429–52), cast bronze, east Baptistery door (Nicolò Orsi Battaglini/Alinari Archives, Florence, photograph by Nicolò Orsi Battaglini [OBN-F-000343–0000]).

created for Orsanmichele; the original is now on display in the Bargello Museum. Although there is some controversy about this relief's being cited as a very early use of perspective,[27] it seems unlikely that the receding orthogonal lines reflect merely a random act. In creating this low relief, Donatello demonstrated the possibility of transferring the techniques of perspective used in painting to the medium of marble sculpture. Because his protean artistic style varied widely from work to work, often with resulting images' lacking conventional perspective rules, it is difficult to know his objective in each instance.

27. Edgerton, *Renaissance Rediscovery of Linear Perspective*, 183n3.

A second example of Donatello's innovative and unique approach to the geometric method is in a roundel, the *Assumption of Saint John the Evangelist* (ca. 1430), in the Old Sacristy in the church of San Lorenzo. Although the roundel is badly deteriorated, it clearly depicts steps on which individuals are positioned. For this work, Donatello altered the traditional rules of perspective in favor of the presentation, but he nonetheless dramatically achieved the feeling of a receding space. He abandoned strict adherence to a single vanishing point; however, the orthogonals, which might otherwise be confusing, are understandable when one considers the placement of the roundel in relation to other significant adjacent works. Clearly, the highly talented and versatile Donatello advanced and expanded the dimensions of the use of geometrical principles.

d. Paolo Uccello

Paolo Uccello (1397–1475) is a notable example of a transitional artist in terms of Albertian principles. His expertise as an artist can be seen in the Green Cloister (Chiostro Verde) of Santa Maria Novella, where the wall frescoes by Uccello and other artists depict stories from the Old Testament. Initiated between about 1425 and 1430, one scene is a particularly interesting example of the way in which Uccello incorporated perspective and color. In this early work, *Deluge*, it is the lightning from the storm that provides the light source, while the tree that is struck serves as the centric point. The turmoil depicted in the scene is both enhanced and focused by Uccello's use of mathematical principles. Some art historians have suggested that *Deluge* includes a portrait of Alberti.

After his work on the Green Cloister, Uccello seems to have made the transition to a more enhanced Albertian methodology. In about 1456 Uccello completed a triptych, the *Battle of San Romano* (fig. 3.6), the center panel of which is now in the Uffizi Gallery. The triptych illustrates the battle fought in 1432 between Florence and Siena. It is the combination of the action of the conflict and the mathematical perspective that gives the panel an aura of tension and grandeur. The chaotic scene is balanced by the geometric arrangement of the perspective grid reinforced by the broken lances and foreshortening of the fallen horse and soldier. Added to Uccello's geometric composition was the portrayal of foreground objects in brighter colors than those meant to recede, an idea suggested by Ptolemy many centuries earlier. In writing about Uccello, Vasari remarked, "Now Paolo was for ever investigating, without a moment's intermission, the most difficult problems of art, insomuch that he reduced to perfection the method of drawing perspectives."[28] Uccello masterfully executed the skill of projecting depth with the use of a grid, thus integrating the science of geometry and art. Fortunately his grid drawing for the *Chalice* (ca. 1450) has been preserved and is part of the collection in the Uffizi Gallery, as is the grid he created for his fresco of Sir John Hawkwood (see 2.2.1.2b).

28. Vasari, *Lives of the Most Eminent Painters*, trans. de Vere, 1:342.

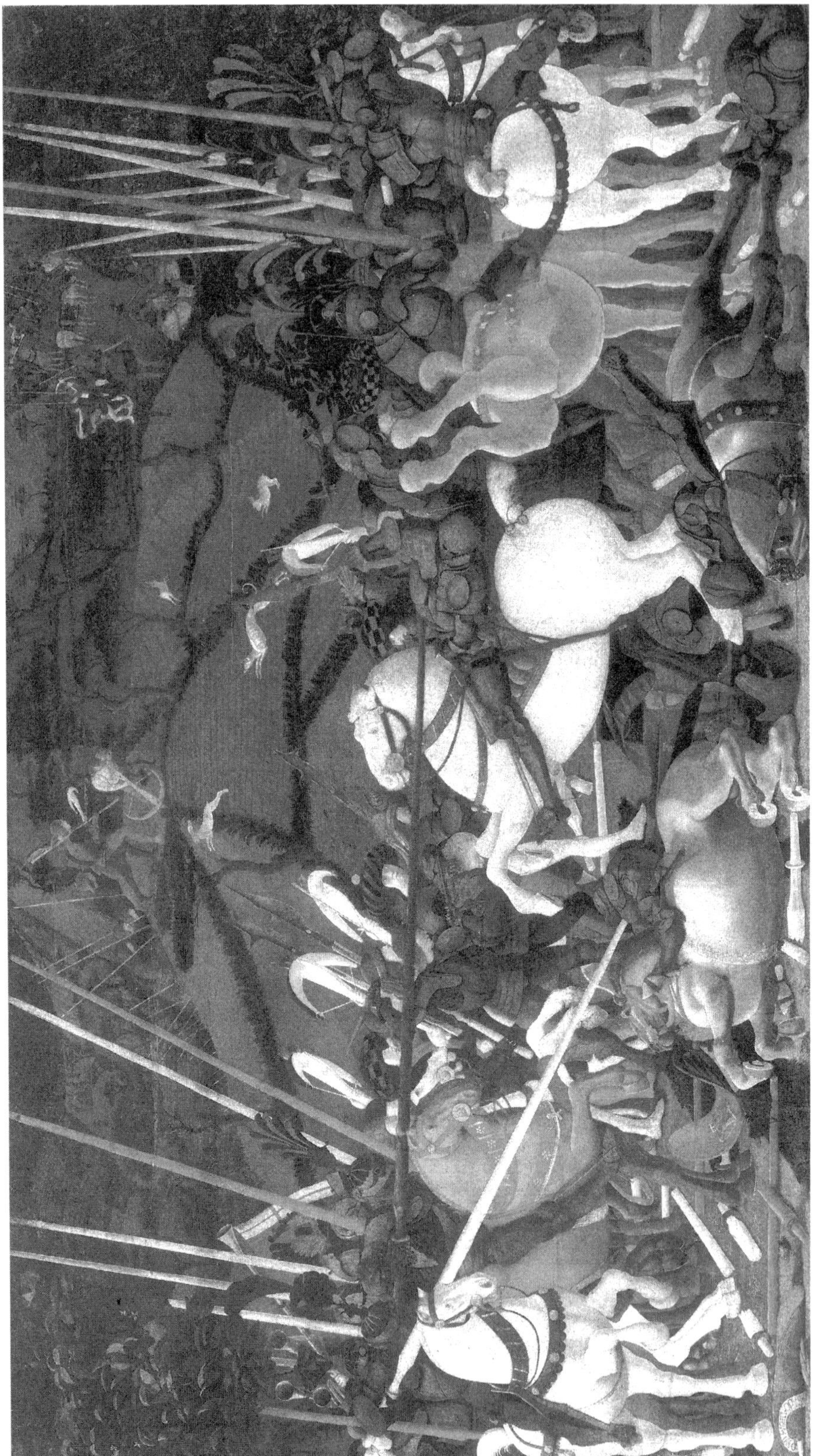

Figure 3.6: Paolo Uccello, *Battle of San Romano* (ca. 1456), fresco, Uffizi Gallery (Alinari Archives, Florence, photograph by Nicola Lorusso, reproduced with the permission of Ministero per i beni e le Attivita Culturali [AGC-F-000112-0000]).

e. Fra Angelico

Born Guido di Pietro, Fra Angelico (ca. 1400–1455) was first person trained as a manuscript illuminator. After becoming a Dominican friar, he joined the community of San Domenico di Fiesole sometime between 1418 and 1421, moving in about 1441 to San Marco in Florence. Fra Angelico did not make major written scholarly contributions to the principles of perspective during his years at San Domenico and San Marco. Recent analysis, however, of his paintings that predated Alberti's treatise and were possibly created even earlier than Brunelleschi's and Masaccio's work offers testimony that Fra Angelico was engaged in accurate geometrical experimentation at about the same time as Brunelleschi.[29]

The Convent of San Marco was in a state of disrepair when the Dominicans began reconstruction in 1437 under the patronage of Cosimo de' Medici. In approximately 1440, Cosimo commissioned Fra Angelico to paint an altarpiece for the high altar. The resulting painting, *Madonna and Child with Saints and Angels*, is one of the first panel pieces to exemplify the rules of perspective. The foreshortening of the carpet gives the altarpiece its sense of depth and provides orthogonal lines that direct the eye to the vanishing point at the midsection of the Virgin Mary. Worthy of note is Fra Angelico's use of symmetry and balance in the diminution in size of the various figures as they recede into the background. Although the panel was damaged during a restoration attempt, it remains an exceptional work of art. By 1443 the individual cells for the friars at San Marco had been rebuilt and were ready for occupancy. Fra Angelico and his assistants undertook the task of creating frescoes to adorn the convent and the cells. He participated in painting the frescoes on the ground floor and supervised the paintings in the cells occupied by his fellow friars. Each cell, although otherwise small, spartan, and austere, contains a fresco portraying a scene from the life of Christ; the scenes are arranged in no particular order as one moves down the hall from cell to cell. At the top of the stairs leading to the cells, Fra Angelico painted the marvelous *Annunciation* (ca. 1438–45), also commissioned by Cosimo de' Medici. Framed by the entranceway to the corridor, the fresco makes a significant impact as the viewer emerges from the head of the stairway into the hall (fig. 3.7). Today the rooms and museum in San Marco stand as testimony to the beauty and depth of Fra Angelico's artistry.

f. Continuing the Process

Andrea del Castagno's fresco of the *Last Supper* in the *cenacolo* of Sant'Apollonia, painted in about 1450, is another interesting example of the use of perspective (see 2.2.1.2e). The artist's technique in painting the tiled floor, the walls, and the ceiling in a mathematically calculated manner gives the illusion not only of the expected depth in the fresco itself, but also of being a continuation of the actual refectory.

Domenico Ghirlandaio (see 2.2.1.2c) was a master of fresco painting, and to that art he applied striking perspective. In his *Last Supper* (1480) in the refectory of the Church of

29. Edgerton, *Mirror, the Window, and the Telescope*, 108–16.

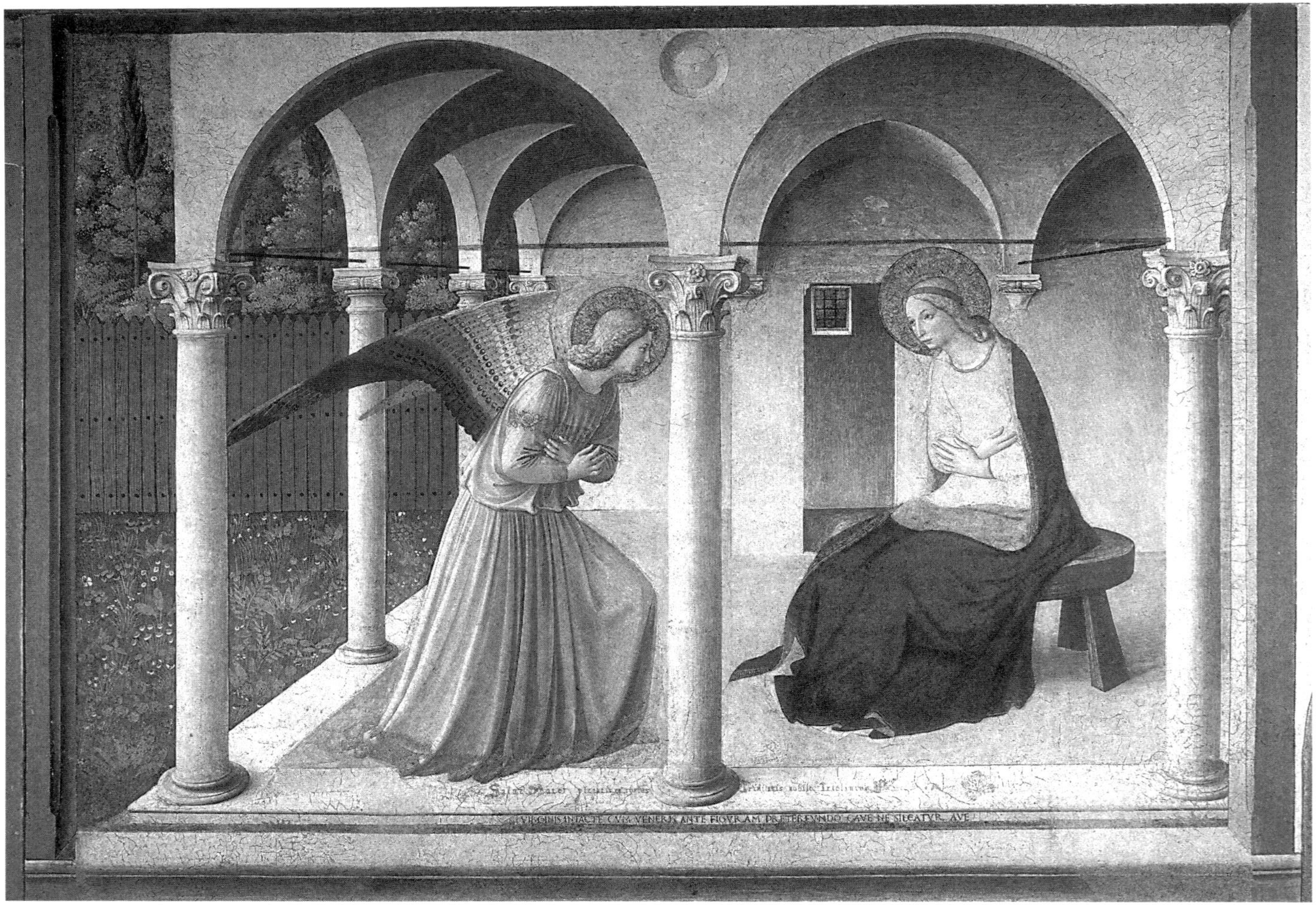

Figure 3.7: Fra Angelico, *Annunciation* (ca. 1438–45), fresco, Convent of San Marco (Nicolò Orsi Battaglini/Alinari Archives, Florence, photograph by Nicolò Orsi Battaglini [OBN-F-000057–0000]).

Ognissanti, Ghirlandaio's depiction of the arches, the tiled floor, the table, and the opening to a garden in the background work harmoniously in concert to draw the viewer into the space occupied by Jesus and his assemblage.

As the fifteenth century progressed, some artists began to modify Alberti's rules in order to achieve an even more dramatic effect. One of these was Domenico Veneziano (ca. 1410–61). In about 1444 he completed his *Saint Lucy Altarpiece* (also referred to as *Madonna and Child with Saints*), now in the Uffizi Gallery. The figures in the painting occupy a symmetrical space that is clearly perspectival, with the tiles on the floor containing a pattern that is foreshortened and arches that appear to recede. However, the remarkable painting does exhibit variations on Alberti's rules. Most notably, the figures are oversized when compared to the architecture in the painting.

Acclaimed artist and mathematician Piero della Francesca was born in Borgo San Sepolcro and made major contributions to the intersection of art and mathematics discussed at length earlier in this chapter (see 3.1.2.2). He painted in Florence for a brief period of time, and one of his important works is displayed in the Uffizi Gallery: a diptych of the *Duchess and Duke of Urbino* (ca. 1472). This painting is an interesting pairing of the oversized portraits of Battista Sforza and Federigo II da Montefeltro against very distant receding perspectival landscapes.

Leonardo da Vinci (1452–1519), a true genius of the Renaissance, was the son of a Florentine notary and his mistress. Showing an aptitude for the arts, he became an apprentice in the workshop of Verrocchio and joined the painter's guild in 1472. After completing his first major works, *Annunciation* (ca. 1472/74) and *Adoration of the Magi* (1481), both now in the Uffizi Gallery, Leonardo left Florence to work for the Duke of Milan, Ludovico Sforza. In 1490, Leonardo began to formally learn the mathematics required for using perspective and optics from Fazio Cardano of the University of Pavia and soon began filling notebooks with his mathematical formulas for geometrical drawings, a practice he continued throughout his life. He meticulously studied many aspects of the use of perspective and mathematics, applying them to his models, scientific drawings, and designs, as well as his art. For example, in his famous depiction of the Vitruvian Man drawn for the new edition in 1511 of Vitruvius's opus, the height of the man divided by the height of his navel yielded an irrational number termed the golden ratio, a numerical value approximately equal to 1.618. In view of the frequency with which it appears in early geometrical constructions and in nature, many contend that the golden ratio depicted in art is simply a normal representation of objects as they naturally appear or as something pleasing to the eye.[30] The three-dimensionality of his invention sketches was innovative, not only in the originality of their designs, but also the manner in which he presented those designs on a flat drawing surface.

30. Of interest, related to the golden ratio is the Fibonacci series, formulated by Leonardo Fibonacci di Pisa (Leonardo Pisano, ca. 1170–1240). It was derived by considering a pair of rabbits and their subsequent reproduction, with certain stipulations. Interestingly, the number of rabbit pairs each month can be predicted by adding the numbers in the previous two months. Moreover, the ratio of rabbit pairs in a given month to the number in the previous month rapidly approaches an irrational number that can be approximated as 1.618, i.e., the golden ratio.

Leonardo was adamant that knowledge of science, including optics and the theory of vision, was vital to those wishing to become accomplished painters. Noted Leonardo scholar Paolo Galluzzi surmised that this Renaissance polymath believed verbal description was inferior to drawing as a means of depicting man and the world.[31] Addressing optics as it related to painting, Leonardo wrote, "The first thing in painting is that the objects it represents should appear in relief, and that the grounds surrounding them at different distances shall appear within the vertical plane of the foreground of the picture by means of the three branches of Perspectives, which are the diminution in the distinctness of the forms of the objects; the diminution of their magnitude; and the diminution in their color."[32]

These are just a few of the many famous artists who graced the Florentine landscape, but what about the others? Not all of the outstanding painters who became adept at using the geometric principles of perspective were involved in the early adaptation of mathematics and geometry to artistic endeavors. The artists discussed here began or advanced a process that continued and evolved for future generations. Exemplifying the spirit of the Florentine Renaissance, they melded art and science so seamlessly that their paintings can be appreciated for their artistic merit and emotional impact without the viewer's having any knowledge of the mathematics behind the art. One notable example is the *Doni Tondo* (ca. 1504) by Michelangelo Buonarroti (1475–1564), another excellent work of art by a master for whom the use of mathematical principles was not so much a learned discipline, but rather an integral part of who he was as a painter.

Some scholars argue that applying rigid mathematical principles of perspective served to restrict artistic freedom, but one has only to look at the works of those painters who mastered the art to see how these new techniques enhanced the artist's ability to create a dynamic painting that, through its apparent physical depth, draws the viewer into the scene. Rather than restricting artistic freedom, the mathematics of perspective provided new avenues by which painters could express their unique artistic talents.

3.2 Cartography: Florentine Maps, a Blending of Art and Science

3.2.1 Early Maps

Modern maps use geometric principles to show exact relationships between areas portrayed, often obtaining data from satellite imaging. In contrast, maps used by ancient cultures were neither drawn to scale nor mathematically derived, but served simply as a way to provide directions to and from various locations using recognizable landmarks. Those maps, however simple, fulfilled their function.

Evidence of early maps has been found around the world, from the pre-Columbian cultures in North America to those of the early peoples in the Arctic and the Pacific Islands. A bronze engraved map from the fourth century BCE was discovered in the Heibei province

31. Galluzzi, *Art of Invention*, 84.

32. Suh, *Leonardo's Notebooks*, 15.

in China, while in China's Hunan province, silk maps have been found that date from the second century BCE. Clay maps dated to 2300 BCE have been uncovered in Mesopotamia (northern Iraq), and there is archaeological evidence that clay maps were also used in Egypt. The Babylonians created what may have been the first map of the world in about the sixth century BCE, but the scant knowledge they possessed of the earth's surface made their clay representations more of a political statement than a useful cartographic instrument.[33] Centuries later, Roman historians spoke of placing maps on public display to celebrate the conquests of their rulers and generals.

Although none of their maps are extant, early Greek natural philosophers had an active interest in cartography. In Miletus, a city in an area on the southwest coast of present-day Turkey, Greek scholars such as Thales (sixth century BCE) and Anaxagoras (500–428 BCE) tried to explain the shape of the earth by relating it to the spherical bodies seen in the sky. Trade fostered the exchange of knowledge and ideas, and led to an increased interest in depicting the earth with some degree of scientific accuracy. Another Milesian, Anaximander (ca. 610–545 BCE), is credited as being the first to construct a map of the known Greek world. Later, in the fourth century BCE, Aristagoras, the former tyrant of Miletus, was attempting to convince King Cleomenes of the Spartans to lead an effort to defeat the Persians and thus liberate the Ionian Greeks. According to Herodotus, he had with him a bronze map of the world, attributed to another Milesian cartographer, Hecataeus (ca. 500 BCE).

In Hellenistic Alexandria, Eratosthenes of Cyrene (ca. 275–195 BCE), a scientist serving as the chief librarian, drew a map of the known world using parallels of latitude and meridians of longitude. Once natural philosophers began to present mounting evidence that the earth was in fact spherical, it became important to estimate its size as accurately as possible. Using a reasonable scientific approach, Eratosthenes noted that during the summer solstice at Cyrene, south of Alexandria, the sun was overhead at noon, while in Alexandria it produced a shadow equivalent to 1/50 of a circle. Making certain assumptions, he concluded that the earth's circumference was 250,000 *stades*, which was fifty times the 5,000 *stades* distance between the two cities. Unfortunately we do not have a correct conversion of *stades* to miles or kilometers, but Eratosthenes's approach provided a means of estimating the circumference of the earth and demonstrated the application of logical reasoning to important problems.

By the second century, the mathematician, astronomer, and geographer Claudius Ptolemaeus (ca. 100–170), known as Ptolemy, was working in the intellectual center of Alexandria, where he assembled his theoretical and practical text on mapmaking, *Geographia.* Composed of eight books or chapters and containing a world map and twenty-six regional maps, *Geographia* moved from the broad view to the more specific: from the world as a whole to the three known continents—Europe, Africa, and Asia—and then to particular areas within those continents. Ptolemy's maps were drawn to scale and used longitude and latitude to project the image of landmasses and seas as accurately as possible. In addition to being a collection of

33. Wilford, *Mapmakers*, 6–11.

maps, *Geographia* was a guide for creating those maps, with portions of the treatise devoted to delineating the actual procedures involved. Ptolemy employed principles governing optics to translate the earth's spherical shape to one that could be depicted on a flat two-dimensional surface. It is not surprising that Ptolemy, an astronomer, emphasized the use of coordinates based on celestial observations and then translated the data to a grid. Keenly aware that it was impossible to accurately portray a spherical surface on a two-dimensional grid, Ptolemy proposed various possible methods of projection. The one he selected as optimum was to retain the parallels (the east-west lines) and have the meridians (the north-south lines) converge at the earth's northern extremity. For small surface areas of landmass or water, the parallels and meridians were to remain at right angles one to another, or nearly so.

Ptolemy also used *Geographia* as a forum to critically examine the projections applied by contemporary cartographers, including Marinus of Tyre, who used a rectangular projection with orthogonal meridians and parallels. In his treatise, Ptolemy presented four methods of map projection: (1) orthogonal meridians and parallels (as per Marinus), (2) linear converging meridians but with curved parallels, (3) curved meridians and parallels with the meridians converging, and (4) a projection of the spherical earth as observed from a distant point.[34]

Discussing his own methodology, Ptolemy distinguished regional cartography—a visual framework of restricted areas—from world cartography, where the relative positions of major landmarks and bodies of water were to be portrayed accurately. Ptolemy stated in book 1, "regional cartography has no need of mathematical method, but here [in world cartography] this element takes absolute precedence . . . It would be well to keep the lines representing the meridians straight, but [to have] those that represent the parallels as circular segments described about one and the same center, from which (imagined as the north pole) one will have to draw the meridian lines. In this way, above all, a semblance of the spherical surface will be retained in both its actual disposition and its visual effect, with the meridians still remaining untilted and still intersecting at that common pole."[35]

Since Ptolemy's model of the universe was incorrect, some, viewing his work two millennia later, consider his contributions to be suspect. But this could not be further from the truth. Ptolemy had a very sophisticated perspective of astronomy and the earth. Moreover, he was firm in his correct interpretation of the shape of the earth. For example, in book 1 of the *Almagest*, Ptolemy wrote, "That the earth too, taken as a whole, is sensibly spherical can best be grasped from the following considerations. We can see, again, that the sun, moon and other stars do not rise and set simultaneously for everyone on earth, but do so earlier for those more towards the east, later for those towards the west. For we find that the phenomena at eclipse, especially lunar eclipses, which take place at the same time, are nevertheless not recorded as occurring at the same hour (that is, at an equal distance from noon) by all observers. Rather, the hour recorded by the more easterly observers is always later than that recorded by the more westerly . . . Hence one can reasonably conclude that the earth's surface is spherical."[36]

34. Neugebauer, *History of Ancient Mathematical Astronomy*, 879–959.

35. Berggren and Jones, *Ptolemy's* Geography, 58, 83.

36. Ptolemy, *Almagest*, trans. Toomer, 40.

3.2.2 Florentine Mapmaking

Road maps had been in use in Italy since Roman times, and during the Middle Ages were used extensively by pilgrims on their travels throughout Europe and the Middle East. By the trecento, an increase in trade and exploration by sea meant that sailors and navigators needed maps that showed reasonably precise locations of coastlines and landmasses, thereby enabling them to undertake more distant and more daring voyages. This need provided an added impetus for cartographers to prepare maps that applied the scientific principles of mathematics (including trigonometry and geometry for surveying), draftsmanship, and the analysis of geological data. As Florentine sailors began charting sea routes for trading purposes, they were able to provide valuable information to the mapmakers and, in turn, benefited from new maps that were more accurate and frequently updated. After centuries of simple, somewhat primitive maps of limited use, mapmaking rapidly assumed a broader importance, becoming a commercial venture in seaports throughout Italy.

Almagest, Ptolemy's astronomical and mathematical treatise, continued to be an influential work over time but, surprisingly, his *Geographia* attracted little attention during the centuries following its first introduction. Florentine humanists such as Francesco Petrarch (1304–74) and Giovanni Boccaccio (1313–75) had shown an interest in geography and mapmaking, but it was the arrival of *Geographia* in Florence that radically altered cartography. In 1397, Manuel Chrysoloras (ca. 1350–1417) brought a copy of the Greek text, along with a world map and twenty-six regional maps, with him from Constantinople. He began translating the book from Greek into Latin under the patronage of Palla Strozzi, a wealthy Florentine book collector. Following Chrysoloras's death, the project was continued, completed by Jacopo Angeli da Scarperia, and dedicated to Pope Alexander V[37] in the hope that it would be looked upon kindly by the church. As word spread of this important work, demands for copies of *Geographia* increased. Strozzi met with several Florentines, including Paolo Toscanelli, Leonardo Bruni, the uncle of Amerigo Vespucci (Giorgio Antonio), and Niccolò Niccoli, in Ambrogio Traversari's cell in the Camaldolese monastery of Santa Maria degli Angeli,[38] which housed one of the major scriptoriums in Renaissance Florence. It was there that they arranged to have the book copied. After the invention in Europe of printing using movable type by Johannes Gutenberg in the 1430s, printed editions of *Geographia* became the standard text for Florentine cartographers.

As multiple printed copies of *Geographia* became available, Florentines expressed an interest in developing maps of the city. The 1469 map by Piero del Massaio presented a reasonably scaled drawing of the major structures and depicted the walled city of Florence both north and south of the Arno (with south oriented at the top of the map), and a few years later the painter and engraver Francesco Rosselli made a more detailed drawing of Florence, referred to as *Veduta della Catena* (*Chain Map*).

37. Edson, *World Map*, 114.

38. Ibid., 132.

By the quattrocento, the development of maps for navigators and traders was facilitated by the application of scientific data to cartography, which was becoming better understood and more sophisticated. Toscanelli, who had helped Brunelleschi understand the mathematics underlying perspective, also had a significant influence on cartography. He clarified the geometrical system of grids for those interested in creating maps for the purpose of visualizing the world. This more quantitative approach soon attracted navigators from other parts of Europe who took advantage of his expertise. Because of the latest advances in cartography, the ocean was no longer thought to be a river enclosing the continents, but was seen as a passage to travel both east and west to all of the earth's landmasses. Although the original correspondence no longer remains, Toscanelli is believed to have written to the king of Portugal explaining his theory that one could shorten the distance of sea travel from Europe to Asia by sailing west, rather than first sailing south to the tip of Africa and then turning east after rounding the continent. While Toscanelli underestimated the circumference of the earth, his theory attracted the attention of a Genoese navigator named Christopher Columbus. Apparently shown a copy of Toscanelli's correspondence with the king, Columbus asked Toscanelli to provide more details. Possibly using the letter and the chart, Columbus, with a scholar's endorsement, began to plan in earnest the voyage of exploration for which he would become famous.

But the importance of mapmaking was not limited to navigational pursuits. The science of cartography became integrated with art when wealthy families realized that maps could be used not only to expand their knowledge of the world, but also to visually connect themselves, both economically and politically, to the world beyond their own city. Printed maps proliferated, and painted depictions of geographical areas began to be used as wall pieces for the palaces and villas of the well-to-do, including those in Renaissance Florence. A display of maps served two purposes: one was to portray the area of land over which the owner had dominion, and the other was to grace one's rooms with decorative murals that provided both an artistic and a scientific adornment. Such maps combined mathematically accurate representations of land and water areas with perspectival paintings of botanical, zoological, and geological information, as well as natural resources and text, all on a single surface.

3.2.2.1 Palazzo Vecchio

An extraordinary example of such a mural cycle located in Florence is one by the Dominican friar christened Pellegrino Rinaldi Danti (1536–86), a cartographer, astronomer, and mathematician—a true polymath who was also trained in the visual arts. He was born in Perugia to a family of goldsmiths, architects, mathematicians, astronomers, writers, and artists. His grandfather Pier Vincenzo Rinaldi, while a trained goldsmith, also dabbled in poetry, astronomy, and architecture. Following an outbreak of the plague in Perugia in the 1490s, Pier Vincenzo moved his family to the countryside, where he studied astronomy and translated Johannes de Sacrobosco's *Sphere* (*Tractatus de Sphaera*) into Italian. It was Pier Vincenzo's cleverness that inspired his friends to begin calling him "Dante," a nickname that evolved to "Dante de Rinaldi" and then simply to "Danti," which became the family's surname. Although both of Pier Vincenzo's children, Giulio and Teodora, studied astronomy, Giulio became an architect and Teodora a painter who was an apprentice to Perugino. Giulio's oldest son, Vincenzo (b. 1530), became an accomplished sculptor who also authored a major work on art. In time Vincenzo would assist his younger brother, Pellegrino (by then known just as Danti), in securing commissions from Grand Duke Cosimo I.

As a boy, Danti had received instruction from his family members, including his aunt Teodora, in astronomy, mathematics, literature, art, and architecture; moreover, he visited the studios of various artists, learning the mathematical basis of perspective. In 1555, after completing his education at the University of Perugia, Danti entered the Dominican order and took the name Egnazio. Continuing with his study of mathematics, philosophy, and theology, he spent some time teaching, but mainly pursued his interests in astronomy, mathematics, and cartography/geography.

At that time, Danti's brother Vincenzo was employed by Grand Duke Cosimo I, who had decided to install maps in the Guardaroba in the Palazzo Vecchio. This building, which had served as the seat of government, was being renovated under the direction of Vasari and converted to a personal residence for Cosimo I and his wife, Eleanora of Toledo. Upon the recommendation of Vincenzo, Cosimo invited Danti to his court and commissioned him to prepare maps of the areas of the world to decorate the Guardaroba. Danti received permission from his superiors to relocate to Florence and was transferred in 1562 to Santa Maria Novella. During the ensuing twelve years, he was assigned numerous important projects, most notably designing and painting the maps in the Guardaroba, designing instrumentation and conducting astronomical experiments in Santa Maria Novella (see 5.2.3), and serving as tutor to the princes, Ferdinando and Francesco, as well as other undertakings.

The Guardaroba, also known as the Sala delle Carte, was designed to be used for the repair and storage of tapestries, pottery, paintings, books, small pieces of furniture, and special holdings collected from around the world by the Medici family. Along the walls of the room were large wooden cabinets to store those possessions, and Danti's maps were commissioned to serve as cabinet decorations. In addition, Danti was asked to design and install a large terrestrial globe for the center of the room.

For Danti, this map cycle and one he later completed for Gregory XIII in Rome were innovative cartographic undertakings. In the Palazzo Vecchio, where Danti was invited to reside, he created thirty-one maps between 1563 and 1575, and the remaining twenty-three were made by Olivetan monk Stefano Bonsignori from about 1577 to 1586. The completed maps, done on a large scale and including all areas of the known world, resulted in a unique and incredible collection. Danti's cartography project for the Palazzo Vecchio could be considered the first comprehensive atlas of the Western world. Nearly two centuries earlier, in 1375, Abraham Cresques, one of the Catalan school of cartographers, was credited with compiling maps comprising a worldview (called the Catalan Atlas). Although it was a large step forward in the field of cartography,[39] the word "atlas" in the title implies an inclusiveness not found in his work. Danti, on the other hand, added present-day atlas-type information such as political, economic, social, and historical notations. The maps also included hills,

39. Wilford, *Mapmakers*, 65.

lakes, cities, people, animals, forests, and other informational drawings to depict features indigenous to many of the particular areas.

Danti's maps for the Palazzo Vecchio represented a bold enterprise accomplished by collecting, comparing, and integrating in a unified form maps made by different people in varying projections and at different times, providing a most complete and accurate worldview. As noted by art historian Francesca Fiorani, the maps in the Guardaroba were basically Ptolemaic in that they were created by applying the grid of parallels and meridians used by Ptolemy and were similarly ordered. There were, however, notable differences. Danti's maps were modified to use the "trapezoidal projection" that appeared in about 1460 as a means of incorporating Ptolemy's scheme of the meridians converging at the poles, with the parallels

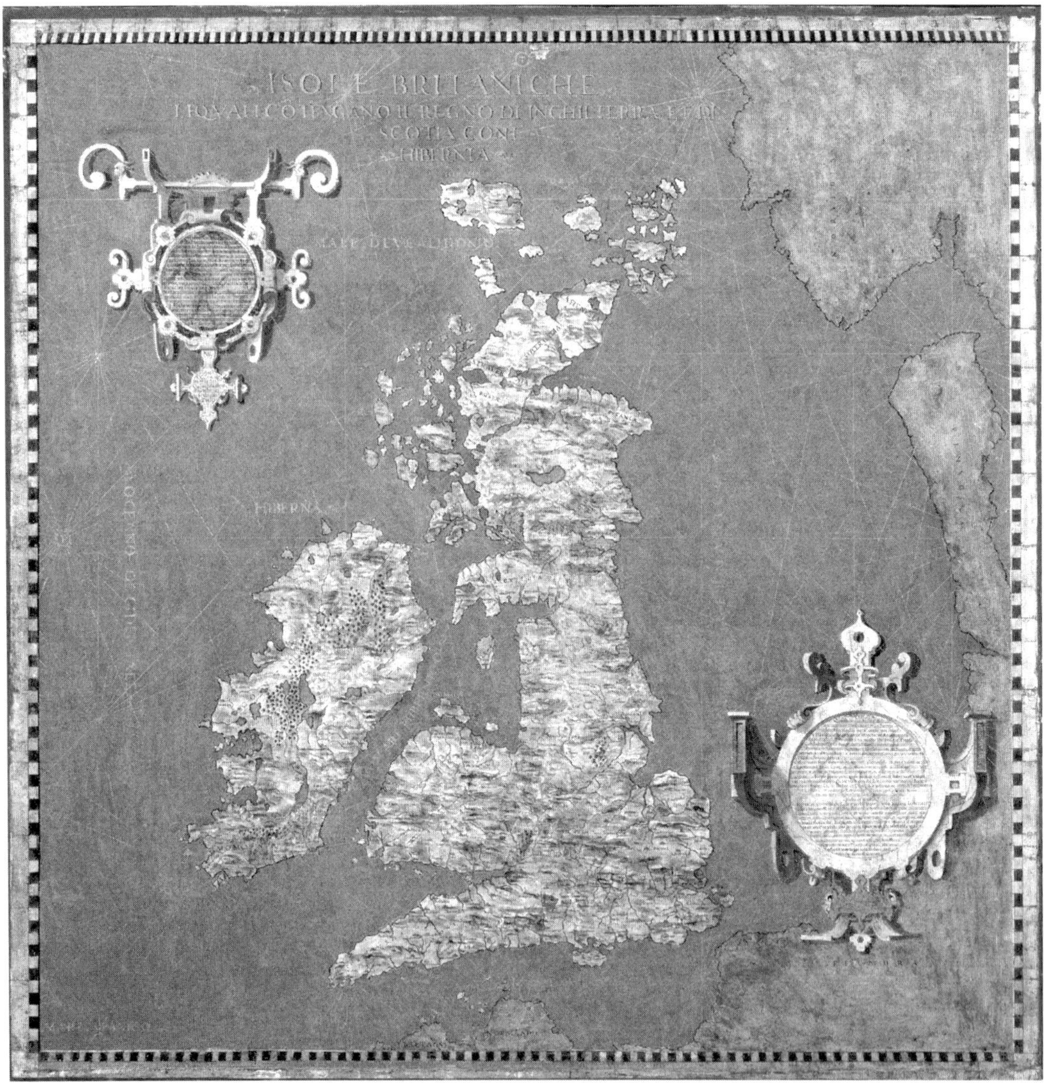

Figure 3.8: Egnazio Danti, Map of the British Isles (1565), Guardaroba of the Palazzo Vecchio (photograph by QwHdenMWIhNP8g, Google Cultural Institute, retrieved from Wikimedia Commons).

remaining equidistant.[40] These contained illustrations that were not sized in relation to each other or to the maps on which they were situated. Superimposed on the maps in corners or in areas lacking landmasses, Danti painted framed inserts to contain the legends that described political divisions, historical facts, and societal customs, as well as economic, geographic, botanical, and zoological data. The cycle excelled as a unique and seamless blending of art and science, presenting in concert geometric plots, representational figures, and didactic information (see fig. 3.8).

To begin this massive undertaking, Danti was assigned space in the Palazzo Vecchio in which to design his cartoons for the maps. Although scant information remains concerning the specifics of the process, we do know many of the sources he used. Danti chose as his initial guide Giacomo Gastaldi's *Cosmographia universalis* (created in Venice in 1561). In addition, Gerard Mercator's *Map of Europe* (1554) was a valuable resource, as was his *Map of the World*. Mercator's 1569 map used his celebrated "Mercator projection" method to represent a spherical globe on a two-dimensional surface. Although still widely used today, the cylindrical projection scheme inherent in this technique leads to considerable distortion, particularly of areas at latitudes well above and below the equator. We do not know the extent to which Mercator's projection influenced Danti's work.

Abraham Ortelius's *Map of the World*, made in Antwerp in 1564, proved especially helpful to Danti for information on the North American continent, and he used the nautical charts of Bartholomeo Velho from 1561 for South America, parts of Africa, and the East Indies. Further sources included charted materials from Spain and Portugal, as well as travel descriptions from voyages of numerous navigators from an extensive collection owned by Cosimo I. To augment areas not represented in other sources and to update earlier maps and charts, Danti requested additional reports, descriptions, navigational charts, and cartographic materials from a variety of individuals who had visited wide-ranging parts of the world. Compiling all of this information, Danti had to accurately redraw maps of different sizes and projections to make a cohesive pictorial representation of the earth's surface. Using tempera, Danti personally painted some of the maps following the design of his cartoons, but others were probably completed by his assistants.

The terrestrial globe, constructed by Danti from about 1563 to 1568, measures slightly more than two meters in diameter (fig. 3.9). The structure holding the globe uses iron arms and was built with a mechanism that allowed the globe to rotate easily. Unfortunately the globe sustained weather damage when it was placed outdoors for a period of time and no longer has its original colors and detail. Yet, standing in the center of the Guardaroba, the globe was and still is a dramatic addition to the area.

In 1571, as work on the maps was still progressing, Cosimo, pleased with the evolving project, appointed Danti the official cosmographer of the court. Cosimo's death in 1574, however, created a climate in which Danti was no longer held in high esteem by Cosimo's son and successor, Grand Duke Francesco. The historical record lacks specific information as to this sudden reversal of court favor, although there have been suggestions of some

40. Fiorani, *Marvel of Maps*, 94.

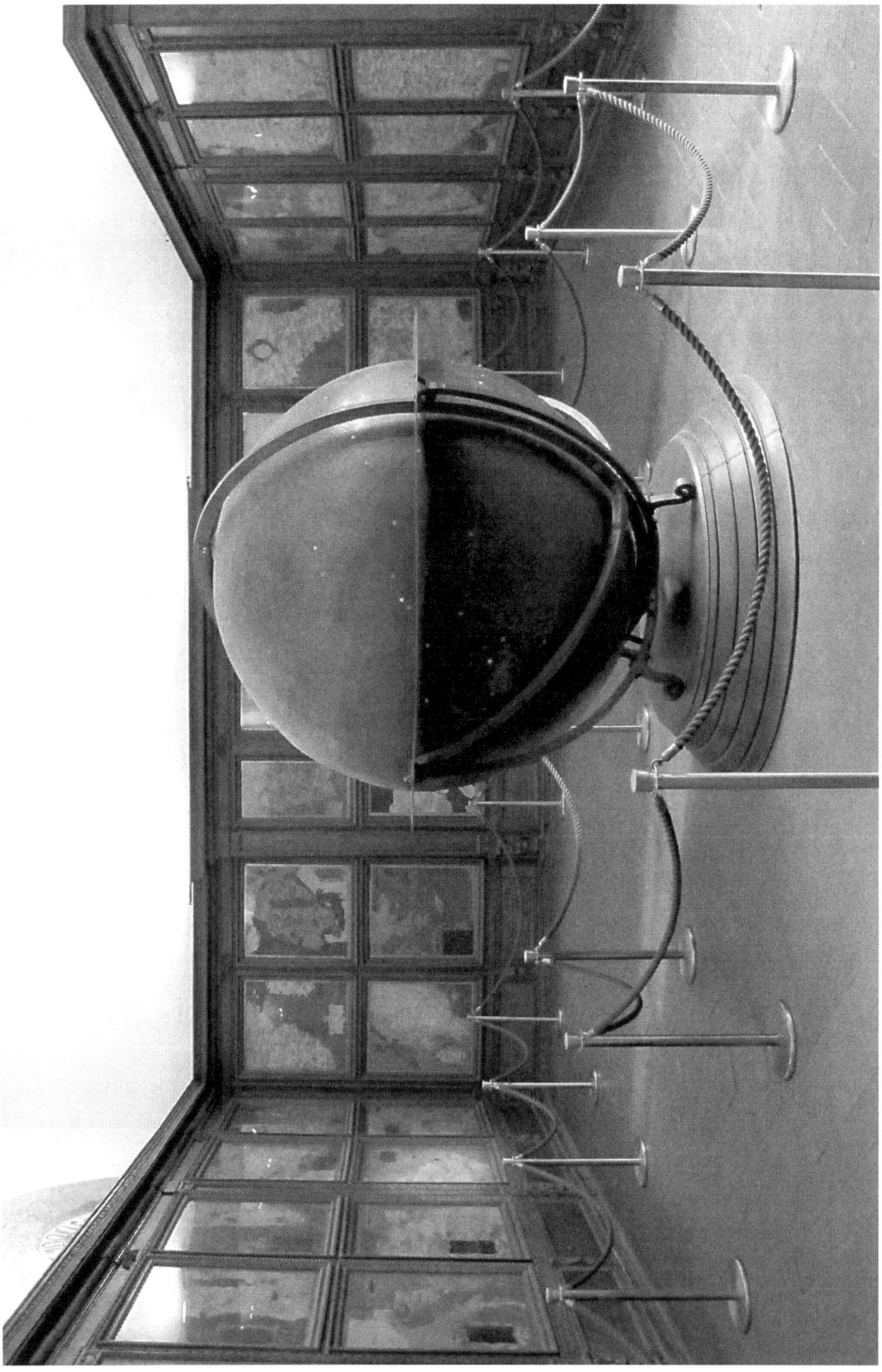

Figure 3.9: Danti, Terrestrial globe (ca. 1563–75), Guardaroba of the Palazzo Vecchio (photograph by Sailko, retrieved from Wikimedia Commons, GNU Free Documentation License).

questionable improprieties on the part of Danti. Francesco exiled Danti from the ducal court and from Florence in 1575. This left Danti's work on the Guardaroba incomplete. Also unfinished were his astronomical installations in Santa Maria Novella, a bold project suggested by Danti and approved by the church in response to Cosimo's request on behalf of Danti (discussed in detail in 5.2.3). Sadly, he was forced to abandon the projects in which he had invested so much of his time, knowledge, and creativity. Assigned by the church to resettle in Bologna, Danti continued his astronomical work unhindered in an atmosphere sympathetic to, and encouraging of, his proposed studies. Later in 1580, as a result of his outstanding cartographic work in Florence, he was commissioned by Pope Gregory XIII to design and install a series of maps for a magnificent gallery in the Vatican, where they remain on display.

Danti's massive undertaking did not remain in its unfinished state; Stefano Bonsignori was awarded the commission to conclude the project. One aspect of the original plan included celestial constellations on the ceiling and a celestial globe. That sphere was to be suspended over the terrestrial globe, probably symbolizing the dominion of the cosmos.[41] The constellations and the celestial sphere, however, never became part of the room. Despite these omissions, the completed Guardaroba, with the majority of the wall space covered with maps, presents the visitor with a unique atlas and remarkable and encompassing image of the world as it was recognized at the time of Cosimo I.

Bonsignori also created the *Map of Florence* (1584), which was dedicated to Grand Duke Francesco. This map is significant for the amount of detail showing the streets, structures, gardens, city walls, and of course, the Arno. That Bonsignori was proud of his masterpiece is documented by his inscription that appears on the map, calling it "worthy on account of its beauty and its magnificence, to be seen by all mankind."[42] Despite Bonsignori's boast, while it remains part of the holdings of the city of Florence, it is no longer publicly displayed.

Mapmaking was also among Leonardo da Vinci's many interests. His notebooks contain numerous examples of his studies of the earth and the ways in which the land could be depicted. His use of relief in the drawings of hills and mountainous areas, the movement of rivers, and the accurate aerial views of cities, including Florence, speak to his expertise in geography, geology, projections, and cartography. Special Leonardo exhibits in Florence frequently include examples of his remarkable maps.

In less than a century, Florentine cartographers and artists used the knowledge and techniques presented in Ptolemy's rediscovered *Geographia* to create a range of extraordinarily impressive utilitarian navigational aids and atlas-type decorative maps, elevating them to an accepted art form. As information about new landmasses and more specific details concerning the three continents became available during the fifteenth and sixteenth centuries, mapmakers depicted the additions and refinements as expansions of the Ptolemaic maps, finally adding a fourth continent, North America, after 1492. As stated by Francesca Fiorani, "It is this Ptolemaic order integrating ancient geography with modern discoveries

41. Cecci and Pacetti, "La sala delle carte geografiche," 142.

42. Levey, *Florence: A Portrait*, 279–80.

that was widely used in the sixteenth century, both in editions of Ptolemy's *Geography* as well as in cosmography books, atlases, and in the Guardaroba Nuova."[43] James McClellan and Harold Dorn noted that "Cartography—the art and applied science of making maps—may well have been the first modern scientific technology."[44]

In addition, it can be argued that cartography is one of the most important examples of the blending of art and science. The mapmakers of Renaissance Florence, by adding their own artistic creativity, elevated the mathematics and geometry of mapmaking to a unique and exceptional art form, producing the splendid cartographic masterpieces that remain extant in Florence.

43. Fiorani, *Marvel of Maps*, 84.

44. McClellan and Dorn, *Science and Technology*, 267.

Chapter 4

Renaissance Medicine

Physicians, Hospitals, Apothecaries, and the Artistic Dimension

> "The importance of these devotional roles [of the hospital church and ward chapel] helps to explain why hospitals became significant centres of artistic patronage."
>
> —John Henderson, *The Renaissance Hospital*

> "It is necessary to a painter that he should know the intrinsic forms of man."
>
> —Leonardo da Vinci, *Leonardo's Notebooks*

> "If you wish to fatigue yourself with it, you will find plenty of recipes [for red, cinnabar], especially collecting them among the monks. But I advise you, that you may not lose your time in the many variations in the methods, to get what you we want from the apothecary's and pay for it."
>
> —Cennino Cennini, *The Book of Art*

An interest in the human body and its physical and emotional connectedness led physicians and artists to more intensely explore ways in which the body operated on a functional as well as visual level. In addition to studying the operational mechanics of the human form, the same artists and physicians promoted the manipulation of natural products for producing medicinals and pigments. Again, it was the humanistic and observational approach that led to advances in art and medicine in an interconnected manner.

4.1 Foundations of Western Medicine

As with other areas of knowledge that entered Florence during the Middle Ages and Renaissance, Western medical thought and practice originated in ancient Greece and the Middle East. The beginnings of early Greek medicine can, in turn, be attributed to influences from Egypt, Mesopotamia, and India.

Records remain from the Mesopotamian civilization that date to the third millennium BCE. These document the existence of Sumerian medical practitioners and list healing concoctions based on plant and animal sources. The doctors were of two types: an *asu* was a practicing physician who used mixtures of natural products to treat patients, and an *ashipu* treated patients with magic. The Code of Hammurabi mentions surgical procedures used by practicing physicians and the instruments alluded to are supported by archaeological finds. The Assyrians and Babylonians continued to use these earlier medical texts, and the library in Ninevah of the ruler and scholar Ashurbanipal contained information on medical diagnosis, rituals, and omens, along with a large collection of works of literature. We know from surviving cuneiform scripts that physicians could distinguish hundreds of different diseases.

Medicine in Egyptian civilization, like that in other early medical practices, was based in part on spirituality and religion, involving rituals and incantations and practiced mainly by priests. Imhotep, the Egyptian god of medicine, played a role similar to that of the Greek Asclepius (Aesculapius in Latin). In Egypt, however, medical care was rooted to some extent in a rudimentary understanding of physiology and anatomy, using knowledge of anatomy gained from the practice of embalming. Moreover, Egyptian physicians used certain forms of surgery, as well as herbal medications.

The original contributions from the Greek civilization occurred primarily in Asia Minor in the fifth century BCE and were followed by the rise of Alexandria, Egypt, during the Ptolemaic dynasty that began in the fourth century BCE when tremendous advances were made in all aspects of medicine, science, and mathematics. Asclepius is credited as the founder of Greek medicine, although it is uncertain if he actually existed or is simply a legend. He appears as a physician in Homer's Iliad, but was later deified as a son of Apollo. The image of Asclepius holding a staff coiled with a single serpent (not two, as commonly portrayed today) has come to symbolize the medical profession. Asclepian temples flourished, and patients treated in them were given a purifying bath followed by days of isolation, fasting, and rest. Greek physicians used natural compounds of plant origin for medical treatment and for magical divinations. In many cases, and much to their credit, the early practitioners empirically determined appropriate dosages and collection methods for the plants.

The first Greek medical school was founded by the Pythagorean Alcmaeon of Croton at Cnidus in about 700 BCE. Alcmaeon considered that good health arose from a balance of forces in the body, as he explained in an early anatomical text. By the fourth century BCE, the major Greek medical schools were at Cnidus and nearby Cos; the latter was started by Hippocrates (ca. 460–ca. 370 BCE). The Hippocratic corpus consists of scores of treatises on various aspects of medical diagnosis and practice. Practitioners at Cos further developed the early Greek concept of humors (referring to bodily fluids, i.e., blood, phlegm, yellow bile, and black bile), coupled with the realization of the importance of keeping them in balance.

Hippocrates moved from a reliance on magical potions and supernatural causes, advocating a more rational basis for diagnosis and treatment. Although Aristotle (384–322 BCE) also wrote on human anatomy, his works were, unfortunately, lost.

Greek medicine reached its zenith in Alexandria during the Ptolemaic dynasty (305–283 BCE), when scholars and students flocked to the city from many parts of the civilized world to study and record their knowledge. The city flourished for several centuries and in that intellectual and cultural center of the ancient world, numerous physicians taught and practiced. New medications were developed and significant contributions to the fields of anatomy and physiology continued up until the city's destruction and the loss of its museum and library.

The association of good health with the four humors continued; in the second century Galen extended this concept, and it survived in a modified form into the Renaissance. Galen of Pergamum (ca. 131–201) was certainly one of the best known and most influential of the Greek physicians. Extraordinarily prolific, his writings influenced medicine for nearly one and a half millennia. After studying in Pergamum, Corinth, Smyrna, and Alexandria, Galen began practicing medicine by treating gladiators in Pergamum, thus gaining firsthand experience in human anatomy. He then relocated to Rome where he served as personal physician to Marcus Aurelius and later emperors. Galen's contributions to anatomy and physiology were highly influential, although he introduced a number of erroneous concepts, many of which were corrected by the Islamic physicians and later by Vasalius and William Harvey.

As today, early medical treatment relied heavily on medications, which were primarily plant-based and to a lesser extent animal- and mineral-based. Many of these are included in a major pharmacopeia written by Pedanius Dioscorides (40–90), who was born in Anazarba, Cilicia, in southeastern Asia Minor. Between about 50 and 70, he prepared a five-volume study, *De materia medica*, that focused on the preparation and properties of medicinals. Dioscorides was also a proponent of testing and modifying preparations to achieve optimum responses. Unfortunately, many translations reorganized his treatise into an alphabetical listing, rather than the grouping of classes and properties originally presented. In addition, as is often the case with multiple copying and translations, a number of errors appeared as later editions were compiled. Nonetheless, this multivolume opus formed the basis for pharmacopeias for some fifteen to sixteen centuries.

With the decline of Alexandria as a leading center for research and training, many of the advances in medicine made by the early Greek physicians were continued in the Middle East. Prior to the seventh century, pre-Islamic medicine included hospitals and apothecaries and was practiced at several major sites, including Baghdad, Cairo, and Damascus. As the Islamic culture rose in the Middle East, physicians integrated the Hippocratic and Galenic traditions of Greek medicine with those of Persia and India. Islamic medicine not only benefited from the strong base of inherited knowledge, but also added substantially to diagnosis, treatment, surgery, dentistry, pharmacology, and physiology. In Baghdad, the noted physician Hunayn ibn Ishaq al-'Ibadi (Joannitius, ca. 808–73) collected writings of the Greeks and, along with his students, translated them into Arabic and Syriac. Included in the translations were works by Hippocrates, Galen, and Dioscorides on medicine and apothecaries, as well as writings by Aristotle, Euclid, and Ptolemy. Hunayn's own writings, along with the translations, were instrumental in launching Islamic science and medicine. Many physicians and apothecaries contributed significantly to the rise of Islamic medicine, including Ali ibn Sahl Rabban al-Tabari in circa 860, Abu Bakr Muhammad ibn Zakariya al-Razi (Rhazes, ca. 854–930), and Ibn Sina (Avicenna, 980–1037), to mention but a few. Avicenna later moved to Persia and practiced medicine there. His writings and discourses covered numerous topics and, along with those of Rhazes, were translated into Latin and later widely used in the medical schools of Italy, thus influencing generations of Italian physicians.

The Moorish conquest of the Iberian Peninsula ensured that much of the medical knowledge from the Greeks and the Middle East entered Europe, where Muslim physicians and apothecaries continued the rich tradition of healing, research, and teaching. As students from Italy and elsewhere came to the leading Iberian medical centers for training, scholars began translating important works from Arabic into Latin. In the tenth century, Cordoba emerged as a major area for medical education, and it was there that Dioscorides's *De materia medica* was translated from Greek into Arabic and then into Latin, becoming widely available and thus having an important impact on the training of physicians and apothecaries for many centuries.[1] Hence, the Muslims can be rightly credited with preserving the early Greek medical knowledge and adding substantially to it, thereby providing a strong base for the further development of medicine in Florence and elsewhere in Europe. In addition to this route of entry into Europe, some manuscripts and ideas were brought by returning crusaders, and some texts from Byzantium were translated directly from Greek into Latin.

In Florence, libraries supported by wealthy patrons benefited greatly from this influx of books on all aspects of medicine and apothecaries. For example, the earliest surviving codex illustrating surgical methods appears to have been written and illuminated at the imperial scriptorium in Constantinople during the reign of Constantine VII. This codex, produced for the Byzantine physician Nicetas (ca. 900), contains images and descriptions of techniques used for over a millennium, with many of those practiced by Apollonios of Cyprus in the first century BCE. The Greek scholar Janus Laskaris, searching for Greek manuscripts under the sponsorship of Lorenzo de' Medici, obtained the codex in Crete. Laskaris and others made copies, and the original is preserved in the Laurentian Library in Florence.

4.2 Doctors and the Practice of Medicine

4.2.1 Training, Health Care, and Regulation

Florentine physicians were members of the Guild of Doctors and Apothecaries, which evolved from the Arte dei Medici e Speziali, formed in 1266. This was one of the seven greater guilds recognized by the Ordinances of Justice in 1293, along with fourteen lesser guilds (see 1.3). The diversity of professional interests embodied in the Guild of Doctors and Apothecaries was not particularly unusual in Florence during the late medieval period. Indeed, this particular guild began as an association of merchants specializing in various

1. Freely, *Aladdin's Lamp*, 108–9.

drugs and spices, and later expanded to include physicians and, in time, painters, saddlers, chandlers, gravediggers, barbers, and others. Although it was one of the larger guilds, the number of physicians enrolled represented a relatively small percentage of the total membership.

In the early fourteenth century, the Guild of Doctors and Apothecaries was split into subgroups representing particular occupations. A statute introduced in 1349 that regulated the practice of medicine for centuries delineated more specific duties, obligations, and responsibilities of the physicians, as well as the requirements and qualifications to be recognized by the guild. One portion of this statute was the clarification of training required for the physicians, surgeons, and empirics who comprised the profession of doctors. Physicians (*fisici, physici*) were those who had a degree in medicine awarded by a university or medical school. Some surgeons, likewise, were university educated, but also took courses and practical training in surgery. A number did not obtain a university degree, but combined their university courses with practical experience under the direction of surgeons. It is probable, however, that many surgeons lacked any level of academic training.[2] Empirics (*empirici*), for example, had no formal university courses and instead received their knowledge from self-study and from observing and assisting licensed doctors. Some of the empirics were highly specialized and often undertook procedures that many surgeons would not attempt. In all cases, aspirants to the guild had to successfully complete an examination before membership was bestowed.

High-quality medical schools were established in Bologna, Salerno, Padua, Pisa, and Rome and attracted students from a broad geographical base, both within and outside of Italy. In order to meet the growing need for doctors in Florence, a university (*studium*) with a medical school was established in 1321 and enlarged after the first outbreak of the plague in 1348. Several doctors comprised the faculty and had the responsibility of giving lectures and practicals, and leading debates and discussions. With some exceptions, obtaining a university medical degree in Italy entailed specific general education and training, which could vary from one medical school to another. Students were expected to enter with appropriate courses in grammar, logic, and mathematics, although some medical schools included these topics as well. It was believed to be necessary to integrate natural philosophy and the liberal arts into a medical education. Indeed, several physicians were adept at drawing and incorporated their talents in depicting various structures of the human body in their publications. At the better medical schools, students would study four years, typically with morning and afternoon lectures and readings based on the writings of Hippocrates (more likely the Hippocratic school), Galen, Avicenna, Averroës (Ibn Rushd), and a few others; evenings were spent in instruction and observation in the more practical aspects of medicine. There were also discussions or disputations, as well as dissections, at which students and practicing doctors were present. At the University of Florence, students were expected to attend two dissections per year, one of a male and one of a female.[3] The cadavers were unclaimed bodies,

2. Park, *Doctors and Medicine*, 65.

3. Ibid., 60.

either prisoners condemned to death by hanging or those who had succumbed in hospitals. Although the field of medical education has changed dramatically over the past five to six centuries, the medieval and Renaissance training paradigm that integrated medical training with other university-level courses continues in modified versions in many countries today.

Since the Florentine Guild of Doctors and Apothecaries maintained strict control over the education and training of individuals authorized to practice medicine, the guild ensured an overall higher competence of its members than was found in northern Europe. For example, in Florence and the greater district of Florence, the practice of medicine, including surgery and dentistry, required licensure by one of the following two mechanisms for aspiring physicians: (a) earning a doctorate from an established university or medical school (e.g., the ones at Bologna, Florence, Padua, Perugia, Siena), or (b) successfully passing a practical examination administered by four doctors. (Other city-states in Italy also required appropriate training or experience for licensure.) Once approved for membership in the guild, which also entailed payment of a fee (often waived or paid only in part), the physician had the title of maestro. As one might expect, however, in actual practice, licensed physicians formed a rather diverse group, composed of individuals with varying levels of expertise who had been trained at somewhat dissimilar universities, often with different standards, or who had passed what could occasionally be categorized as a substandard examination.

The plague (Black Death) of 1347 to 1348 placed a major burden on all Florentines and resulted in a questioning of the capabilities of those in the medical profession, who clearly could not control a disease that reduced the population of Florence, estimated to be over 100,000 in the late 1330s, by more than half. As is now known, the bubonic plague was caused by the gram-negative bacterium *Yersinia pestis*, which originated in or near China[4] and was carried by fleas (*Xenopsylla cheopis*) on rats via the trade routes. It now seems probable that the plague reached Italy from Kaffa, Crimea, carried by Italians escaping a siege by the Mongols in the fourteenth century, during which plague-infected cadavers were thrown into the city, an early example of biological warfare.[5] Succeeding major outbreaks in 1449/50, 1457, 1464, 1468, 1478/79, and 1527–30, with lesser outbreaks throughout this period,[6] continued to add to the devastation caused by the initial entry into Florence. The ranks of the physicians were also reduced by the plague, although many had sufficient means to escape to the countryside during the worst times. Such departures alluded to by Boccaccio, if factual, would have greatly diminished respect for the profession; however, recent studies refute those claims of physicians abandoning the city.[7]

The years during and immediately after the plague of 1347 to 1348 negatively impacted the practice of art. Not only were there artists among the victims, but funds available for commissions were greatly diminished. Although there is a paucity of art related to the Black

4. Morelli et al., "*Yersinia pestis* Genome Sequencing."

5. Wheelis, "Biological Warfare."

6. Morrison, Kirshner, and Molho, "Epidemics in Renaissance Florence."

7. Wray, "Boccaccio and the Doctors."

Death, Florentine painter and sculptor Andrea Orcagna (d. 1368) created a fresco for the church of Santa Croce entitled the *Triumph of Death*. All that remains is a fragment of this fresco, one of three that was originally painted. It depicts forlorn figures facing Death and includes an inscription begging Death to take those for whom there is no hope.

In the aftermath of the initial outbreak, there was an influx of immigrants to Florence, many of whom called themselves physicians but, although licensed to practice medicine in Florence, were woefully unprepared to do so with any degree of competence. In response to the laxity in granting licensure, beginning in 1351 a number of physicians introduced amendments to the 1349 statute, ostensibly tightening the entrance examination for non-university trained physicians and in 1392 formed the College of Doctors. The motives for these actions were several: a genuine attempt on the part of several guild members to improve the quality of medical care, an effort to restore the respect of the medical community to the status it held before the Black Death devastated the city, and possibly a desire to reduce competition. The number of physicians was never large, fluctuating between twenty-six and seventy-one in the century following the plague; however, the number of physicians per 10,000 citizens increased from five in 1338 to a range of seven to fourteen between 1352 and 1470.[8]

The practice of medicine in Florence could involve any one, or a combination, of the following: private practice to patients who paid for the services rendered; service to the city, including any number of secular organizations; staffing of hospitals and apothecaries; or service to religious institutions. In 1427, those who practiced medicine had the third highest wealth in the city, less than the bankers and merchants/wool dealers and greater than lawyers, apothecaries, and others.[9] But a number of physicians also donated their services to the poor and to certain secular and religious organizations.

Complementing the standard medical treatments (e.g., medications, generally herbal-based, and surgery when required), Florentine doctors also prescribed certain diets and rest to facilitate recovery. In addition to the writings of Galen and others that influenced medical practice in the Renaissance, one book in particular was widely used: *Medical Sermons* by Niccolò Falcucci da Borgo San Lorenzo, who practiced from the end of the fourteenth to the beginning of the fifteenth century. According to medical historian Katherine Park, the book was "a discussion of the causes, symptoms, and cure of all known illnesses, ordered from general to particular and according to the organ affected."[10] Borgo San Lorenzo's influential book gave a rationale for the dietary, medicinal, and surgical practices, providing doctors with cause-and-effect guidance that melded practice and theory.

Women working in the hospitals primarily filled nursing roles, including assisting physicians and surgeons. In the process, many women became adept at performing medical procedures and dispensing medications. In addition, physicians' daughters might learn patient care from working with their fathers. Less common, but nonetheless

8. Park, *Doctors and Medicine*, 57.

9. Ibid., 121.

10. Ibid., 210–11.

present, were a very limited number of female physicians who were members of the guild and contributed to overall health care in Florence.[11] It was not until September 13, 1788, however, that the first woman surgeon in Florence, Maria Petrocini Ferretti, was approved to practice. She was trained in a long apprenticeship with her husband, Francesco, who practiced in Anghiari, and also attended lectures at the Florentine School of Surgery, located in Santa Maria Nuova (see 4.3.1). After completing her training and presenting her request to Grand Duke Pietro Leopoldo d'Asburgo Lorena of Tuscany, she was granted permission to take the examination on August 16, 1788, and approved to practice shortly thereafter.[12]

Although efforts were made to ensure adequate training for physicians, there were incidents in which remedies were useless and their treatment techniques proved to be dangerous to their patients. For example, it was not uncommon for women to develop puerperal fever, a form of septicemia, after giving birth or after a miscarriage or abortion. The causative agent was generally a strain of Streptococcus leading to an infection of the upper genital tract that could prove fatal for the mother. One famous case involved the birth of Caterina de' Medici, the future queen of France. Her mother, Madeleine de La Tour d'Auvergne (1501–19), who was the wife of Lorenzo, the Duke of Urbino and grandson of Lorenzo de' Medici, developed a fever and diarrhea. For treatment the court physicians gave her syrup to aid in the removal of the afterbirth and later ordered that she consume what is presumed to have been a chicken broth.[13] She died about two weeks after delivery. Not only were physicians extremely limited in their ability to treat infections, they were in many cases probably the source of the infection, which could arise from their procedures carried out under unsanitary conditions. Ironically, women who lacked resources to be attended by a physician during childbirth may have had better outcomes; those attended only by midwives or female relatives were less likely to develop an infection. In the latter part of the nineteenth century a Hungarian physician, Ignaz Semmelweis, reported that mortality rates were lower in deliveries assisted by midwives because they tended to wash their hands more effectively and with greater frequency than did the physicians.

Florence was not recognized as one of the major medical centers in Italy during the Renaissance; however, some interesting and important advances in understanding human physiology and the treatment of diseases occurred there. For example, the Danish doctor and physiologist Niels Stensen (known as Steno) was brought to Florence by Grand Duke Ferdinando II as the court physician. While there, based on dissections of dogfish and rays, he correctly hypothesized that human ovaries also contained eggs that were then transported to the uterus.[14] Dissections became more sophisticated and yielded new information that contributed to better medical practice.

11. Ibid., 71.

12. Lippi and Vannucci, "Maria Petrocini Ferretti."

13. Lippi, "Strange Destiny."

14. Cobb, *Generation*, 99–101.

4.2.2 Understanding the Human Body for Medicine and Art

Renaissance physicians in Italy, including Florence, used autopsies and dissections both to determine causes of death and for teaching purposes. Although the Catholic Church in principle prohibited desecration of the human body, it did permit dissection of certain unclaimed cadavers. Artists too had an interest in understanding anatomy in order to more realistically represent the human form in sculptures and paintings. Dissection was not a sanctioned endeavor for artists; however, hospitals were known to occasionally release an unclaimed body for them to personally study. In *On Painting*, Alberti discussed the importance of anatomical studies as a means of obtaining knowledge of the human body. Vasari described the painter and sculptor Antonio del Pollaiuolo (d. 1498) as one of the great Florentine artists who performed dissections with the aim of understanding the muscularity underlying the outward appearance of the human form. Art was becoming more focused on accuracy in portrayal as it was transformed from the rigid stylized painting to a more realistic representation, a trend driven by the humanistic appreciation of the classical, often nude, Greek and Roman sculptures.

Leonardo da Vinci was also one who deemed the study of anatomy as vital to artistic success. In addition to performing dissections in Rome and Milan, he was also permitted to carry out some in Santa Maria Nuova, where they were already being conducted by doctors. In his notes on illustrating the human body, Leonardo wrote, "Therefore it is necessary to perform more than one dissection; you will need three in order to have a full knowledge of the veins and arteries . . . and three more for a knowledge of the tissues and three for the tendons and muscles and ligaments and three for the bones."[15] Leonardo certainly broke new ground, not only in the repeated use of dissections, but also in the manner in which he drew the systems of the body in exquisite detail, meticulously labeling each diagram and adding notes to annotate his sketches (fig. 4.1). He also filled his notebooks with instructions to artists describing how to draw each particular part of the body. Leonardo had great reverence for the precision with which the human body was constructed and the way in which it operated, as well as an appreciation for the opportunity to study and learn from the corpses made available to him. On a notebook page containing his drawing of the neck with the vertebrae and muscles, Leonardo included this thought: "O speculator! Concerning this machine of ours, let it not distress you that you impart knowledge of it through another's death, but rejoice that our Creator has ordained the intellect to such excellence of perception."[16]

According to Vasari, Michelangelo also performed anatomical studies in order to understand the workings of bones, veins, muscles, and so forth, and he dissected corpses to learn about structure and the various changes associated with movement. He is known to have conducted his dissections in a room in the Monastery of Santo Spirito with the blessings of the prior, Fra Niccolò Bichiellini, to whom Michelangelo had presented a wooden crucifix for the church. The corpses were likely obtained from the hospital affiliated with Santo

15. Galluzzi, *Renaissance Engineers*, 79.

16. Suh, *Leonardo's Notebooks*, 144.

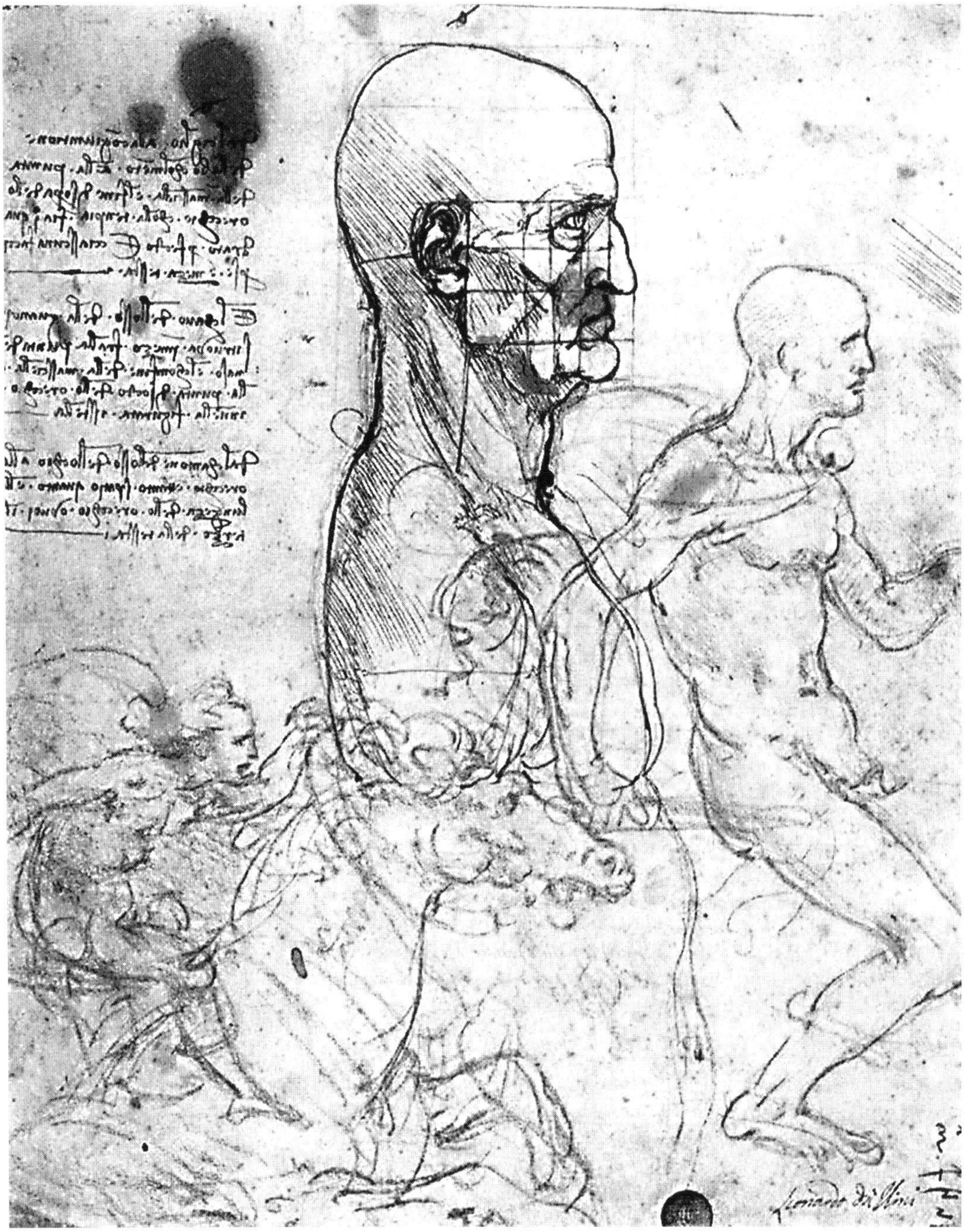

Figure 4.1: Leonardo da Vinci, Anatomical sketch: Profile of Man and Two Riders (1490–ca. 1504), red chalk, pen and ink on paper, Accademia of Venice (Web Gallery of Art, retrieved from Wikimedia Commons).

Spirito (cadavers were also acquired from the common graves in Florence). Although it can be presumed that Michelangelo made many anatomical drawings, only a few have survived. Those that do are remarkable in their vivid detail, especially in relation to the depiction of the skeleton and musculature. One has only to view Michelangelo's extraordinary sculpture of the David, carved from 1501 to 1504 and now in the Galleria dell'Accademia, to understand his mastery of human anatomy (see 2.3.1.2c).

Several proposals concerning Michelangelo's possible anatomical inclusions in painting have been suggested. For example, some art historians speculate that he incorporated human organs into his adornment of the Sistine Chapel. The background to *God Separating Earth from Waters* has been interpreted as a bisected kidney,[17] and it has been theorized that in *The Creation of Adam*, God and the angels are superimposed on a midsagital section of the brain.[18]

Viewed from the twenty-first-century mind-set, where dissections are now conducted by students and pathologists in well-lighted rooms with ample air circulation, we can only marvel at the commitment and tenacity of Leonardo, Michelangelo, and others who explored the human body for the purpose of enhancing their art. Their laboratories were poorly lighted with only candlelight or torches, and the stench, particularly during the warmer months, must have been nearly unbearable. Public dissections for students and interested laypersons were generally performed in larger rooms under somewhat better conditions.

While artists in Renaissance Florence studied human anatomy to enhance the quality of their art, some later artists created pieces designed specifically to depict elements of human anatomy. One example is the marvelous wax models that are housed in La Specola. This technique of making wax models was introduced earlier in Florence, but it was not until the seventeenth century that the pieces were used to teach anatomy. Opened in 1775 by Pietro Leopoldo di Lorena along with Felice Fontana, La Specola is the oldest public museum in Europe. At one time it housed a large collection of Medici holdings, some of which are still on display. The wax models on view in La Specola are exquisitely done and document a close connection between art and the human body.

4.2.3 The Role of the Divine in the Healing Process

Complementing the medical treatment of disease in Florentine hospitals was the importance of curing the soul. Leon Battista Alberti (1404–72), noted architect, writer, and mathematician, stated that hospitals were places where the secular and sacred could be reconciled.[19] Christ was understood to be the divine physician who, together with the medical doctor, created an environment for healing. To this end, the hospital church was positioned in an area of the compound that provided easy access to patients and their families wishing to attend services or in need of solace. Another important area set aside for contemplation was the

17. Eknoyan, "Michelangelo: Art," 1191.

18. Ginn and Lorusso, "Brain, Mind, and Body."

19. Henderson, *Renaissance Hospital*, xxxiv.

internal cloister, an open space within the hospital. In addition, the wards for men and for women contained altars for prayer and for the invocation of the healing powers attributed to God, Jesus Christ, the Virgin Mary, and an array of saints.

Art was used to enhance the environment in the portions of the hospital devoted to religious meditation. Patrons of hospitals played a large part in arranging for the adornment of areas meant to bring comfort and hope to the healing process. The paintings and sculptures remaining today in extant hospitals and their churches, and in galleries to which many of the works have been moved, clearly attest to the fact that prominent painters of the period accepted those hospital-related commissions. Examples of such art are highlighted in the following discussion of two representative Florentine hospitals.

4.3 Florentine Hospitals

Medieval and Renaissance hospitals provide splendid examples of the intersection of health care, charity, piety, and art. As in many Italian cities, the first hospitals in Florence catered to a broad constituency, often providing lodging to travelers in addition to caring for the sick. Later hospitals were often founded expressly to attend to a specific need, such as caring for the poor, for invalids, or for orphaned or abandoned children, as well as providing medical care for members of the general population.

In the eleventh century, when the first set of Florentine city walls was constructed, five hospitals served a population of some 20,000. By the fourteenth century, when the last walls were built, there were forty-nine hospitals administering to a population of about 100,000. Of course, the plague that ravaged Florence in 1348 spurred the opening of additional accommodations in spite of the large population decline of over 50 percent. Toward the latter part of the fifteenth century, Florence had regained a portion of its pre-plague population with an increase in the number of hospitals to sixty-four. The reason for the growth was multifactorial and reflected not only the population expansion, but also the emergence of Florence as one of the most powerful city-states in Italy, indeed in Europe. In this vibrant culture, individuals of means, lay organizations such as confraternities, and civic and religious groups were expected to provide charity to the needy, as well as facilities for medical services. These groups responded and by the middle of the sixteenth century, the number of hospitals was sixty-eight, although several centuries were required for the population to fully recover to the level recorded in the early fourteenth century.[20]

In many cases, the exterior structures of the hospitals that opened between 1000 and 1550 are still standing. Two of these Renaissance hospitals, Santa Maria Nuova and Ospedale degli Innocenti, remain in use and serve as excellent examples of the integration of art, architecture, and medicine.

20. Henderson, *Renaissance Hospital*, 6, 341–55.

4.3.1 Santa Maria Nuova and Sant'Egidio

4.3.1.1 History and Mission

Santa Maria Nuova combined beauty and functionality exceptionally well, so much so that other cities emulated its plans and appearance when they built their own hospitals. Located in Piazza Santa Maria Nuova, the hospital of Santa Maria Nuova was founded in 1286 and began admitting patients several years later. The father of Dante's beloved Beatrice, the merchant Folco di Ricovero Portinari, established the hospital near the center of Florence to serve the medical needs of the poor. The hospital was built on land that became available after the destruction of a portion of the twelfth-century city walls and was near the convent of the Frati Saccati and the church of Sant'Egidio. Following suppression of the Frati Saccati by Pope Gregory X (1274), its land and structures were purchased by Portinari and Messer Lorenzo di Jacopo da Bibbiena, who was to become director of the hospital. Male and female patients were treated and housed in the original structure, which at first held seventeen beds. The hospital later acquired land south of the piazza on the corner of Via Sant'Egidio and Via delle Pappe, where a separate women's ward was constructed in the early part of the fourteenth century.

Growing in size and quality, the hospital had developed a stellar reputation by the fifteenth century, and in the sixteenth century had expanded to between 100 and 160 beds for male patients and about half that number for female patients. As in most other Renaissance hospitals, the wards were either rectangular (much longer than wide), L- or T-shaped, or in the form of a cross. Patients were generally placed two to a bed, and sometimes even more. In the hospitals, doctors administered to the poor who were ailing and to a few others, including priests and those of means who chose not to be treated at home. The more privileged patients and those from the church, along with a few other groups of patients, including the mentally ill, were placed in separate areas, isolated from the general wards. In the early 1500s, there were nine doctors totally dedicated to or affiliated with Santa Maria Nuova, six of whom were more experienced, plus three assistants who lived on the premises and gained valuable training from the doctors. Over the centuries, various specialists were paid (or donated time) to fulfill particular needs such as eye care and cataract removal, accomplished by lowering the lens behind the pupil with a needle.

The statutes and ordinances of Santa Maria Nuova between the fourteenth and sixteenth centuries illustrate the admission process, which included, at some point, taking patients to the chaplain to confess their sins. The ordinances described how the hospital staff was to accept the sick:

> Diligently and charitably, tend to and comfort, and refresh, nourish and clean and administer to all their needs, and treat them with all their strength and charity . . . In the middle of the infirmary there is a little bed covered with a cloth on which the sick are laid when they first arrive. As soon as they are admitted the infirmarer comes to them to determine the nature of their illnesses. He assigns the feverish and those with skin lesions or wounds to empty beds . . . and he sends one of the head nurses to wash the sick person's

> feet and another sub-infirmarer to get [their clothes] . . . Once the sick person has been assigned to a bed, he [the sub-infirmarer] comes, collects all the clothes taken off by the patient and wraps them up. He notes his name, father's name, place of origin, family name and the amount of money he is carrying, if any, and he writes all this on a label attached to a bundle, which he places in a storeroom called the Pergola . . . This depository is lined with chests or cupboards identified by the letters of the alphabet and the bundle is placed in the cupboard corresponding to the patient's initial. Thus, when a patient recovers and wishes to leave, it is easy to find his effects once he tells his name.[21]

The ordinances give the impression of a caring and compassionate medical and religious staff assigned to the hospital, but there was no cure for many illnesses and the sheer inconvenience and discomfort of two or more patients to a bed cannot be overlooked. Despite these difficulties, hospital records and a variety of documents by several authors show that Santa Maria Nuova apparently fulfilled its original mission of caring for the poor with (for the time) competent personnel at all levels. The hospital was relatively self-contained with its own garden, kitchen, laundry, apothecary, and chapels, and, of course, its own church. It, nonetheless, depended on external sources for many of its needs.

Santa Maria Nuova has continued as a respected hospital for over seven centuries. While undergoing extensive remodeling and expansion over time, it remains a legacy of the Florentine commitment to health services for all classes of citizens, and particularly those in financial need.

4.3.1.2 Healing through Art

Sant'Egidio, the church adjacent to Santa Maria Nuova, was founded almost three decades prior to the building of the hospital, which then assumed control of the church. The church's dedication to Saint Giles was appropriate because of its role in the life of the hospital. Born in Athens in the middle of the seventh century, Saint Giles renounced his possessions, giving them to the poor, and went to southern France, where he lived until his death early in the eighth century. He was known as a patron saint of the poor and crippled because he cured his own wound after being shot by an arrow, and then cured a beggar of his illness. Saint Giles first sought solitude in the forest where his only companion was said to have been a hind, but emerged and founded an abbey in France prior to his death. Important to its reputation, Florence's church of Sant'Egidio claimed to own various relics of Saint Giles.

The art commissioned for Sant'Egidio, some of which can be seen today in the Uffizi Gallery, the Bargello Museum, and the refectory of Sant'Apollonia, represents important examples of the spiritual component of healing. Notable works include an altarpiece, the *Adoration of the Magi* by Lorenzo Monaco (1370–1425), created in about 1422 and now in

21. Ibid., 162–64.

the Uffizi. This painting depicts the Magi paying homage to the Virgin Mary and the Christ Child, appropriate subject matter for those in need of the curative powers attributed to the Virgin. The use of vibrant colors and gold leaf adds to its splendor. Four artists, Domenico Veneziano (ca. 1400–1482), Piero della Francesca (d. 1492), Alessio Baldovinetti (ca. 1425–99), and Andrea del Castagno (ca. 1422/23–57), worked to complete a six-part fresco for the new choir. The resulting cycle, *Life of the Virgin*, was one of the most impressive to be executed during the Renaissance. The first portion, painted between 1439 and 1445, consisted of three scenes, *Meeting at the Golden Gate*, *Birth of the Virgin*, and *Marriage of the Virgin*. From 1451 until 1453 the last three scenes in the cycle were completed, *Presentation in the Temple*, *Annunciation*, and *Death of the Virgin*. As was true with the altarpiece, it was the story of the life of Mary that served as a reminder of her healing grace. Unfortunately for the art world, the fresco was covered and perhaps destroyed during a reconstruction of the church in the sixteenth century. Fragments of the sinopia are conserved in Sant'Apollonia (see 2.2.1.2e). Also now located in the Uffizi, the *Coronation of the Virgin* was painted between 1435 and 1443 by Fra Angelico (ca. 1387–1455). There is some debate about its exact location in Sant'Egidio, although it was most certainly meant as an altarpiece. This magnificent painting speaks to the wondrous impact of the powers of Mary and Christ with their location among the heavenly clouds surrounded by a court of thirty-eight angels and an additional forty-eight holy figures.

In addition to the hospital church, the internal cloisters (courtyards) of Santa Maria Nuova provided space for contemplation and prayer, having a therapeutic value of balancing body and soul. For patients and staff alike, providing an area away from the wards was an important component of the healing process. Religious art that promoted spiritual reflection and renewal were often placed within a cloister. Occasionally the walls of the hospital cloisters were frescoed, but more commonly those types of major artistic endeavors were found in convents and churches. The walls of the cloisters of Santa Maria Nuova did not contain monumental frescoes; rather, small devotional works were the norm. In Santa Maria Nuova, the cloister, Chiostro delle Medicherie, was enhanced by a glazed terracotta *Pietà*, probably created by Giovanni della Robbia (1469–1529), that speaks to the compassion of the Virgin Mary. Devotional art also played a role in the decoration of the area above doorways leading into the cloisters, underscoring their roles as transitional spaces from the activities of the hospital to the quiet of the courtyards. Works such as altarpieces and crucifixes adorned the walls of male and female wards. These wards, long and lined with beds, each had a chapel-like area with an altar that was often surrounded by devotional frescoes. The altar, visible from any of the beds, was meant to provide solace to the infirm.

The powerful imagery of the Christian theology in the original elegant art and divine frescoes speaks to the mission that was considered key to the hospital's role in assuring patients and their families of the spiritual powers of Mary, Christ, and the heavenly Father, clearly capable not only of healing, but also of the redemption of the soul. Because of Santa Maria Nuova's size and commitment to combine the spiritual and physical healing process, by the middle of the 1400s the hospital complex had become one of Florence's most important locations for commissioned works.

4.3.2 Ospedale degli Innocenti

4.3.2.1 History and Mission

In 1294, the Florentine Republic charged the prosperous Arte della Seta (Guild of Silkworkers) with responsibility for the care and education of abandoned children. At least two such facilities were already in existence: the Spedale di (Santa Maria) San Gallo (located just outside the northern city walls) and Spedale di Santa Maria della Scala. Both were small and unable to meet the needs of the city. An impetus for the new *spedale* was a provision in the will of Francesco di Marco Datini providing funds for the creation of such a facility.[22] By the early part of the fifteenth century, the Guild of Silkworkers had purchased land for a new building in the Piazza Santissima Annunziata, located perpendicular to the church of the Santissima Annunziata. This was an area where pilgrims congregated to worship the Virgin Mary, and it seemed appropriate that the role of the Mother Mary be associated also with caring for abandoned infants. The guild hired Filippo Brunelleschi (1377–1446) to design and oversee construction of the facility. Work began in 1419 and, although it was unfinished, the Ospedale degli Innocenti was officially opened January 25, 1445, with the first child accepted on February 5.

The orphanage was staffed at all times and infants could be left in a basin located beneath a window on the portico. Later, in 1660, a wheel was constructed at the opposite end of the portico; infants could be placed anonymously on the outer side of the wheel, which was then turned through the opening in the hospital wall. The Innocenti has retained an interesting array of identification markers, for example, half of a coin and/or one member of a pair of earrings, that were placed with children left at the facility by a parent, relative, or caretaker so they might later claim that child if they desired. In the latter part of the nineteenth century, use of the wheel entry was discontinued, but the facility remains open today with its charitable work focusing on the needs of children and their families.

The large number of babies, abandoned because of plague, poverty, and famine, forced the guild to hire wet nurses from both the city and the countryside. Although some of the children were adopted, most of the babies and youngest children were placed in foster care until they were seven years of age, after which they were returned to the orphanage for basic schooling. Following that, the girls were either married, employed by the guild, became nuns, or were accepted to work for the orphanage. Boys, on the other hand, were readily placed into one of the workshops of Florence. In the first year of operation, the Ospedale degli Innocenti accepted 90 babies; during its first three years, 260 children entered the orphanage. In 1534 the facility housed 550 children and was responsible for an additional 1,036 children in the care of wet nurses and foster families; in 1681, 861 children were in residence (756 girls and 105 boys) and 2,618 children were in the care of families.[23]

22. Saalman, *Filippo Brunelleschi: The Buildings*, 36–37.

23. Piccini, *Spedale degli Innocenti*, 6–12.

The Silkworkers Guild elected a head administrator (prior) by secret ballot, and that head was assisted by three elected directors. The orphanage was funded by a combination of taxes collected from the silkworkers, bequests, funds from the City of Florence, and money raised by the hospital itself. Additional income was generated from the on-site garden where produce was grown not only for the consumption of the children, but also for sale to the people of Florence. With the resources available, the hospital was able to provide numerous services to the children under its care. For example, at the time of the Innocenti's opening, an infirmary, supervised by a doctor and perhaps an apothecary, was in place and staffed by visiting physicians. Unmarried women provided permanent nursing and charitable support.

Some interesting, and at times original, medical, nutritional, and public health initiatives and applications were used at the Innocenti. During the plague, for example, the children were washed with strong vinegar.[24] There are no controlled studies on such a preventive measure, and it is questionable that this simple action could retard the infection. Yet, if one reads a contemporary description of someone infected with the plague (e.g., Boccaccio's account in the introduction to *The Decameron*), it is reasonable that someone viewing these symptoms would deduce that making the skin less hospitable might provide some protection to the individual.

> It began in both men and women with certain swellings either in the groin or under the armpits, some of which grew to the size of a normal apple and others to the size of an egg (more or less), and the people called them gavoccioli. And from the two parts of the body already mentioned, within a brief space of time, the said deadly gavoccioli began to spread indiscriminately over every part of the body; and after this, the symptoms of the illness changed to black or livid spots appearing on the arms and thighs, and on every part of the body, some large ones and sometimes many little ones scattered all around. And just as the gavoccioli were originally, and still are, a very certain indication of impending death, in like manner these spots came to mean the same thing for whoever had them.[25]

Other health initiatives that proved to be effective were initiated. In 1577, Grand Duke Francesco de' Medici visited Ospedale degli Innocenti and learned of the shortage of wet nurses; he reported that in Spain, children were given cow's milk. This method of nourishment was then introduced at the Innocenti with great success, the milk being dispensed from small glasses with teats. In another illustration of the Innocenti's innovative techniques, workers observed that when infants in poor health were taken to the Baptistery for baptism, their illnesses often became worse. Theorizing that this was due to contact with infectious agents in the unsanitary city streets, the baptismal water, and the general population, the Innocenti erected a baptismal font in the hospital—a most unusual action for a secular site—so that sick infants did not have to be temporarily removed from their safe environment.

24. Ibid., 10.

25. Boccaccio, *Decameron*, trans. Musa and Bondanella, 4.

Health programs continued over the following centuries. In 1756, in response to a high mortality rate of entering children, mainly due to smallpox, staff at the Innocenti performed an experimental treatment (variolization) in which children were inoculated using human material from patients suffering from smallpox and hospitalized in Santa Maria Nuova or San Matteo. This approach opened the way for the formation of the Istituto Vaccinogeno Toscana at the Innocenti in 1822. These few examples illustrate the Florentines' inventiveness and creativity in meeting numerous health challenges.

The Ospedale degli Innocenti also used innovation in the design and decoration of the orphanage, demonstrating the Florentine commitment to provide a comfortable and nurturing environment. For the staff and the citizens of the city, the Innocenti served as a place of beauty and inspiration in which the children in its care were safe, nourished, and enriched.

4.3.2.2 Importance of Art

Construction of the Ospedale degli Innocenti, exceptional in its beauty and geometric proportions, was begun in about 1419. Its designer, Filippo Brunelleschi, was destined to become one of the foremost architects of his period (see 6.3.1.2). As *operato* (director of the works), he spent eight years supervising the building of the main parts of the structure, managing even the minute details. Most of the documentation has survived, providing many details about the construction.

In Florence, hospital façades were considered important: they were expected to be beautiful and often to emphasize the type of care given at the facility through the art that embellished the design. Hospital façades frequently featured a loggia that unified the frontage of the building and provided a buffer between the street and the hospital complex. The art and architecture of the façade of the Innocenti are exceptionally exquisite, working in concert to provide a welcoming entrance to the hospital. Brunelleschi's loggia runs the length of the building, with its arches proportioned such that the height of Corinthian columns supporting each arch is approximately the same as the distance between the columns and the depth of the portico (figs. 4.2 and 4.3). This balance provides both a harmonious and supportive effect for the observer, be it a visitor or a person delivering an abandoned child. Henderson states, "the façades of hospitals built in Florence from the later fourteenth century onwards have been hailed . . . as part of the Renaissance aesthetic . . . Brunellseschi's graceful design for the Innocenti . . . is often characterized as a milestone in the development of architecture and one of the major achievements of the early Renaissance."[26]

In 1487, the adornment of the portico was completed by the addition of blue and white terracotta roundels placed between each of the arches, the work of Andrea della Robbia (1435–ca. 1525–28). A member of the family famous for its terracotta masterpieces, Andrea produced some of Italy's most noteworthy terracotta pieces. Each roundel on the hospital portico portrays an infant wrapped in swaddling clothes (fig. 4.4) and beautifully proclaims the building as one created for the protection and nurturing of the young. The overall impact

26. Henderson, *Renaissance Hospital*, 73.

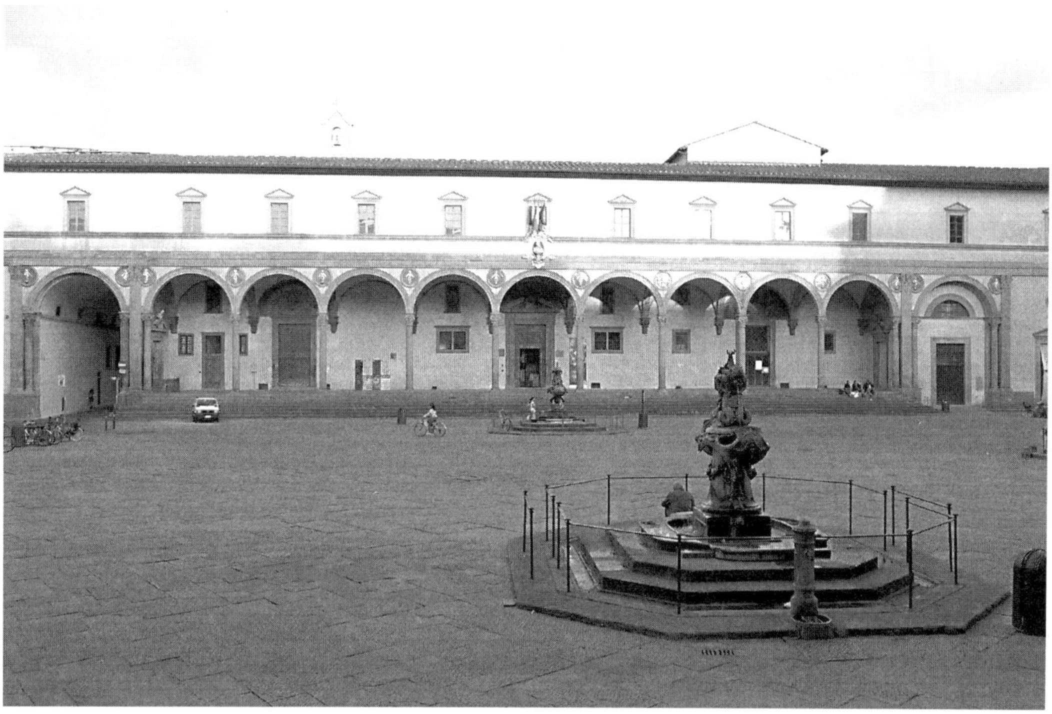

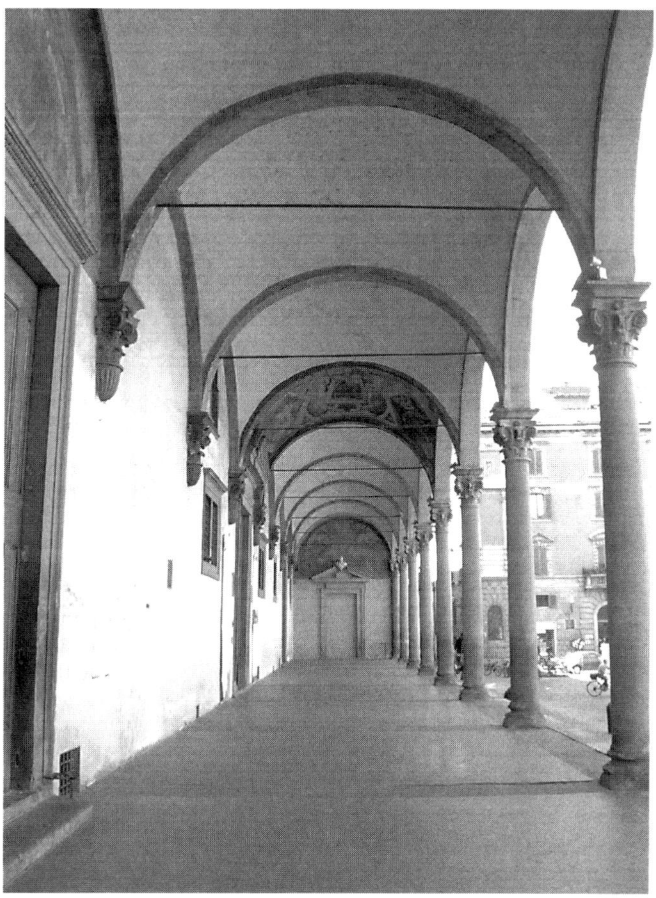

Figure 4.2: Filippo Brunelleschi, Ospedale degli Innocenti façade arches (photo by Warburg, retrieved from Wikimedia Commons, GNU Free Documentation License).

Figure 4.3: Filippo Brunelleschi, Arches at Ospedale degli Innocenti (photograph by authors).

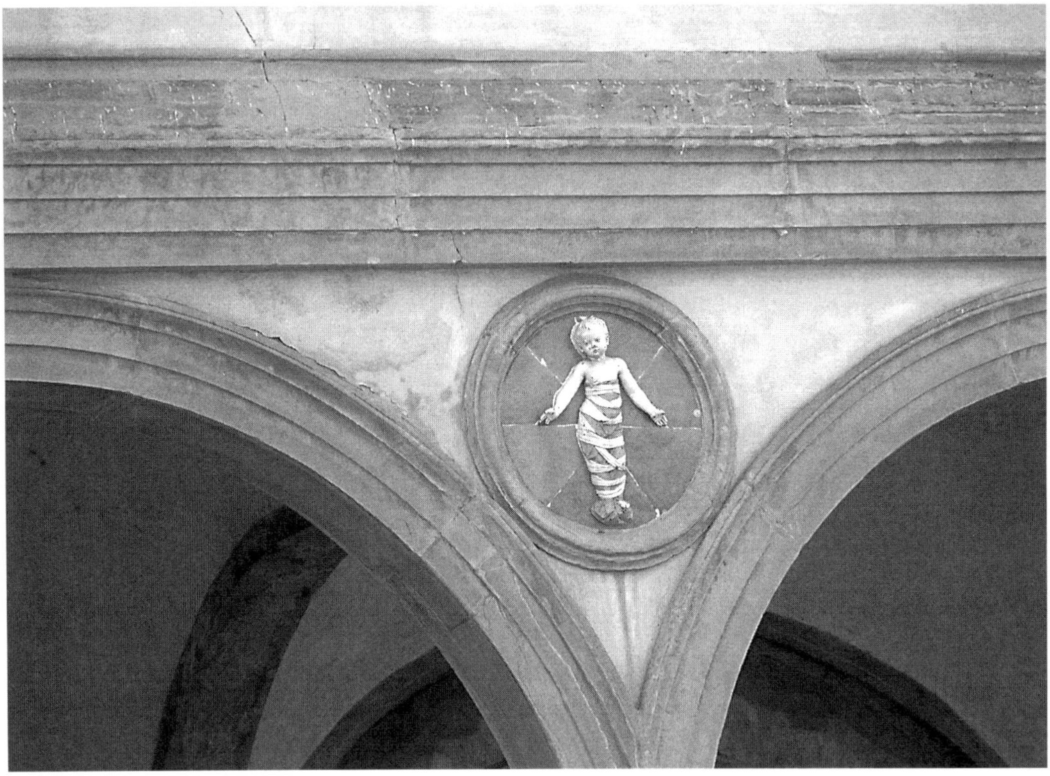

Figure 4.4: Andrea della Robbia, Roundels added to Ospedale degli Innocenti (1487) (photograph by Bruno Barral, retrieved from Wikimedia Commons, Creative Commons License).

of the proportions and the art of the façade is remarkable. Andrea della Robbia followed the Innocenti installation with a second similar work, a series of glazed terracotta medallions that can be seen on the Brunelleschi-inspired loggia (built 1489–96) of the former Ospedale di San Paolo in the Piazza Santa Maria Novella. The roundels depict saints and contrasting wealth. As with the Innocenti, the loggia provided a buffer between the street and the hospital while creating an artistic and welcoming entry.

The design of the interior of the Ospedale degli Innocenti itself is similar to that of contemporary convents, consisting of a central cloistered courtyard surrounded by the various structures that made up the hospital complex. The central cloister was enclosed by a portico supported by columns, as was a smaller second cloister. The cloisters served as quiet spaces meant to provide curative serenity to staff and children alike. The main floor also included an office, a reception area, a space that was perhaps a pharmacy, a foundling ward, women's refectory, infirmary, men's refectory, and kitchen. Above these rooms and the front portico were the dormitories for the children. The wood ceiling of the boys' dormitory, designed by Brunelleschi and not intended as a public space, is especially beautiful and meant to evoke a feeling of strength, order, safety, and tranquility for the boys it housed.

The church of the Innocenti, located north of the central cloister, was architecturally simple, consistent with the rest of the structure, and was adorned with many notable works of art. The guild commissioned a panel of the *Adoration of the Magi* (1488) for the chapel's high altar. For this work, the guild selected Domenico Ghirlandaio (1449–95), one of the leading fresco artists of the early Renaissance (see 2.2.1.2a). Ghirlandaio often combined religious themes with the secular world, depicting Florentines in scenes from the Bible. He used this combination of the spiritual and earthly in the *Adoration of the Magi*; in addition to *Mary and the Christ Child*, Ghirlandaio included the *Massacre of Innocents* and portraits of adults who are thought to be merchants belonging to the Guild of Silkworkers. The artist also painted Rome in a background that employs perspective extremely effectively.

Above the main street entrance of the church, where it could be seen by children, staff, and parishioners who attended services, was a fresco by Giovanni di Francesco, *Eternal Father in Glory Surrounded by Angels and Martyrs*; Andrea della Robbia created an *Annunciation* for one of the chapels (now placed over a doorway in the entry courtyard). These works lent themselves to the reiteration of the connection between the care of the body and the soul by invoking images of God and the Virgin Mary, both caring healers.

Over the centuries the church has undergone extensive renovation, but the original and impressively beautiful timber-beam ceiling created by Brunelleschi is still visible through a window above a more recent vaulted ceiling. Throughout its many years of operation, additions and alterations have also been made to the physical structure of the Innocenti itself. For example, following the 1966 flood, an excellent museum (Picture Gallery) was established to preserve, in a second floor area previously used as the boys' dormitory, the original works of art from both the church and the orphanage. Ghirlandaio's *Adoration of the Magi* was removed from the high altar in 1786 when the church underwent renovation and is now displayed in the Picture Gallery. Also on view in the Picture Gallery are numerous paintings and sculptures that were part of the original decoration scheme and were related to the building's mission/purpose. Among these are a *Pietá with Saint Joseph and an Angel* painted by Giovan Battista Naldini (1537–91), who was one of the children raised in the Innocenti and trained as an artist; a painting by Domenico di Michelino entitled *Madonna degli Innocenti*, which was probably used as a processional banner because of its depiction of the Virgin Mary sheltering children who had been abandoned; and an altarpiece for the church created by Piero di Cosimo entitled *Madonna and Child Enthroned* (1493) that, with its brilliant colors, caring facial expressions, and infant putti, speaks to the protection of Jesus by his loving mother Mary.

Clearly, the art discussed above, including Andrea della Robbia's roundels depicting children wrapped in swaddling clothes and Ghirlandaio's *Adoration of the Magi* in which baby Jesus is being lauded and admired, worked in combination with the design of the Innocenti and was an important component of the mission of the hospital. Brunelleschi was the ideal architect to conceive of and implement the plan in order to achieve the desired effect. It was Brunelleschi's intent to reduce architectural forms to a minimum, unifying rather than differentiating distinct areas of a building, even arranging the placement of windows to bring light uniformly throughout the interior of the structure.[27] Upon entering a Brunelleschan designed space, one has a sense of harmony and symmetry, especially effective in a hospital

27. Saalman, *Filippo Brunelleschi: The Buildings*, 76.

setting such as the Innocenti where comfort and solace were of particular importance. His architectural aesthetic, one of simplicity, seems ideally suited for a building meant as a substitute home for Florence's abandoned children. The Innocenti today stands as a testament to the commitment of the Florentines to aid the underserved.

4.4 Apothecaries

The most common medical intervention was the prescribing of medications. Much as is the practice today, physicians prepared a regimen of medications that were filled by trained hospital apothecary staff. In addition to the hospital pharmacies, small apothecary shops were located throughout Florence, and many paid a doctor to be present for a certain number of hours during the week to examine patients and write prescriptions that the staff would fill. Those apothecaries also sold some spices and standard medications that people could buy without having a doctor's order. Some physicians owned their own pharmacies and treated their patients at those locations, and a number of Florentine physicians personally invested in the drug and spice trade. In 1558, about forty-six apothecaries were registered in Florence serving some 60,000 to 70,000 inhabitants in the area.[28] Monastic pharmacies also emerged as purveyors of medicinals starting early in the thirteenth century.

4.4.1 Monastery and Convent Apothecaries

In the fourteenth century, the friars at Santa Maria Novella started a small infirmary, and in 1508 a spice and drug shop (*spezieria*) was erected on the site to prepare herbals for the use of the friars. After being closed for about twenty years, it was reopened in 1612. Santa Maria Novella employed a trained herb specialist who worked under the direction of Fra' Angiolo Marchissi, considered the founder of the apothecary, and the shop was expanded and began dispensing medications to the Florentine public. Interestingly, members of the Medici family were known to be clients.

At present, the pharmacy is the site of the Officina Profumo-Farmaceutica di Santa Maria Novella. Although the business is now a purveyor of perfumes, soaps, and the like, much of the old pharmacy remains. The rooms used for herbal preparations are now open to the public and display early pharmaceutical jars, mixing bowls, mortars and pestles, green glass vessels, the original ovens, and apparati used for distillations dating from the 1600s. The herb garden, currently marked by a plot of grass, is no longer in existence, but a small experimental garden, located outside the distillation room, remains. The entryway and center sales room formerly served as the chapel for the friars, and the frescoed sacristy now functions as a library dealing with herbal recipes.

Nuns, as well as friars, also began opening apothecary shops. Although nuns had a long-established tradition of mixing and dispensing medicinals for members of their own religious communities, beginning in about 1500 they started marketing medications to the

28. Strocchia, *Nuns and Nunneries*, 13.

public. During the period from 1338 to the early cinquecento, the number of nuns in Florence had increased from about five hundred to over two thousand (living within a four-mile radius of Florence).[29] Their entry into the apothecary field provided an important contribution to the city and also ensured a steady source of income to the nuns. In Florence, Santa Caterina, a convent under the directorship of the friars of San Marco, was one of the first, in 1517, to sell its medications. Its infirmary, that also served as the apothecary, prepared and distilled preparations for the public; included in their inventory were oils, powders, pills, purgatives, syrups, and a formulation of distilled rose water for which they became known. Early in the endeavor, the shop's profits were not significant, as the nuns mainly distributed the medications to the poor and charged little, if anything, for the service. Their business continued to grow, however, and in 1570 the nuns expanded their business and moved the shop into a larger structure close to their convent but on a major street (now Via Cavour). The enlarged pharmacy served both those with health concerns and those in need of paint pigments (see below), and the nuns began to realize greater income. Many of the raw materials for the medicinal preparations were provided by the friars of San Marco, who also began dispensing herbal medicines in the mid-quattrocento.

Another convent-operated pharmacy was opened in 1535 by the Annalena nuns in the San Spirito district, which had few other apothecary shops. The nuns supplied the pharmacy with their own herbs, which they cultivated in their large gardens and farms. Nuns sold their preparations to the general population, but convent pharmacies also became a special source of medicinals for Florentine women who were more comfortable dealing with female apothecaries. Lay community apothecary shops were more likely to be staffed and patronized mainly by men. Within the confines of the convent, nuns learned preparation techniques from a lead nun well versed in medicinals, and the skills were passed from sister to sister. Nuns also acquired information and training from printed pharmaceutical materials and learned from physicians and lay herb suppliers[30] who supplemented plants grown by the sisters.

4.4.2 Ingredients and Recipes

The preparation and application of medicinal curatives became a common subject of medical writing with texts focused on the uses of simple medicinals and the mixing procedures and proportional amounts for compound medications. Some physicians, like those who attended the Medici families, left records of medical conditions and medicinal plants used in treatments, which provide a rich source of information on medicinal recipes. These increasingly refined botanical/pharmacological records of preparations, including hundreds of recipes, were readily available in Florence from the late eleventh century, but the use of spices and herbs was clearly not an exact science. Although the formulas for many prescriptions were relatively uncomplicated, prescription dosages that proved to be efficacious for

29. Ibid., 13.

30. Strocchia, "Nun Apothecaries."

some individuals worked differently, minimally, or not at all in others. This was attributable as much to variation in the degree of strength of many of the spices and herbs as it was to differences in tolerance and metabolism among individuals. Another difficulty lay in plant nomenclature, which was often confusing; moreover, where manuals did not include drawings of the actual herbs, problems of identification arose. Accurate botanical illustration thus became important to successful preparation in apothecaries.[31]

Historically, painters had incorporated plants in their work for artistic purposes. As but one example, consider Sandro Botticelli's *Primavera* (ca. 1482) that is now displayed in the Uffizi (see 2.2.2.3b). In this magnificent painting, interpreted by many to be allegorical for the growth associated with spring, hundreds of plant species and different flowers can be seen. Some of the same plants and flowers shown also appear in simple and compound recipes used by apothecaries. Many other artists, including Fra Angelico and those in the della Robbia workshop, integrated various botanical species in their paintings.

A considerable amount of information has survived on the herbs and concoctions prescribed by Florentine doctors and apothecaries in the Middle Ages and Renaissance. The majority of the recipes used were based on uncomplicated mixtures of common herbs. Some, however, involved multiple ingredients, numerous steps in the preparation, and at times would involve toxic materials. One such example is a family of herbs used in medicines that had its origin in Pontus on the Black Sea. King Mithridates VI (ca. 132–63 BCE), of Persian and Greek Macedonian ancestry, working with an even older medicinal mixture, developed mithridatum, which is believed to have been used as an antidote to poison. The concoction was modified by Celsus and later by Galen, who suggested that theriac, as it was known in the Roman world, was not only useful in countering ingested poisons, but also in treating digestive problems. Various versions of recipes to produce theriac had emerged by the Renaissance and contained anywhere from sixteen to sixty-four ingredients. Most of these constituents were from a wide source of plants, including ginger and cardamom of the Zingiberaceae family, St. John's wort, anise, wild poppy, myrrh, cinnamon, saffron, and others representing many families, all of which have physiological effects.[32] Other recipes contained animal parts, often from lizards or poisonous snakes. Difficult to prepare and requiring time for maturation, theriac was expensive and thus available only to the rich. Because of its supposed effectiveness as an antidote, theriac (or the earlier version, mithridatum) was recommended to treat plague victims. But other, simpler compounds utilizing ingredients such as camphor, Armeniun bole, red sandalwood, and aloe were also used as treatment for plague. During the Renaissance, Florentines conducted research, often supported by the Medici family, in order to identify compounds that were efficacious, less complicated to prepare, and involved readily available ingredients.[33]

Since the Middle Ages, monasteries had been planting herb gardens to ensure a steady supply of plants used in medicinal remedies. One notable example of such a garden is the

31. Siraisi, *Medieval and Early Renaissance Medicine*, 141–47.

32. Norton, "Pharmacology of Mithridatum."

33. Lippi, "Diseases of the Medici Family."

Orto Botanico, also called the Giardino dei Semplici (Garden of the Simples), associated with San Marco and still in existence. The garden was established by the Dominican nuns in Florence and was an important source of the raw or simple ingredients used by Florentine apothecaries to prepare curative herbal medications. In 1545, the garden was acquired by Cosimo I de' Medici, who placed Luca Ghini in charge. Ghini had studied at the University of Bologna and later taught medicine and the use of herbs in Bologna before his calling to Florence. Under Ghini the garden evolved into an experimental laboratory for the study of plants and their pharmacological properties. The Giardino dei Semplici, one of the oldest herbal gardens in Tuscany, now houses plants from many parts of the world and can be visited by the public.

4.4.3 Standardization of Recipes

With pharmacies located within the hospital complexes, private apothecary shops, and monastic orders all dispensing medicinals, the scope of the apothecary trade became very diverse. The burgeoning numbers of herbs and mixtures being used necessitated some form of standardization to ensure consistent recipes throughout Florence and to maintain high quality. To that end, in 1498 (modern date 1499) the College of Physicians and the Guild of Apothecaries published the *Ricettario fiorentino*, often referred to as the *Nuovo riceptario*, to be used by the apothecaries. The complete title, translated into English, was *New Formulary Compiled by the Most Renowned College of the Distinguished Doctors of Art and Medicine of the Magnificent City of Florence.*[34] This volume, the first official pharmacopeia in Europe, was written by the physician Master Hyeronimo, son of Ludovico dal Pozzo Toscanelli,[35] with the support and guidance of Savonarola, the powerful Dominican prior. Later in 1515, Santa Maria Nuova published a collection, the *Ricettario*, prepared by Hectorre di Lionello di Francesco Baldovinetti. This collection included more than a thousand simple and complex recipes for medicinal cures, drawn from the experiences of the numerous doctors treating patients in the hospital. The doctors observed patients who had been given a particular recipe to determine if the product had the beneficial effects desired; if not, they altered the formulation to increase its efficacy. This trial-and-error approach combined accepted preparations with observation of outcomes and efforts to improve medications. Many of the medicines described in this opus were originally drawn from Greek and Roman (e.g., Galen), as well as Islamic sources (e.g., Avicenna, Mesue [Yuhanna ibn Masawaih], and Rhazes).[36] A second edition of the 1515 publication appeared in 1550 entitled *El ricettario dell'Arte, et universita de medici, et spetiali della citta di Firenze. Riveduto dal Collegio de medici per ordine dello illustrissimo et eccellentissimo signore duca di Firenze.*

In the mid-sixteenth century, Cosimo I formed two magistracies to oversee convent pharmacies as part of the effort to achieve an even better degree of standardization of the

34. Henderson, *Renaissance Hospital*, 415n28.

35. Huguet-Termes, "Islamic Pharmacology," 233.

36. Henderson, *Renaissance Hospital*, 297–301.

medicines being sold in Florence. During the same period, the College of Physicians was given regulatory power to inspect all operating Florentine apothecary shops to ensure the consistent quality of all pharmaceuticals dispensed. To further this aim, in 1597 the College of Physicians deemed it important to produce a new pharmacopeia, the *Ricettario fiorentino di nuovo illustrato*, which they dedicated to Grand Duke Ferdinando (fig. 4.5). This volume is divided into three sections. Part 1 contains an alphabetical listing of the plant, animal, and mineral components and discusses, with illustrations, the processes of preparation, including distillation. Part 2 describes, in an alphabetical listing, the different amounts (weights and volumes) to be used in individual recipes. Part 3 lists the myriad recipes in common use.

Although raw ingredients such as plants, minerals, and bones were processed with relatively straightforward and unsophisticated techniques, such as grinding with mortar and pestle, Florentine apothecaries also used the more sophisticated method of distillation, where flowers and plants were boiled in water with the volatile components escaping the liquid and being collected in a vessel attached to the boiling container.

4.4.4 Vital Role of Apothecaries in Art

Apothecaries, as well as being sources of medical curatives, were the places from which artists most often obtained the pigments they used to prepare painting materials. Some substances had both medical and artistic uses, but apothecaries also stored materials specifically for those involved in art. Perhaps because of the close connection between apothecaries, artists, and the ingredients common to them both, painters, who did not have a guild of their own, formed a subgroup within the Guild of Doctors and Apothecaries.

Thanks to Cennino Cennini's 1390 treatise *Il Libro dell'arte*, much is known about the kinds of materials used and about the role of apothecaries in supplying those materials. In *Il Libro dell'arte*, Cennini describes the practice of painting from creating colors, to preparing the surfaces to be painted, to techniques for applying the color (see 2.2.1). From this work, we learn that artists purchased some of the pigments they used in lump form at apothecaries, while others were purchased in the form of a prepared powder. In some instances, artists would seek their own sources for pigments, usually either natural mineral deposits, by-products of glass production, or colors from animal and plant sources.

In most cases, artists obtained pigments from apothecaries, then prepared the paint in their workshops, combining the pigments with a medium to transform them into the type of paint they desired. For a pigment that came in a lump form, the artist needed to grind it to a powder. To accomplish this the artist placed the pigment on a hard flat stone (preferably a slab of porphyry), then using a small stone cut to fit in the palm of the hand at one end and flat on the other, ground the pigment between the two surfaces. Pigments differed in the amount of time required for grinding. Cennini wrote that the process took skill in that some pigments ground for an excessive length of time could lose color while other pigments improved the longer they were ground. The artist had to possess an intimate knowledge of the characteristics and parameters of each ingredient in order to achieve an excellent final product. Once the pigment was in powder form, water could be added to remove any residual lumps. Medium was usually not added to the ground pigments until immediately before the actual painting was to begin.

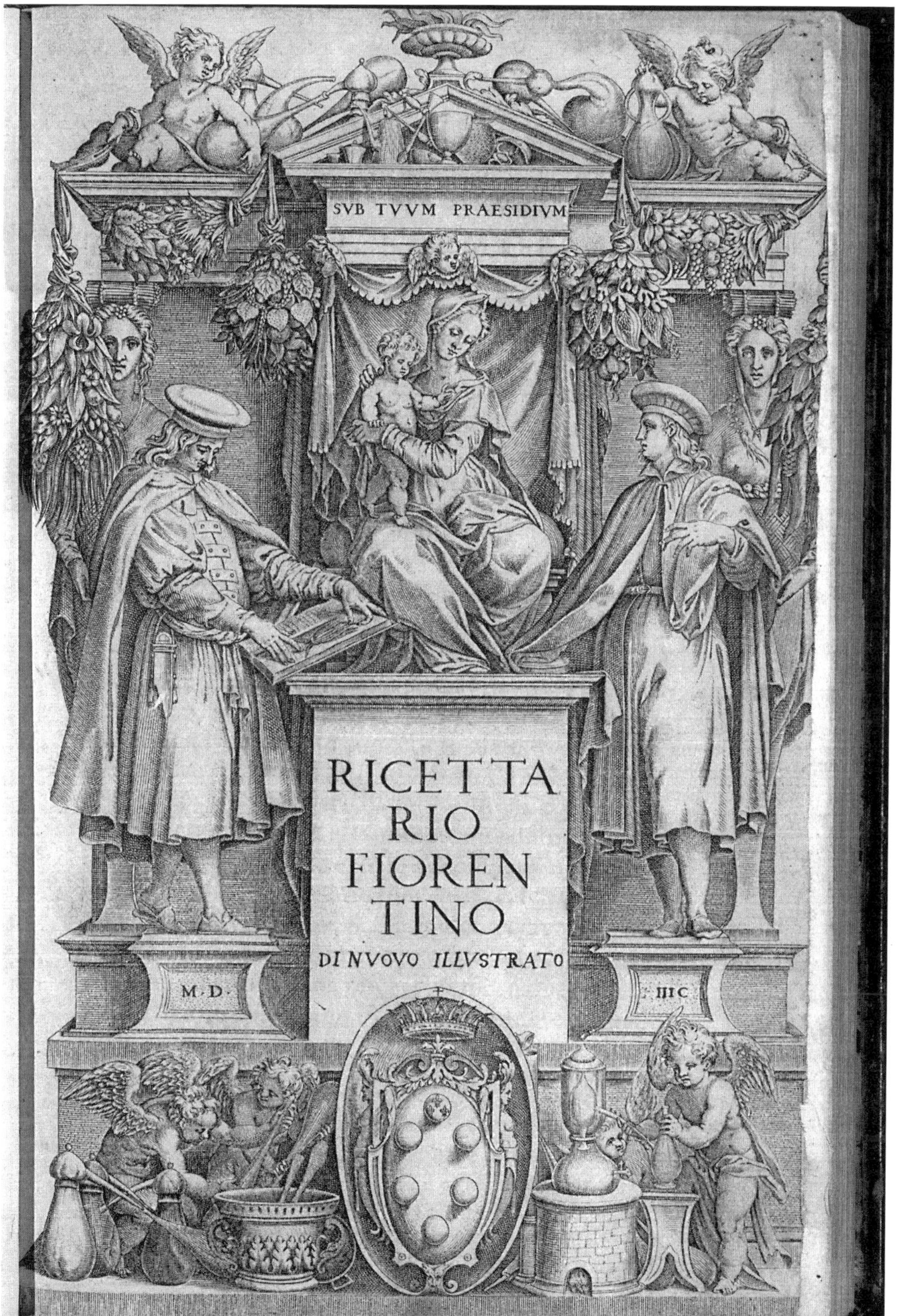

Figure 4.5: Title page, *Ricettario fiorentino di nuovo illustrato* [Fiorenz]: [Marescotti], 1597 (ETH-Bibliothek, Zurich, Alte und Seltene Drucke; EDIT 16 CNCE 3199 [http://dx.doi.org/10.3931/e-rara-9313]).

There were some pigments that were created using complicated recipes; for those, Cennini advised that artists should, for the most part, rely on the expertise of the apothecaries rather than attempt to perform the task in their workshops. Cinnabar, for example, required a chemical process to produce the color red, and the production of malachite green (verde azzurro) involved artificial preparation that Cennini did not explain, recommending instead that artists purchase the pigment from an experienced apothecary. Cennini also included information regarding the ideal properties of various pigments and the qualities artists should look for when selecting them from an apothecary. For the innovative Florentines, the science of color became a key ingredient in their mastery of artistic technique. Clearly, artists placed paramount importance on the colors, the materials from which they were derived, and the sources and the selection of pigments offered; apothecaries were central to that process.

Pharmacies and art also intersected in the creation of containers used to store herbs, spices, and syrups. The most common vessels were terracotta jugs (for oils); glazed containers, having more inert surfaces (to hold syrups and oils); and storage boxes made of lead (for unguents), glass (for distilled water), copper (for syrups), or wood (for dried, minced, or lump materials). The container chosen depended on the type the apothecaries found most suitable for preserving a particular ingredient. The earthenware jars and jugs became valued for their artistry as well as for their practical uses.

From ancient times, people have used earthenware pots to store various items, including herbals used by healers. Greek, Egyptian, and Roman civilizations left evidence of medicine-related pottery, as did various cultures from the Near East and western Asia. By the Middle Ages the range of ingredients in medicines was extensive; pharmacopeias and inventories document the use of various species of plants, including their stems, leaves, fruit, and seeds, and a variety of animal-derived ingredients, such as fat, ground horns, and various organs, most of which needed to be stored in appropriate containers. To meet the demand for the large number of different types of jars and jugs in which to store an ever-increasing array of ingredients and prepared formulations, pottery workshops were expanded in the fourteenth century. Commissions for vessels from hospitals, pharmacies, and monastic apothecaries accounted for a high percentage of the output of many of the workshops in central Florence and the surrounding area, primarily in Montelupo.

Pottery craftsmen worked to develop vessels that could safely contain liquids, have sufficient durability for frequent use, and also be decorative. As early as the eighth century, artisans from the Middle East were formulating glazes that not only enhanced the beauty of ceramic vessels, but improved their ability to hold fluids without leaking or affecting the fluid being stored. Their vessel coating found to be most effective was a mixture of water and ash from tin, lead, and melted sand and potash, which made it possible to fire pots without losing the integrity of the painted decoration. Traders from the Iberian Peninsula acquired pots in western Asia that used a tin coating, and local artisans refined the technique that became known as maiolica (see 2.6.1). They exported the vessels to the Italian peninsula, where the formula was adopted by workshops in Orvieto in the thirteenth century and later in Florence, as well as in Deruta, Siena, Faenze, and the Umbrian centers of Gubbio, Pesaro, Castel Durante, and in Venice. During the fifteenth century, Italian workshops began to inscribe some medicinal vessels with the names of the intended contents, as well as a designated symbol for a particular hospital or pharmacy. As was true for pottery in general, artistic innovations continued as workshops enhanced the decorative quality of the pharmaceutical jars.

An examination of the vessels in their variety of shapes and configurations speaks eloquently of the integration of art, science, and utility. It was during the Florentine Renaissance that workshops began to take an even greater interest in the ornamentation of pharmaceutical jars and jugs. Interestingly, the themes that dominated were flora, especially pine cones (identified in particular with Florentine vessels); stylized birds, fish, and other animals; and an occasional coat of arms or classical motif. But rarely did designs relate to the medicinal contents. The principal colors were blue, purple, and green; other commonly used colors were yellow, brown, and ochre. Cylindrical jars without handles (*albarelli*), pots with double handles (*orcioli*), jugs (*boccali*), spouted jars with either one or two handles, and covered shallow pots were remarkable for their artistic and intricate designs. By the mid-sixteenth century, artisans in the Florentine workshops in Montelupo often made the pot handles to resemble braided ropes or the heads of dragons.

A number of Western museums have collections of Florentine pharmaceutical jars, most notably the Victoria and Albert Museum in London, while in Florence a few examples can be viewed in the Bargello Museum, which displays works from Montelupo made between the fourteenth and seventeenth centuries. Some of the showcased items are small apothecary jars and *boccali* from the Santa Maria Nuova pharmacy, clearly identified by their distinctive emblems. The Officina Profumo-Farmaceutica di Santa Maria Novella (see 4.4.1) also has a collection of particularly interesting medicinal vessels.

4.5 Medicine Depicted in Art

Many works of art were placed in hospital settings for their curative value. There are also examples of paintings located in various areas of Florence that were created for the purpose of depicting some particular ailment or medical practice, or where a medical condition was shown, although not intended as the central theme of the painting.

4.5.1 Campanile (Duomo)

There are several interesting examples of artists using the practice of medicine as the subject of their work. One early depiction is found on the Campanile (Bell Tower) located next to the Florence Cathedral. Andrea Pisano (d. 1348) and his workshop created hexagonal tile reliefs for the first level of the tower and another cycle of lozenges on blue backgrounds for the second tier. The tiles in the first level portray a narrative of mankind, starting with the biblical stories in Genesis and continuing with tiles focused on the range of man's skills, including the creative arts. One of the hexagonal tiles located on the south side of the Campanile illustrates the practice of medicine (fig. 4.6). It depicts four Florentines in the office of a physician. Two patients are carrying a urine specimen in a glass flask encased in a straw sack, while another patient is having his urine specimen visually observed by the physician.

The relief draws attention to a common practice in the life of a Florentine doctor. Physicians were expected to be able to distinguish the variations in urine color, clarity, odor, and sedimentation in order to diagnose a malady. This type of examination was repeated often during a patient's illness to monitor changes in the patient's condition and/or adjust the treatments. The technique was so important to the practice of early modern medicine that the urine container, called a matula, was often depicted in paintings and drawings with a physician/patient theme.

4.5.2 Bargello Museum

Another portrayal can be seen in a small tabernacolo on the exterior wall of the Palazzo del Bargello at the corner of Via Ghibellina and Via del Acqua. This painting, executed by Fabrizio Boschi in 1589, shows Saint Bonaventure visiting plague victims. Florence experienced numerous outbreaks of the plague from the mid-fourteenth century until the late seventeenth century. Because plague was highly contagious, those who were willing to risk their lives visiting and treating its victims were held in high regard. As a child, Saint Bonaventure was believed to have been cured of a life-threatening illness by the prayers of Saint Francis of Assisi. During Saint Bonaventure's lifetime, he was greatly appreciated for visiting those who were extremely ill and often contagious, making him an appropriate subject for this painting. Because of the outside location of this work, damage was extensive, and thus it was restored by Silvestra Bargellini in 1996. Unfortunately even with the relatively recent restoration, the outside location, coupled with the condition of the painting and the protective covering, which has deteriorated and become heavily fogged, render it difficult to observe.

4.5.3 Santa Felicitá and the Palazzo Vecchio

Paintings depicting medical conditions, although not meant as the central theme, can also be seen in Florence. One is an unusual portrait of Saint Matthew located in the Brunelleschi-designed Barbadori Chapel in Santa Felicitá. Frescoed by Jacopo Pontormo (1494–1557), possibly with input from his student Agnolo Bronzino (1503–72) between 1525 and 1528, a roundel in the upper right-hand corner shows what could be interpreted as additional breasts on his chest and an obvious extra growth on his neck with what appears to be a nipple. The extra breast on the chest, a condition known as polymastia, was known to occasionally occur and was even claimed by some to be a sign of increased fertility. The latter depiction could be interpreted as a goiter, an enlargement of the thyroid gland often caused by a lack of iodine in the diet or by a thyroid tumor. A similar medical condition is also visible in a fresco in the Palazzo Vecchio. The small chapel in the private quarters of Eleonora di Toledo, wife of Grand Duke Cosimo I, boasts a beautiful frescoed scene executed by Bronzino from about 1541 to 1545. This painting of the story of Moses includes a woman, not central to the theme, but with an obvious swelling of the neck, another possible representation of a goiter.[37]

37. Lippi and Rotella, "Bronzino or Pontormo?"

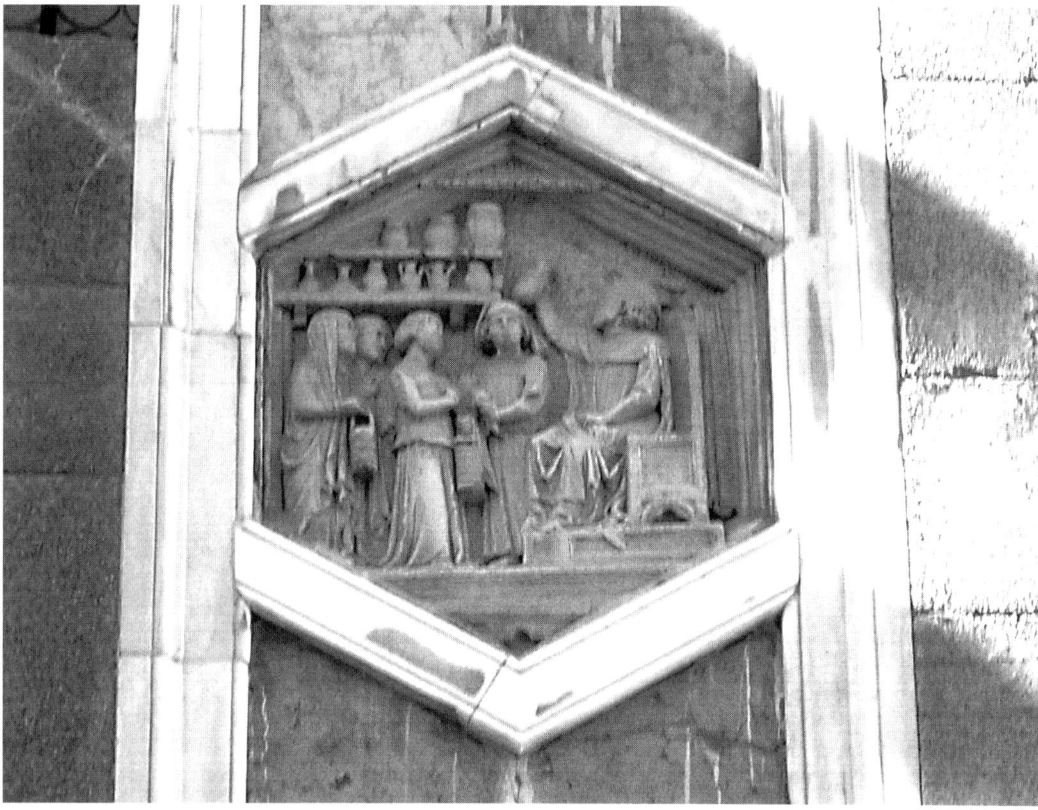

Figure 4.6: Andrea Pisano, Hexagonal tile depicting patients and physician examining urine (ca. 1340s), Campanile (photograph by authors).

Although a goiter is fairly rare today and relatively easy to treat, the condition was prevalent during the early modern period, especially in inland areas. As had been true for many centuries, physicians during the Renaissance considered medical disorders to arise from an imbalance of the four humors, phlegm, blood, black bile, and red or yellow bile, or an imbalance of hot, cold, wet, and dry conditions within the body. Treatment of disorders such as those mentioned above was centered on the maintenance of a correct balance of these factors.

4.5.4 Oratorio Buonomini di San Martino

In the Piazza San Martino, off Via Dante Aligheri, is the Oratorio Buonomini di San Martino. This structure houses a charitable organization founded in 1441 by a prior of the Dominican order of San Marco, Antonio Pierozzi. It was composed of men, two from each district of Florence, whose mission was to unobtrusively provide emergency funds to citizens of means who, for whatever reason, found themselves in financial need. The frescoes in the small intimate meeting hall, attributed to Ghirlandaio and his workshop, show members and non-member citizens performing charitable work. Of interest to this discussion, the first painting to the left of the front entrance depicts a child-birthing scene. The woman in bed has just given birth to an infant shown lying next to her. Present in the scene are two individuals attending the woman, possibly her husband and a family friend or midwife, and a man, who may be a member of the charitable organization, presenting a chicken and a flask, probably containing wine, to another attending woman. This is but another example of the ways in which art permeated much of the Florentine medical and pharmaceutical arena, each enriching the other in important and interesting ways.

Chapter 5

Astronomy and Time Reckoning

> "Finally, there are the instruments themselves, constructions both beautiful and useful, conduits of light through vast dark spaces, defunct sites of science, living objects of wonder . . . There is something romantic, even sublime, in witnessing the faithful rendezvous of sun and rod arranged centuries ago."
>
> —J. L. Heilbron, *The Sun in the Cathedral*

In their increased desire to better understand the cyclical movements of the heavenly bodies and to precisely translate their observations to years, months, days, and hours, the Florentines built on astronomical knowledge of the past, utilizing monumental structures, advanced apparati, and mathematical principles. Artists and scientists portrayed those visual observations in accurate drawings, thus making the astronomical information more accessible. In addition, some painters used celestial phenomena and instrumentation as components in their works, rendering dramatic artistic and scientific statements.

5.1 Ancient Origins of Renaissance Astronomy

Early civilizations throughout the world observed and recorded the relative movements of the sun, moon, planets, and stars, and developed solar and lunar calendars for secular and religious purposes. As they increased their understanding of the periodic motion of celestial bodies, many cultures came to believe that those orbs influenced, or even dictated, human life and the environment, and that therefore they could predict events based on the movements of the stars. Over time, this system of astrology became quite detailed and sophisticated in its structure, but while astrology may have been more familiar to most people, astronomers continued to study celestial bodies from a scientific and mathematical perspective. Much of the astronomical knowledge to which Renaissance Florence had access was derived from Greek and Islamic scholars with an interest in astronomy and mathematics, the latter based on the foundations developed from geometry. Those early astronomers, benefiting from empirical developments in Mesopotamia and Egypt, provided the Europeans with substantial astronomical information that established a foundation for the Scientific Revolution to follow.

The Sumerians, as early as 3000 BCE, had described and named the constellations that we recognize today. By the eighth and seventh centuries BCE, the Babylonians, taking advantage of Sumerian astronomical knowledge, added their own observational and mathematical findings that were codified on clay tablets. Their empirical observations of periodic planetary movement and detailed records of the moon enabled them to predict lunar eclipses with considerable accuracy. In the sixth and fifth centuries BCE, the city of Miletus (in modern-day Turkey) spawned a number of Greek physicists and astronomers—most notably Thales (ca. 624/25–547/46 BCE), Anaximander (ca. 610–546 BCE), and Anaximenes (ca. 585–528 BCE)—who contributed greatly to the scientific information that would reach Florence some two millennia later. Sadly, much of our knowledge from those early astronomers is based on quotes and attributions from later natural philosophers. But we do know that the prevailing model of the universe, as championed by Aristotle (ca. 384–322 BCE) in early Greece, placed Earth at the center of the universe with the sun, the moon, and the planets orbiting around it.

This geocentric view of the universe was widely accepted for many centuries, although the work of Aristarchus of Samos (ca. 310–230 BCE) foreshadowed an idea that would be considered revolutionary seventeen centuries later. Aristarchus probably studied at the Lyceum in Athens with the physicist Strato of Lampsacus (ca. 335–ca. 269 BCE), who was a student at Aristotle's school in Athens, then tutored Ptolemy (Claudius Ptolemaeus, ca. 100–170) in Alexandria before returning to Athens to head the Lyceum. Aristarchus suggested that planetary motion could be explained if Earth rotated daily on its axis and revolved annually in a circle around the sun. His ideas were rejected in favor of the prevailing geocentric view advanced by Aristotle and later Ptolemy, but his works were not entirely lost. The Polish astronomer Nicolaus Copernicus (1473–1543) had access to the work of Aristarchus, and although he acknowledged his work in the original manuscript of *De revolutionibus orbium coelestium*, it is interesting that for some unknown reason he neglected to mention Aristarchus's heliocentric model in the version that was published. Two other influential Greek astronomers were Apollonius of Perge (ca. 262–190 BCE) and Hipparchus of Nicea (ca. 190–120 BCE). Apollonius is best known for his study of planetary motion, and Hipparchus, a preeminent astronomer, applied mathematical precision to his studies and is credited with discovering Earth's precession and for the invention of the astrolabe (discussed later in this chapter).

As Alexandria, founded by Alexander the Great (356–323 BCE) in the third century BCE, matured into a significant international center of research and higher learning in science and philosophy, it became the focal point of Greek scientific inquiry, boasting dedicated scholars including Ptolemy. Alexandria's comprehensive library collected the writings of the early Greeks, later adding Ptolemy's *Mathematike syntaxis* (*Megale syntaxis*, or the *Great Compilation*, but usually known by its Arabic title, *Almagest*) and *Tetrabiblos*. *Almagest* represented the first comprehensive treatise on astronomy, and *Tetrabiblos* became a major source of information on astrology. Ptolemy, like Aristotle, favored the geocentric view of the solar system. The troublesome observation that some planets appear to orbit backwards around Earth (planet retrogression) was cleverly, albeit incorrectly, explained by Ptolemy. He introduced the concept of epicycles, which theorized that planets move in a small circle around a point that in turn moves in a circular orbit around Earth. Despite the fact that Aristarchus's view of the universe turned out to be more accurate, Ptolemy's writings on astronomy were much more widely known and accepted throughout the ancient and medieval world.

When, after war and natural disaster in the third and fourth centuries, Alexandria was no longer the epicenter of scientific study, Islamic scholars inherited the astronomical knowledge and theories of the Greeks. They made numerous modifications to the work of Ptolemy and other Greek astronomers and added their own astronomical observations and calculations. Following the Islamic conquest of the Iberian Peninsula in the eighth century, Cordoba and Toledo developed into major research and educational centers for astronomy, medicine, mathematics, and philosophy. These centers sponsored great libraries that attracted Christian scholars for study, many of whom were responsible for translating a number of the works from Arabic to Latin. Other Greek scientific treatises were translated into Latin by scholars in Sicily. Eventually the books in Latin were copied or purchased for the developing libraries in Florence, such as the Laurentian Library; a number of the Greek treatises also made their way to Florence, where they were translated by a few of the intellectuals of the time. Manuel Chrysoloras (ca. 1350–1417), a Byzantium humanist, taught in Florence from 1396 to 1400. While in Florence, he translated Greek manuscripts, making them far more accessible to many Florentines. The Belgian Dominican monk William of Moerbeke was also responsible for translating numerous books and manuscripts from Greek to Latin. A number of the books that had been translated from Greek to Arabic and then to Latin were translated back into Greek from the Latin for the libraries of Constantinople. What a tour through time and space!

By these rather convoluted modes of entry into Florence, the Renaissance astronomers gathered an impressive amount of data from careful observations made over the course of two millennia. For many years, Renaissance astronomers interpreted those data based on an incorrect Aristotelian-Ptolemaic model of the universe. Even after Copernicus published, in 1543, *De revolutionibus orbium coelestium*, in which he explained his heliocentric view of the universe, many scientists considered his ideas to be merely theoretical until Galileo Galilei (1564–1642) presented supporting evidence from his observations made with a telescope. Although the idea had been suggested earlier, it was this Renaissance Florentine scientist who was central to the paradigm shift from an Earth-centered to a sun-centered understanding of the solar system.

Renaissance astronomers were also very interested in measuring, again with high accuracy and precision, the obliquity of the ecliptic (the angle between Earth's equatorial plane and its orbital plane, or the tilt of Earth), a major scientific problem that astronomers had pursued for about two thousand years. The early Greek astronomers knew that Earth's axis was tilted in relation to the sun, resulting in variation in the seasons. They theorized that the projection of Earth's north pole defined a celestial pole about which all bodies seem to rotate daily. They determined the tilt of Earth's axis to be about 23.5 degrees. The Greeks observed that despite their best efforts to make precise calculations, there seemed to be variations they could not completely explain or predict. Hence, the Florentine astronomers were tackling a fundamental problem inherited from earlier scientific deductions. Today we know that the tilt varies between about 22.1 and 24.5 degrees with a periodicity of about 40,000 years.

From a modern perspective, astronomy's focus on understanding the nature of the solar system may seem more important, but to early modern Europeans, its relationship to timekeeping was much more central to their daily lives. The problem of developing an accurate timekeeping instrument was more prosaic, but was a highly difficult problem to solve. The church had a desire for an accurate solar-based calendar in order to determine the equinoxes and solstices with reasonable precision and to establish correct dates for Christian holidays. Equally, cities desired a calendar that was consistent year after year.

Throughout history, one challenge that all astronomers faced with regard to a solar-based or lunar-based system was that a solar year consists of 365.2422 days and a lunar month of 29.5306 days, both numbers being rounded. A culture that followed a lunar-based calendar would quickly find itself over time, for example, observing the beginning of spring in a different month. Various societies found ways to make corrections for these awkward numbers of nature, but whatever system was used, the calendar would invariably drift to some degree unless appropriate corrections continued to be made.

The Catholic Church, and indeed the Western world, had been using the Julian calendar since it was introduced in 46 BCE by Julius Caesar, who enlisted the aid of Sosigenes of Alexandria to overcome the major problems associated with the mixed solar/lunar system of the old Roman calendar. Sosigenes took 365.25 days for the solar year, a relatively close value; moreover, he attempted to make corrections by having three years consist of 365 days and one year of 366 days. This system worked reasonably well for short periods, but it drifted appreciably over time; in the 1500s it was some ten days off, which meant that the vernal equinox was falling on March 11 instead of its decreed day of March 21. During the Florentine Renaissance, this inaccuracy was particularly annoying to the Catholic Church, which needed to calculate the correct date for its celebration of Easter and thus was willing to accommodate astronomers in their quest to make better measurements for reckoning time. Several of the sites used for solar time determinations in Renaissance Florence were in fact located in churches.

In the 1570s in Florence, astronomer and mathematician Fra Egnazio Danti (1536–86) designed an astronomical quadrant and armillary sphere that were placed on the façade of Santa Maria Novella, as well as a gnomon in the church interior to determine equinoxes, solstices, and to improve calendrical time. Because of its height, length, and stability, a church was an ideal location to place a gnomon, which consisted of a circular opening in the roof or façade through which sunlight cast its rays on a meridian, a strip on the floor marked to indicate dates and zodiacal times.[1]

Danti's extensive astronomical and mathematical studies conducted in Florence are discussed in detail later in the chapter, but it is interesting to note that when Pope Gregory XIII

1. Strano, "Origins of Gnomonics."

decided, in response to a recommendation from the Council of Trent (1545–63), to correct the Julian calendar, he invited Danti to Rome to serve as a member of the commission. The resulting Gregorian calendar (1582) solved the problem of the Julian calendar, which gained three days every four centuries. It was decided that years divisible by four should have one day added, unless the year was exactly divisible by one hundred.

In addition to instruments used for calculating dates, scientists also developed non-mechanical devices for measuring the hours. Sundials had been in use in the Western world for at least five thousand years, but they obviously worked only during periods of sunshine, so early cultures also experimented with candle clocks and hourglass-type instruments. Babylonian, Egyptian, and Greek inventors, seeking more accurate methods of timekeeping that could be used both day and night, developed water clocks (*clepsydrae*). The Romans devised intricate gearing that enhanced these clocks' capability, and Islamic cultures advanced the accuracy of the instrumentation. After Islamic water-clock technology arrived in Spain, more complex designs evolved.

Although the ideas that led to the development of mechanical clocks had been germinating for some time, it was not until the latter part of the thirteenth century that inventors in Europe began to use escapements (pendulums or balance wheels), weights and balances, and gears and ratchets to transfer energy and advance gear trains in fixed increments, thereby moving indicators that marked the time. It was probably in northern Europe that the first mechanical clocks appeared. These were operated by a weight that moved a set of geared wheels as it descended. One of the earliest mechanical clocks for which the workings can be documented is the Dover Castle Clock, which is dated to the first half of the 1300s; other clocks also appeared in England during that time period. Henri de Vick's clock, now located at the Palais de Justice in Paris, was constructed in the 1370s, and the medieval clock in the Salisbury Cathedral in England was built in or sometime before 1386. The Salisbury clock is today probably one of the oldest European clocks still in working condition.

In Italy, mechanical clocks were first mentioned in the early to mid-fourteenth century in cities such as Milan, Florence, and Padua. About 1348 Giovanni de'Dondi (1318–89), a pioneering clockmaker from Padua, began constructing an ingenious astronomical clock (*astrarium*) that was weight-driven with an escapement. By the beginning of the fifteenth century, use of this method had spread to other parts of the Italian peninsula and clocks were found on public buildings in most European towns, with their faces showing either twelve- or twenty-four-hour segments. If the clocks were reset nightly, they were fairly accurate.

5.2 Solar Time Measurement: The Sites, the Scientists, and the Art

In Florence's Baptistery of San Giovanni, there was an opening in the roof through which sunlight struck an inlaid disk on the floor. Probably dating to the fifth century, this early astronomical site is believed to have served as a measurement of annual time, namely, the summer solstice. Later, during the Florentine Renaissance, astronomers installed instruments such as armillary spheres, gnomons, and disks to track and measure the movement of the sun and assist in calculating solar time in additional sites throughout the city. These locations were selected for their height, orientation, stability, and availability. In keeping with the spirit of the Florentine Renaissance, the sites chosen for these measurements were, by and large, buildings of architectural splendor that housed magnificent art.

5.2.1 The Baptistery

Located near Santa Maria del Fiore, known as the Duomo, and Giotto's bell tower, Campanile di Giotto, the Battistero di San Giovanni is one of the oldest buildings remaining in Florence. Based on architectural and archaeological evidence, the structure that originally existed on the site is believed to have been a Roman temple dedicated to Mars. The structure was consecrated in the fifth century to the Christian religion and then replaced or altered in the sixth century. The octagonal Baptistery that one sees today was begun in 1059 and completed in 1128. Over the next several centuries, many generations of Florentine citizens were baptized there, including Dante Alighieri (1265–1321), Amerigo Vespucci (1454–1512), Niccolò Machiavelli (1469–1527), and Medici family members including Duke Cosimo I (1519–74). Today the Baptistery is best known for its bronze doors added by Lorenzo Ghiberti and Andrea Pisano in the fifteenth century.

The historical astronomical component in the Baptistery consisted of an inlaid marble disk on the floor and an opening in the roof. The disk, with the sun located in the center, contains the following palindrome: "En giro torte sol ciclos, et rotor igne" (Here, I the sun turn obliquely the circles and am turned by fire).[2] Radiating out from the sun are two rows of wedges, the outer one of which has the signs of the zodiac and around the outer circumference is another inscription, which reads in part, "Florence, florid, furnished with all good things / asked that this work be filled with the constellations of the skies / . . . / in the low flooring appear the illustrious things of heaven."[3] The disk was originally placed near the north door so that the rays of the sun, or perhaps the shadow cast by a sphere at the site of the opening in the center of the domed roof, apparently marked the zodiacal disk at the summer solstice. Inlaid in a marble slab, the disk was moved during renovation beginning in the eleventh century, probably because it was no longer functioning due to the precession of the equinoxes over about five centuries rendering the gnomon and the disk in an incorrect relationship. In its current location on the east side of the building, such an illumination is no longer possible. Writing in the fourteenth century, Villani mentioned that the Baptistery was topped in the eleventh century by a *capannucio* (a little hut) consisting of a small roof on four short pilasters, thereby partially closing the hole. The hut was replaced in 1448 by the structure seen today.[4] Historical documents are sparse and confusing as to the actual progression of events and the functioning of this early use of a building for astronomical purposes, but it did serve as a precedent that would continue during and after the Renaissance.

Today the Baptistery is one of the most frequently visited sites in Florence, in large part because of the grand and imposing bronze-paneled doors on the exterior (see 2.3.5.2a), the

2. Innocenti, "Solstitial Sundial," 19n1.

3. Ibid., 19n2.

4. Ibid., 13.

magnificent interior marble-clad walls, and the mosaic ceiling (see 2.4.3.1). The zodiacal disk can be viewed today, but is protected from pedestrian traffic. This represents one of many examples of the joint use of a common space for religious, artistic, and astronomical purposes.

5.2.2 The Florence Cathedral (Duomo), Santa Maria del Fiore

In 1475 Florentine scientist Paolo dal Pozzo Toscanelli (1397–1482) placed a gnomon, presumably to accurately measure the summer solstice and probably the obliquity of the ecliptic, in Santa Maria del Fiore. Toscanelli, son of the physician Dominic Toscanelli, was an outstanding fifteenth-century scientist who worked in the fields of mathematics, astronomy, astrology, and cosmography. Born in Florence, he began studying mathematics under the tutelage of Giovanni dell'Abacco and continued at the University of Padua, where he studied mathematics, medicine, and philosophy. After graduating in 1424 with a doctor of medicine degree, Toscanelli returned to Florence, where he formed friendships with Filippo Brunelleschi (for whom he was also a tutor in mathematics), Leon Battista Alberti, and other prominent Florentines. A plaque commemorating Toscanelli is attached to a building across the street from the Palazzo Pitti, designating the location of his residence. Toscanelli was acclaimed for his accurate and painstaking study of comets and, as noted in the discussion on perspective (see 3.1.2.1), is believed to have influenced Brunelleschi's mathematical approach to depiction of depth in art. In addition, as mentioned in the section on cartography (see 3.2.2), it is probable that Christopher Columbus used Toscanelli's information and chart in planning his 1492 voyage of exploration.

By the time Toscanelli installed the gnomon in the Duomo, he was very well known and respected, not only in Florence but throughout much of Italy. Someone of lesser stature than Toscanelli may have had difficulty in securing permission from the Catholic Church to add scientific devices to a church building, particularly to the Florentine Duomo. His stellar reputation and his prior relationship with Brunelleschi, designer and builder of the dome of the Duomo, certainly aided him in his request. At over ninety meters from the ground, the placement of Toscanelli's gnomon (fig. 5.1) made it the highest one anywhere at the time. Since Santa Maria del Fiore, like most Catholic churches, was positioned with its major axis running in an east–west direction, Toscanelli placed his meridian (marker line on the floor) in the north transept in a standard north–south orientation. Because of the height of the dome, the transept would not accommodate a complete meridian, so Toscanelli installed a relatively short one (ca. ten meters) that permitted measurements at and near the summer solstice when the shadow would be shorter. There is no surviving documentation explaining Toscanelli's overall objective, but it is likely, as indicated by the distinguished mathematician Regiomontanus (1436–76), that he wished to measure the obliquity of the ecliptic, the tilt of Earth's polar axis relative to the plane of its orbit. Using his precise measurements of the summer solstice, Toscanelli was able to accurately determine the length of the year and correct the thirteenth-century Alfonsine tables, which represented an updated version of the earlier eleventh-century Toledan tables. These compilations were made by Muslim scholars in Spain and used to calculate, relative to fixed stars, the positions of the planets, the moon, and the sun.

Toscanelli's gnomon, a bronze disk with a gnomonic hole of about five centimeters, was located in the lantern atop the dome to allow sunlight to penetrate into the Duomo. While the diameter is consistent with the rule that the opening should not be greater than 1/1,000 its height from the ground, the size shows that, for whatever reason, Toscanelli chose to install a gnomon with a smaller opening; based on its height, the opening could have been as much as nine centimeters in diameter. Referring to the size of the gnomic opening and the geometry of the cathedral, it has been pointed out that the use of a small opening results in a less bright image but permits greater contrast and resolution.[5] The bronze plate with the gnomonic hole was slightly repositioned in 1511, but the plate that can be seen today likely has the hole's original bushing, fabricated to Toscanelli's specifications in the fifteenth century.

This still-functional gnomon has had an interesting history. The Opera del Duomo had the plate removed in the nineteenth century during a restoration of the lantern over the dome. However, there was such an outcry in Florence over this act that the Opera decided to have the removed gnomon reinstalled by Father Antonelli in 1866. Because of the restoration, the plate could not be placed in its exact earlier location, but was installed less than a centimeter higher.

The only remaining portion of the original meridian installed by Toscanelli is the marble disk, several centimeters in diameter (fig. 5.2). In 1510 two sections of marble were placed around Toscanelli's original circular section of marble, forming a much larger circle in which Toscanelli's disk is not centered. The diameter of this new circular section of marble corresponds to the stenopaic image of the sun, that is, the image formed on the floor by the sunlight after passing through the five-centimeter opening in the gnomon. Although only marginally legible today, the more recent marble disk contains the wording, "MDX, PRIDIE, ID, IVNII" (June 12, 1510), the day marking the summer solstice in the old calendar.

There is little information available on how Toscanelli used his gnomon or how it was applied in the years after his death, but we do know that a Dominican friar, Egnazio Danti (see 3.2.2.1 and 5.2.3), was using Toscanelli's gnomon in the sixteenth century. It was not until 1757 that the Jesuit priest Leonardo Ximenes (1716–86) published a well-documented book, based on surviving records and his own exhaustive measurements, in which he provided detailed information on the installation of the gnomon and meridian designed by Toscanelli and justified the need for a new meridian in Santa Maria del Fiore. While little is known of Toscanelli's scientific objectives, other than perhaps to accurately determine the summer solstice, Ximenes explained that his own major goal was to evaluate variations in the obliquity of the ecliptic. Toscanelli's very high gnomon was required in order to accurately detect the small angle, thus enabling a comparison of Ximenes's results with those determined in1510.[6] Ximenes's request for a new meridian was approved and after extensive and careful measurements the work was completed in a relatively short time.

5. Mazzucconi, Ranfagni, and Righini, "Gnomon of Santa Maria del Fiore," 49.

6. Ibid., 50.

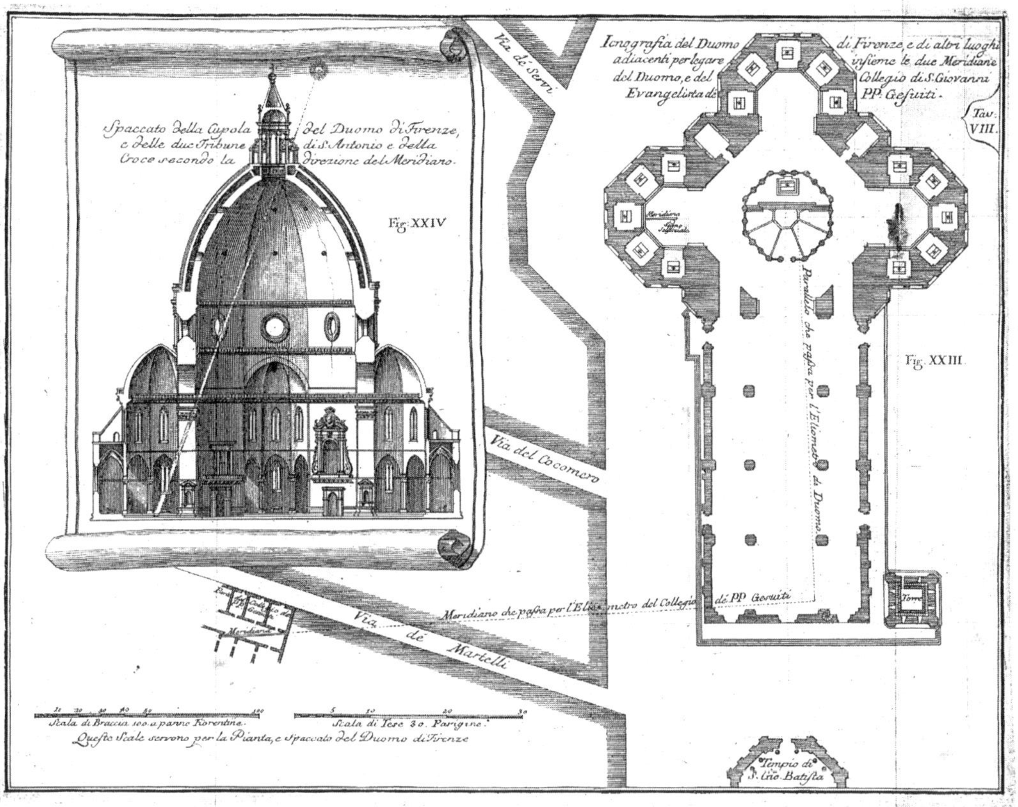

Figure 5.1: Leonardo Ximenes's depiction of beam of light admitted by the gnomon in Santa Maria del Fiore (Duomo) at the summer solstice. Meridian designed by Paolo dal Pozzi Toscanelli (ca. 1475), drawing from Ximenes, *Del vecchio e nuovo gnomone fiorentino e delle osservazioni astronomiche* . . . , 1757 (MED 1809, tav. VIII, photograph by Renato Idra, inv. # 68400, courtesy of Museo Galileo, Florence).

Ximenes, the son of Spanish parents, was born in Trapani, Sicily, in late December 1716. He received his early education at a Jesuit school in Trapani and completed his education at the Roman College. At the age of fifteen he was inducted into the Order of Saint Ignatius and ordained in 1743. Upon entering the Society of Jesus in 1750, Ximenes transferred his property to his mother and paternal uncle and retained only a small allowance. The order sent him to Florence, where he soon requested that he be appointed a mathematician at the University of Florence. This request was rapidly approved, and Ximenes was given the title of imperial geographer. In 1773 Clement XIV suppressed the Order of Saint Ignatius, but Ximenes was able to remain in Florence as a guest at the Casa di San Giovanni and continue his work, which involved surveying, mapping, hydraulics, engineering, and other areas, as well as astronomy. Funds derived from his appointment as professor in Florence and mathematics tutor enabled Ximenes to leave sufficient resources to endow two professorships, one in hydraulics and the other in astronomy. After his death, Ximenes's instruments and library on astronomy were given to the professor of astronomy, who was charged with continuing annual observations of the summer solstice in the Duomo and maintaining the observatory

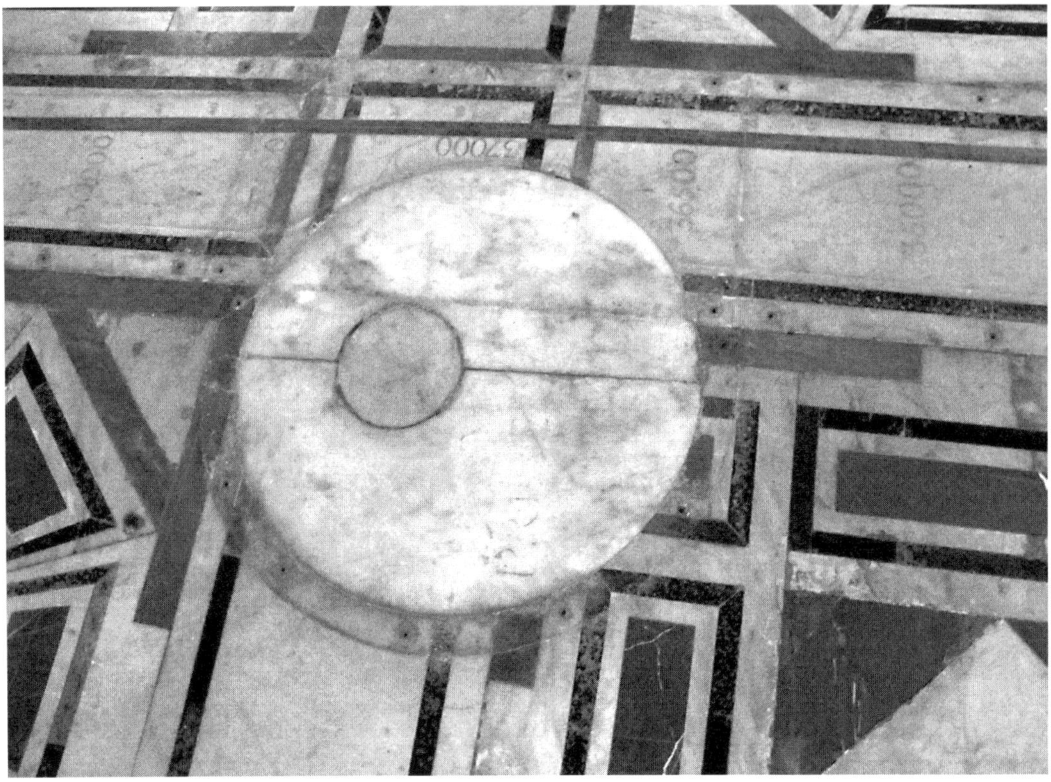

Figure 5.2: Paolo dal Pozzi Toscanelli, Marble disc (ca. 1475), modified by Leonardo Ximenes (mid-18th century), in the floor of the Duomo (photograph by Sailko, retrieved from Wikimedia Commons, GNU Free Documentation License).

Ximenes had established in the Jesuit college (Casa di San Giovanni). In 1859 the observatory was moved to Arcetri, near Florence, where it was dedicated in 1872 in honor of Galileo, who spent the last eleven years of his life in house arrest near the site.

To develop an accurate record and better understand Toscanelli's instrument, Ximenes made meticulous measurements of the vertical height of the gnomon (91.05 meters) and the horizontal distance from the point on the floor immediately below the gnomon to the center of the marble disk (33.69 meters). He also found that Toscanelli's meridian line deviated slightly from true. After completing his astronomical measurements and comparing those he determined in 1755 with those from 1510, Ximenes concluded that the progression of Earth's tilt over time was approximately 30 arcseconds per century, a value now known to be 47 arcseconds per century (0.013 degrees per hundred years). That error notwithstanding, Ximenes did establish unequivocally that the obliquity of the ecliptic was indeed decreasing, as commented on by Nicolaus Copernicus and Egnazio Danti. Moreover, the value Ximenes obtained for the obliquity of the ecliptic was reasonably accurate for the time.[7]

7. Heilbron, *Sun in the Cathedral*, 244–45.

To commemorate his accomplishments, Ximenes placed an inscription on a pier in the south transept acknowledging the Grand Duke of Tuscany for supporting the studies. Around the same time, Ximenes had an elliptical marble plate installed in the floor of the cathedral to mark the perpendicular point and show the units of measurements employed (the Florentine *braccia* and the Parisian foot).

After Ximenes's death in 1786, there was no further interest on the part of the church or the scientific community in continuing astronomical measurements in Santa Maria del Fiore. Since that time, and unfortunately for the visitor today, the cathedral has partially obscured Toscanelli's marble disc and Ximenes's eighteenth-century inlaid marble and brass meridian line with chairs. Moreover, much of the north transept is now closed to the public, although certified guides can be hired to escort visitors to this area. Also, thanks to the efforts of several individuals and the support of the Opera del Duomo, each year, scientists, historians, and a few invited guests are permitted the use of the cathedral for celestial observations.

5.2.3 Santa Maria Novella

In 1574, about a century after Toscanelli installed his gnomon in Santa Maria del Fiore, a Dominican friar, Egnazio Danti, gained the support of Cosimo I de' Medici (1519–74), Grand Duke of Tuscany, for his astronomical studies. Much as Toscanelli had influential friends like Brunelleschi and Ximenes had the support of the Grand Duke of Tuscany, Danti took full advantage of Cosimo I's patronage to use his considerable talents under the grand duke's sponsorship.

Pellegrino Rinaldi Danti was born in Perugia into a multitalented family of artists, mathematicians, astronomers, writers, and goldsmiths (see 3.2.2.1). He followed in his family's footsteps, training as a cartographer, astronomer, mathematician, and visual artist; he also become a Dominican friar and took the name of Egnazio. Danti's older brother introduced him to Cosimo I, who then commissioned Danti to decorate the map room (Guardaroba) in the Palazzo Vecchio (see 3.2.2.1). To the satisfaction and delight of Cosimo I, Danti's output of maps was prodigious and the work beautifully executed.

Ever on the alert for grandiose projects, Cosimo found in Danti a receptive and inventive mind, even though he was at times overly ambitious. One of Danti's plans, proposed to Cosimo but never begun, was the construction of a canal across the peninsula to link the Adriatic and Mediterranean. Another project involved the reform of the calendar. To avoid having Easter coinciding with the Jewish Passover, the early church had defined Easter as the first Sunday after the first full moon of the vernal equinox. The Council of Nicea decreed in 325 that the vernal equinox fell on March 21, but by the sixteenth century the equinox was falling ten days earlier. As the patron of the lead astronomer of Florence, Cosimo consulted with Pius V (who had supported his recognition as Grand Duke of Tuscany) about reform of the calendar, an important but sensitive issue in the church at the time.

Danti's knowledge of astronomy led him to convincingly argue that accurate determinations were needed for the length of the year and the timing of the equinoxes in terms of the calendar. In order to make precise measurements, he proposed to Cosimo that two

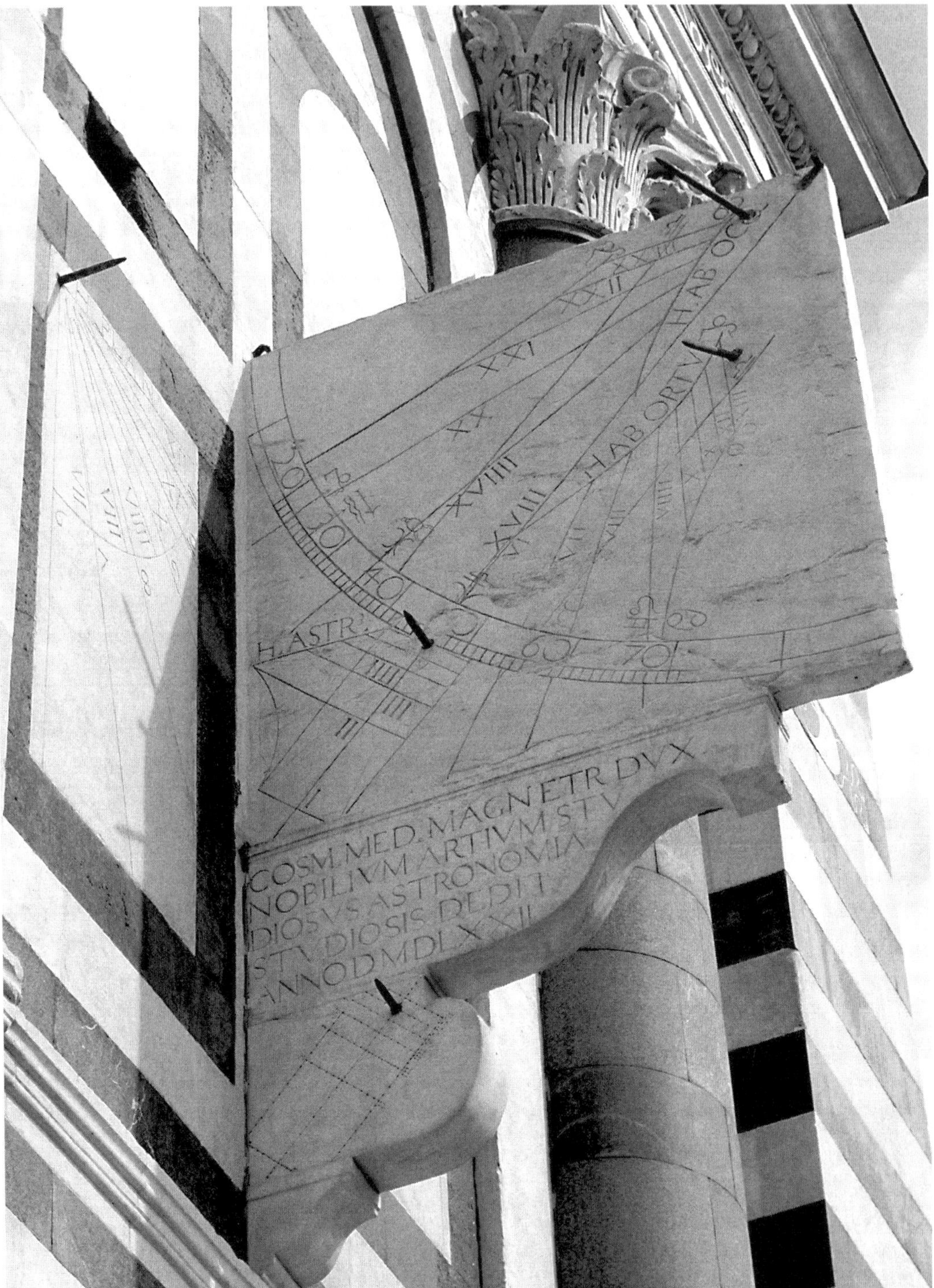

Figure 5.3: Egnazio Danti, Astronomical quadrant (1572) on façade of the Church of Santa Maria Novella (photograph by Sailko, retrieved from Wikimedia Commons, GNU Free Documentation License).

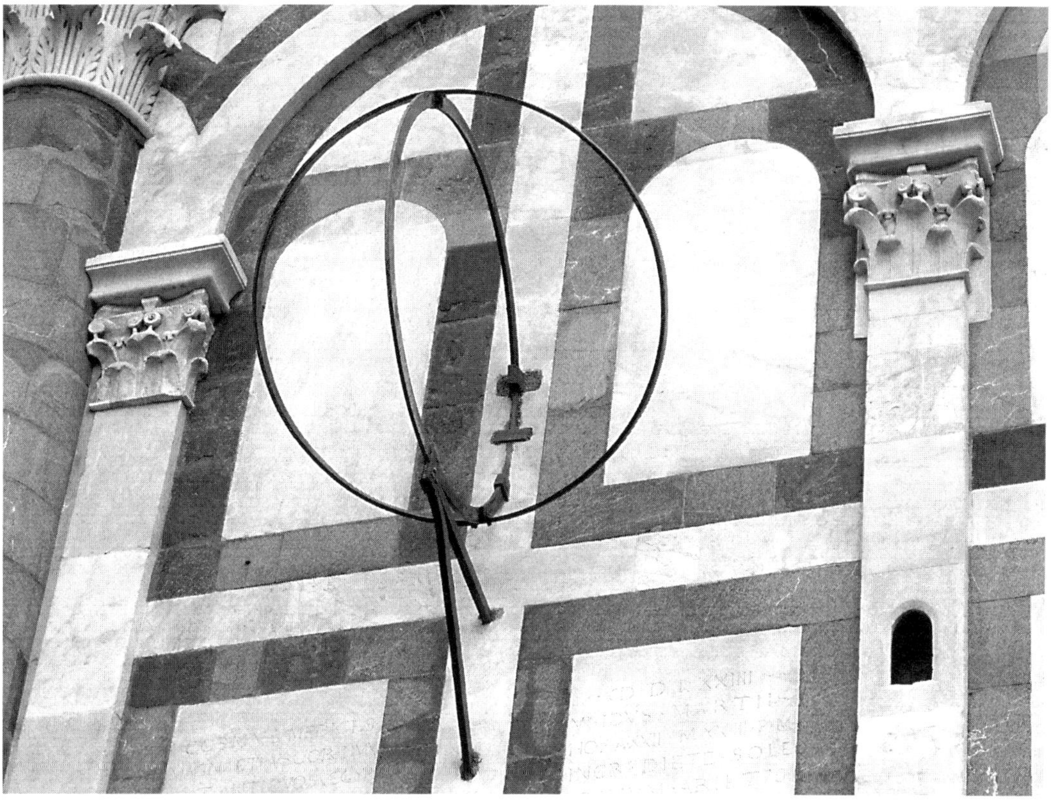

Figure 5.4: Egnazio Danti, Equinoctial armillary (1574), façade of the Church of Santa Maria Novella (photograph by Sailko, retrieved from Wikimedia Commons, GNU Free Documentation License).

astronomical instruments be placed on the façade of Santa Maria Novella. Danti's justification for recommending this particular church had nothing to do with its being the religious home of Danti. Rather, this Dominican church had a north–south orientation with the façade and the main entrance facing a southerly direction. This change in directionality from the standard east–west orientation occurred during the latter part of the thirteenth century when the church was renovated, resulting in a configuration of a north–south nave and an east–west transept. The reason for the change in orientation is not documented, but for Danti it served the useful purpose of permitting measurements throughout the year. This is also the first and only instance of a cathedral façade in Florence being utilized in this way, with the two instruments flanking the front doors.

Cosimo's support enabled Danti to design and install an astronomical quadrant on Santa Maria Novella in 1572 (fig. 5.3) and an equinoctial armillary in 1574 (fig. 5.4), which he planned to use to determine the equinox and solstice. The quadrant contains a short cylinder made of bronze that serves as a gnomon, casting shadows on the gradations on either side of the meridian as the sun passes. Both the solstices and the equinoxes are marked.[8]

8. Settle, "Gnomons and Astronomical Instruments," 77.

In 1574 Danti installed the second of the two devices, the armillary sphere he designed after studying works by Ptolemy and other earlier and contemporary astronomers. The instrument is deceivingly simple in design, consisting of two rings, each 1.75 meters in diameter, with one on a vertical plane and one tilted on the plane of the celestial equator (the plane defined by Earth's equator, which is equidistant from both poles). The finer details of the armillary sphere are unfortunately not readily visible from the ground, but there are gradations on the vertical ring used to mark a ray of sunlight from an opening in the upper portion of the equatorial ring. In a public display on the vernal equinox in 1574, Danti demonstrated the measurements that could be made with the instruments and, more importantly, showed that equinoxes and solstices defined by the church deviated appreciably from the true astronomical days. Justifiably proud of the design, the quality, and the beauty of these instruments, Danti signed both of the exterior astronomical devices.

Both Danti and Cosimo realized that more accurate measurements were required for meaningful scientific investigations, one being the determination of the obliquity of the ecliptic. In a 1571 publication, Danti, surveying values of the obliquity of the ecliptic from Ptolemy in circa 150, noted a diminution in the angle of the tilt.[9]

Accurately determining this seemingly rather esoteric value, and understanding its changes, became the next intriguing problem for Danti to solve. To that end, again with the support of Cosimo, Danti created and installed the first of two gnomons in Santa Maria Novella. With permission granted by the patrons (the De'Ricci family) of the rose window on the south façade of Santa Maria Novella, in 1574 or 1575 Danti made a metal-encased opening in the window, about twenty-one meters from the ground.[10] Unfortunately, the opening was removed during a recent restoration, but based on the recommendation of science historian Thomas Settle, a new opening was installed at or near the site of the original when the window was replaced.[11] Although Danti did not install a permanent meridian for his first gnomon, he engraved markers on the floor corresponding to the equinoxes indicated by the signs of Aries and Libra; the sign for the winter solstice (Capricorn) is barely visible, while Cancer, the sign for the summer solstice, is missing. As the gnomon was designed, the image of the sun ray passing through the gnomon and striking the floor would take a northerly route of just over forty meters in going from the summer solstice to the winter solstice.

During this time Cosimo died (1574) and his son Francesco (1541–87), Danti's former student and, as it turns out, a formidable adversary, became the new Grand Duke of Tuscany. Desiring to place a new gnomon higher up on the façade of Santa Maria Novella and anxious to begin meaningful measurements, Danti petitioned Francesco and the church to permit him to make a new opening. Upon receiving approval, Danti proceeded to place the gnomon about twenty meters from the ground, thus permitting more accuracy in the mea-

9. Heilbron, *Sun in the Cathedral*, 63.

10. Settle, "Egnazio Danti as a Builder of Gnomons," 94; Bartolini, *I Fori Gnomonici*, 36.

11. Settle, "Gnomons and Astronomical Instruments," 81.

surements due to the length of the meridian, particularly during the equinoxes and winter solstice.[12] For this Danti made a hole through the external marble, the inner church wall, and the interior wall of the church and installed a metal-encased plate giving a pear-shaped opening of about eight centimeters long and three and five centimeters in the narrower and wider widths. Danti was anxious to complete his instrument and hoped to begin measurements for the winter solstice, and perhaps even for the autumnal equinox, but his plans were thwarted when he received a letter from Serafino Cavalli, general of the Dominican order, commanding that Danti leave Florence within twenty-four hours and report to the Convent of San Domenico in Bologna. One reason given, not documented, was that Grand Duke Francesco had charged Danti with improper conduct; Danti's close association with Francesco's brother Cardinal Ferdinando (1549–1609) did not help. He was, however, able to receive some temporary reprieve, allowing him to determine the 1575 winter solstice before leaving Florence. Because of this sudden and unexpected change of plans, Danti never had the opportunity to install a meridian proper.

While in Bologna, Danti designed and installed two gnomons, one located in the Dominican convent of San Domenico and the other in the large Dominican church of San Petronio. During this period Danti continued his studies on astronomy and served as professor of mathematics at the University of Bologna. Danti's gnomon in San Petronio would have permitted the kind of accurate measurements of the declination of the sun at or near the solstice he had hoped to make in Florence. Unfortunately, the gnomon was destroyed during a renovation, and the one visible today was constructed by Gian Domenico Cassini in the middle of the seventeenth century.[13] There are no documents of Danti's presumed observations, so we do not know whether he was able to make the measurements he had proposed.

Despite falling out of favor with Grand Duke Francesco and having to leave Florence, Danti was able to continue his career under other patronage. He authored, translated, and edited numerous books; built astronomical, weather, and surveying instruments; and mapped the region of Perugia. He was a dedicated and knowledgeable teacher of mathematics and astronomy, and was appointed papal cosmographer and mathematician in 1580 by Pope Gregory XIII. In that capacity, Danti was assigned the task of mapping the Papal States (the maps are visible in the Gallery of Maps lining the Cortile del Belvedere). He was also made a member of the commission to reform the calendar. Danti supervised the painters in the Vatican, prepared the plans for the repair of the harbor at Portus, and assisted Fontana in relocating and raising the Egyptian obelisk that now stands in front of Saint Peter's. In appreciation of the many achievements of this remarkable man, Gregory XIII appointed Danti bishop of Alatri in 1583, and by all accounts he performed superbly in that position until his death in 1586.

12. Settle, "Egnazio Danti as a Builder of Gnomons," 94.

13. Ibid., 96–97.

5.2.4 The Uffizi

There is evidence from a 1591 description of Florence published by Francesco Bocchi that a gnomon for determining the solstices and equinoxes was installed in the Tribune of the Uffizi. This octagonal room within the Uffizi complex was designed by Buontalenti and designated by Grand Duke Francesco as a gallery to display select items from the Medici family's vast artistic collection. When Ferdinando I became Grand Duke of Tuscany, he had certain scientific items moved to the Tribune, perhaps commemorating his father, Cosimo, and Danti. The document does not describe the gnomon, but it is likely that a small hole was installed in the base of the southern window of the lantern, allowing the winter solstice to be measured at the top of the northerly wall of the Tribune, the equinoxes midway down, and the summer solstice on the floor.[14] Unfortunately, nothing remains of this gnomon, although the room itself is a delight to visit, providing a reminder of the combination of science and art that flourished in Renaissance Florence.

5.2.5 The Pitti Palace

In the centuries after Toscanelli placed his gnomon in the façade of the lantern of Santa Maria del Fiore, the idea of locating astronomical equipment on public and private buildings became more acceptable. The painter Anton Domenico Gabbiani (1652–1726) and the mathematician Vincenzo Viviani (1622–1703) renovated a room in the Pitti Palace that became known as the Gnomon Room (Stanza della Meridiana). A small but lovely and readily accessible gnomon is located in this room in the wing of the palace now denoted as the Palazzina della Meridiana. The room was part of the apartment of Grand Duke Ferdinando II, a sponsor and student of science who, with his brother Prince Leopold de Medici, founded the Accademia del Cimento in Florence in 1657 and hosted its meetings in the palace.

Gabbiani frescoed the room in 1693 and featured Galileo and Amerigo Vespucci being welcomed by personifications of Fame, Glory, Doctrine, and Time. One of Galileo's students, Vincenzo Viviani, who served as court mathematician under Cosimo III, installed the gnomon in 1696 with the advice of Gian Viviani Cassini (1625–1712), arguably one of the leading experts of the time. Formed by an opening in a sheet of metal, the oculus was inserted into the south portion of the room, while the meridian line, made of brass in a marble frame, was set partially in the floor and extended to the northerly wall, marking the summer solstice. The brass line then continued up the wall, the end of which indicated the winter solstice. The various zodiacal signs are inscribed on the brass line, as are the months and days. The gnomonic hole was closed in 1776 to accommodate a new construction project,[15] but the meridian remains. The Gnomon Room provides an excellent opportunity to closely inspect the type of measuring device that was so widely used throughout history. As

14. Camerota, *Line of the Sun*, 93–94.

15. Chiarelli, Barbolini, and Garofalo, "Gnomon in Palazzo Pitti."

with many of the scientific instruments of this and earlier periods, the artistic component can only be realized when the instrument is in use, thus allowing natural sunlight, or the moon, or galaxies, or many other aspects of nature to be as much of the experimental observation as the fabrication itself.

5.3 Measuring the Hours in Florence

Church bells were rung in the Middle Ages to announce the start of religious services, but by the early trecento bells began to be used to inform citizens of the hour. People were able to determine the approximate time of day by listening to the bells, often with patterns of ringing differentiating the hour, quarter hour, and half hour. The sundial was the most common mechanism used by the bellringers to determine the hour. Since sundials were inoperative in darkness and cloudy weather, there was a strong incentive to develop reliable mechanical clocks. As public clocks became more common, they presented an opportunity to display not only factual information but also artistic elements in their presentation.

5.3.1 Ponte Vecchio

A bridge has spanned the Arno at its narrowest point since the first century, replacing a ferry used to connect both banks of the river along the Via Cassia (the second-century Roman road linking Rome and Florence). Floodwaters repeatedly swept away the early bridges, culminating in the destruction of the 1294 stone bridge expanse in 1333. In 1345, a substantial new bridge, designed by Taddeo Gaddi and known as the Ponte Vecchio (or Old Bridge, acknowledging the antiquity of the bridge's history), was built. The thirty-two-meter bridge consists of three base arches resting on two stone piers, and its width accommodates shops and other structures along the sides, with a small plaza in the center section. On the roof of one of the lower buildings adjoining the central plaza, there is a sundial that faces south (fig. 5.5). The sundial, in the shape of a white marble cup, is divided by slender raised divisions marking the hours, these being indicated by a projecting gnomon that casts its shadow onto the cup. The sundial sits on a small marble column and is so positioned that it is visible from ground level. The south orientation is designated by a small sculpted lizard. At the sundial's base, there is an inscription, marginally legible, that dates the history of the current bridge and that of the dial: "In the Thirty Three after One Thousand Three Hundred the bridge collapsed due to floods of water, twelve years later as pleased the Commune it was rebuilt with this ornamentation."[16]

5.3.2 Palazzo Vecchio

The tower of the Palazzo Vecchio housed the first public mechanical clock manufactured in Florence. It was installed in 1353, and is based on a twenty-four-hour day, beginning with the first hour after sunset (I) and ending at XXIV. Constructed by Niccolò di Bernardo, a

16. Magrini, "Ponte Vecchio." http://Brunelleschi.imss.fi.it/itineraries/place/PonteVecchio.html.

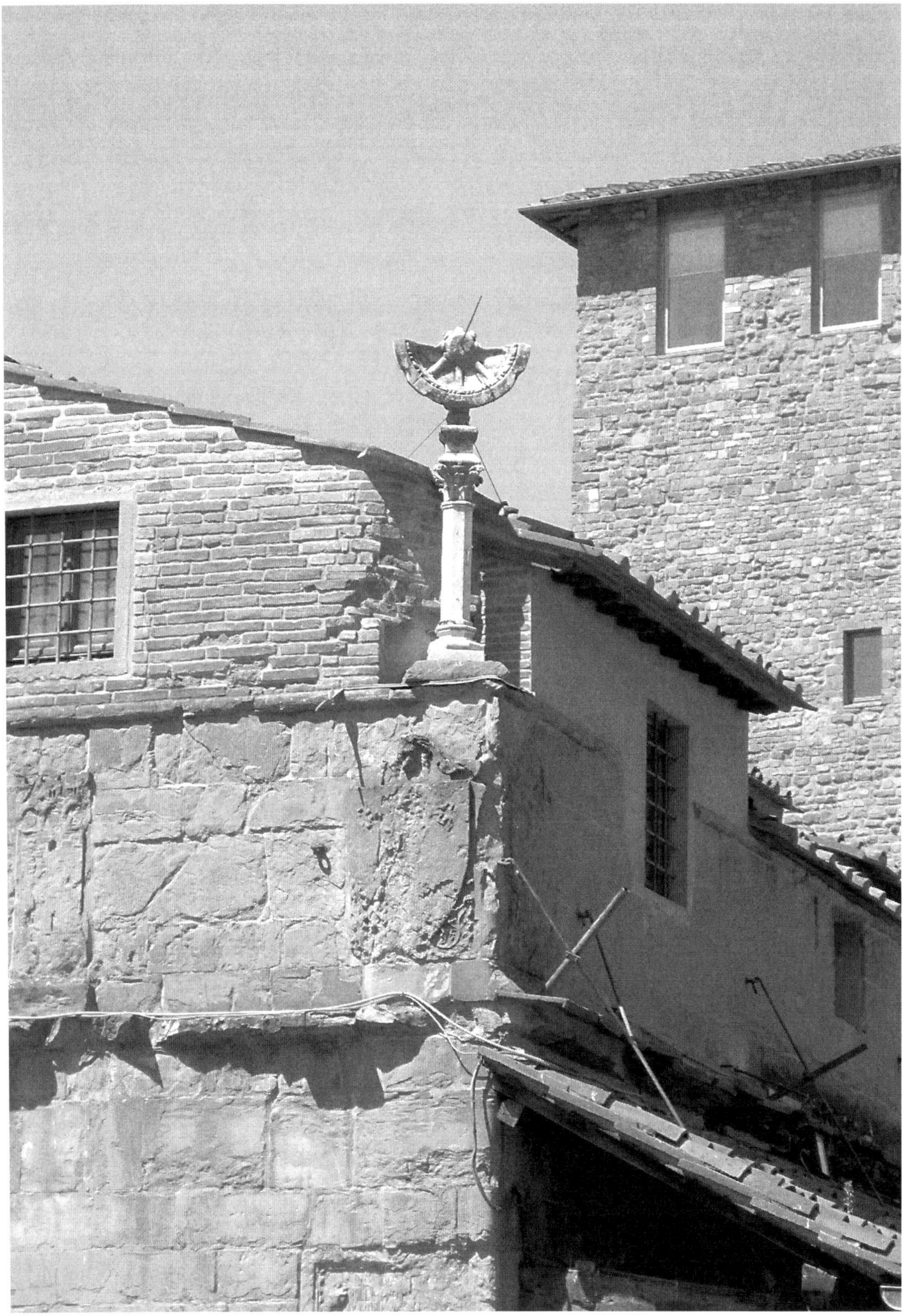

Figure 5.5: Sundial (mid-1300s) on the Ponte Vecchio (photograph by Sailko, retrieved from Wikimedia Commons, GNU Free Documentation License).

Florentine watchmaker better known as Nicolò Berardi, it was made in a workshop near the Palazzo Vecchio on a street later named Via dell'Oriuolo (Street of the Clock). Although no documents describing the exact configuration of the mechanism have survived, a reasonable deduction of the workings can be gleaned from what is known about other contemporary clocks. Probably encased in a wrought iron frame, the clock likely utilized strong cables from which two stone weights were suspended. The weights served to move the geared iron wheels transmitting the force to the escapement (the shaft fitted with notches) to progressively ring a bell each hour. A device called a *ribotta* was undoubtedly installed to ring the bell a second time, ensuring the public would notice. Having no numbered dial, the clock marked time solely by the incremental bell ringing. The clock currently on the façade of the Palazzo Vecchio was installed in the seventeenth century.

5.3.3 Duomo

In 1443 Angelo Niccolai degli Orologi, believed to be a descendant of Nicolò Berardi, was commissioned, probably by the cathedral, to make and install a clock for the interior wall of the Florence Cathedral (Duomo), Santa Maria del Fiore. It is reasonable to assume that any clock to be placed in the Duomo would be expected to be artistically decorated as well as technologically accurate. To that end, also in 1443 the noted Florentine painter Paolo Uccello (1397–1475) was appointed to fresco the face for the clock situated above the main entrance to the cathedral. The resulting fresco was a square containing images of the four Evangelists in the corners against a blue background. Roman numerals used to indicate the hours were placed in a circle, with I at the bottom and moving in a counterclockwise direction to XXIV to the left of the I. Uccello also gilded the single hand, which was artistically designed to represent a star. In 1750, the dial was modified to conform to French clocks bearing numbers from one to twelve, but a 1968 restoration by Umberto Baldini reverted the dial on Uccello's clock to the twenty-four-hour design, which can be seen today.

Sadly there are no records of the original configuration. However, if the mechanism was typical for its time, it would have consisted of two weights and a series of gears that moved the hands. Unlike some other contemporary clocks, this one was not designed to ring a bell announcing the hour. Rather, the hand indicated the time, beginning with the hour following the setting of the sun. A number of restorations were done during the ensuing two centuries, but by 1688 the clock was no longer capable of reliable timekeeping, and the internal mechanism was replaced. This project was undertaken by Antonio Niccodemi, who used a pendulum connected to the escapement, thereby gaining greater precision. The current mechanism was installed in 1761 by Giuseppe Bargiacchi. Work on the timepiece has been done periodically, most recently by Andrea Palmieri in 2004. The frescoed face remains prominently visible today.

5.3.4 Galileo Museum

An excellent example of the integration of art and science is a remarkable and unusual Florentine clock. It was originally commissioned by Lorenzo the Magnificent as a gift for the king of Hungary and Croatia, but there are some reports that two versions were created. Whatever the number, upon completion in 1509/10, one was purchased and then donated by the Capitani di Parte Guelfa to the Signoria and placed in the Palazzo Vecchio. Cosimo de Medici then had it installed in the room next to the Map Room (Guardaroba). It was a planetary clock designed and built by the well-known and highly accomplished clockmaker Lorenzo della Volpaia (1446–1512) to display not only the time, but also the positions of the planets, the moon, and the sun. In order to fabricate this complex mechanism, della Volpaia needed to call upon his extensive knowledge of astronomy, mathematics, and scientific instrumentation, as well as his artistic and clock-making skills. It is known that in 1560, Girolamo della Volpaia (ca. 1530–1614; grandson of Lorenzo della Volpaia) restored the clock, but by the close of the cinquecento it was missing.[17] Fortunately sufficient information was found in notebooks retained by the della Volpaia family to allow a replica to be built in 1994 under the direction of the Galileo Museum, where it can be seen today.

The outer rim of the clock bears twenty-four hourly numbers beginning with I at the top and proceeding to XII at the bottom, then beginning again with I and proceeding again to XII at the top. An inner circle shows the twelve signs of the zodiac, and within that circle, the central portion contains seven circular regions. In the center is a circle displaying the sun and the phases of the moon, and grouped around it are six circles. Five of these depict the planets known at the time—Mars, Mercury, Venus, Jupiter, and Saturn—and a sixth features a dragon displaying the eclipses and lunar nodes (the points where the lunar orbit crosses the ecliptic). There are also two globes mounted to the sides of the clock face, one showing Earth and the other showing the night sky. Included as well is a chime denoting the hours. It is run by a weight-driven system connected to a highly complex set of gears that turn the hands on the various dials on the clock face, creating the relative motions of the planetary bodies. Assuming that the reproduction is accurate in terms of coloration and design, artistically it is a beautiful piece of work.

5.4 Understanding the Solar System: Galileo Galilei and the European Scientific Revolution

Galileo Galilei (1564–1642) was another polymath of the Italian Renaissance, arguably one of its best known and most provocative. His major contributions provided much fuel for the Scientific Revolution, placing him on the leading cusp in astronomy, physics, and mathematics. Galileo, one of six children, was born to Vincenzo Galilei (ca. 1520–91), a famous lutenist, music theorist, and composer, and Giulia Ammannati (1538–1620) in Pisa (then part of the Duchy of Florence). The family moved to Florence when Galileo was a young boy, and he was educated in the Camaldolese monastery in Vallombrosa and then at the University of Pisa, where he studied mathematics rather than medicine as his father had hoped. In 1589, at the age of twenty-five, he was appointed professor of mathematics at the University of Pisa; he transferred to the University of Padua in 1592 to teach mathematics, astronomy, and mechanics, and remained there until 1610.

17. Palmieri, "Ancient Mechanical 'Oriuoli,'" 20–27.

Over his lifetime Galileo made seminal observations in several areas of science and technology, and also applied his vast knowledge to practical inventions. Recognizing the need for a universal instrument that could be used to perform a variety of mathematical and geometric calculations, he designed what became known as Galileo's Compass. This deceivingly simple, but really quite sophisticated, instrument had applications for mathematics, geometry, astronomy, currency exchange, interest rates, and more. This instrument was extremely successful, earning him considerable profit. One of the sixty known to have been fabricated and sold in Padua is displayed in the Galileo Museum.

In 1608 Galileo, learning of the telescope that had recently been invented in The Netherlands, was intrigued with the possibilities it presented to astronomers and immediately embarked on constructing one with a magnification of about three times normal vision (3x). This was the first of several telescopes Galileo built, with later versions achieving higher magnifications that allowed him to make important astronomical observations. Upon looking upwards with his first rather crude telescope, Galileo saw a cratered moon and a multitude of stars, including many clusters encompassing the Milky Way; with each improvement to his instrument, Galileo was able to observe more exciting and previously unknown celestial bodies. Using his telescope with a 30x magnification, Galileo was for the first time able to actually see that Jupiter was circled by four moons, now known as Callisto, Europa, Ganymede, and Io. This far-reaching and radical discovery of orbiting spheres, along with the observable phases of Venus indicating rotation around the sun, led Galileo to challenge the prevailing and long-held view that all heavenly bodies revolved around Earth. This observation eventually led him to advance theories that placed him in opposition to the teachings of the church. He published these new findings in a small work entitled *Sidereus nuncius* (*The Starry Messenger*, or *Messenger from the Stars*) that appeared in 1610. An engaging book that demonstrates Galileo's careful studies and clarity of expression, it has some simple geometrical diagrams and occasionally refers to elementary trigonometry, but is devoid of complicated mathematical and physical formulae. Rather it relies on Galileo's well-documented observations, which he beautifully depicted, calling on the artistic training he had received in his younger years, including his mastery of perspective, drawing, and use of shading. That early training was crucial for the dissemination of his observations, and his artistic and scientifically accurate drawings are still reproduced today. One image that is often duplicated is Galileo's figure of the waxing moon in the first quarter as seen on December 3, 1609 (fig. 5.6). That picture, and others reproduced for *Starry Messenger*, was prepared by an engraver from Galileo's wash drawings. In contrast to the published versions, Galileo's original sketches were remarkable watercolor renditions of the visual field he witnessed. As Samuel Y. Edgerton describes it in *The Mirror, the Window, and the Telescope*, "With the deft brush strokes of a practiced watercolorist, he laid on a half dozen different grades of washes, imparting to his images an attractive soft and luminescent quality."[18] Continuing, Edgerton commended Galileo

18. Edgerton, *Mirror, the Window, and the Telescope*, 162.

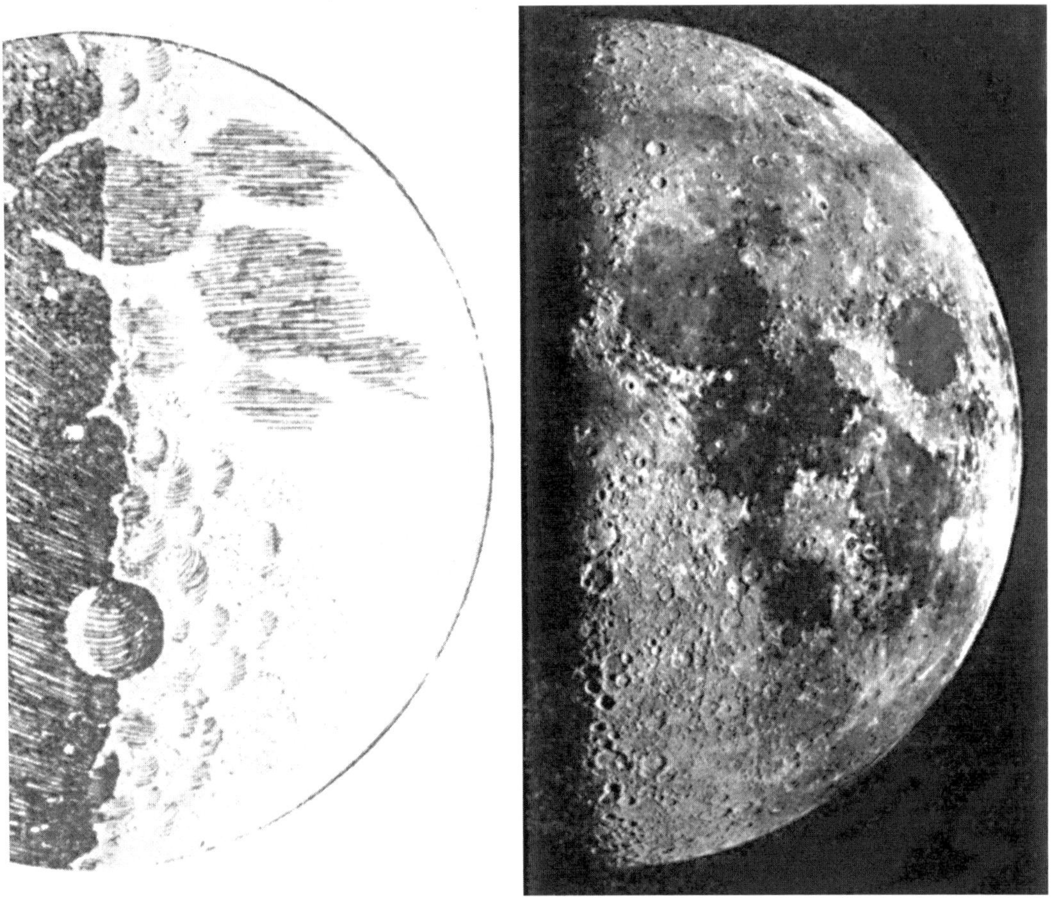

Figure 5.6: *Left*: Galileo Galilei, Drawing of the moon surface (1609–10), published in *Sidereus nuncius* (1610). *Right*: Photographic image of the same view (photograph by ECeDee, composition by ECeDee, retrieved from Wikimedia Commons).

for "his ability to marshal dark and light washes to increase their mutual intensities."[19] Galileo was able to show multiple craters enhanced along the line separating the light and dark fields of the moon. His artistic talents were recognized by his contemporaries, and he was appointed to the Florentine Academy of Art in 1615.

Galileo's ability to artistically render his astronomical studies in his publications was essential, for it allowed his readers to understand the importance of the careful observations being interpreted by a keen and well-trained analytical mind. The combination of Galileo's clear depictions and his analysis of the data provided a firm foundation in support of the Copernican model of the universe and discredited the Aristotelian-Ptolemaic model that had prevailed for over a millennium. Near the end of the book, Galileo, never one to mince his words, wrote,

19. Ibid., 162.

> We have moreover an excellent and splendid argument for taking away the scruples of those who, while tolerating with equanimity the revolution of the planets around the sun in the Copernican system, are so disturbed by the attendance of one moon around the earth while the two together complete the annual orb around the sun that they conclude that this constitution of the universe must be overthrown as impossible. For here we have only one planet revolving around another while both run through a great circle around the sun: but our vision offers us four stars wandering around Jupiter like the moon around the earth while all together with Jupiter traverse a great circle around the sun in a space of 12 years.[20]

Knowing that the ideas he was publishing were radical, Galileo sought political support by cleverly naming the Jovian moons the Medicean stars to honor Grand Duke Cosimo II de' Medici (1590–1621) and his three brothers, and wrote a rather lengthy letter to accompany the copy of *The Starry Messenger* he sent to Cosimo. His letter to the grand duke is at the same time eloquent, loquacious, and effusively laudatory:

> Since under Your auspices, Most Serene Cosimo, I discovered these stars unknown to all previous astronomers, I decided by the highest right to adorn them with the very august name of Your family. For since I first discovered them, who will deny me the right if I also assign them a name and call them the Medicean Stars, hoping that perhaps as much honor will be added to these stars by this appellation as was brought to other stars by the other heroes? For, to be silent about Your Most Serene Highness's ancestors to whose eternal glory the monuments of all histories testify, Your virtue alone, Great Hero, can, by your name, impart immortality to these stars.[21]

To the surprise of no one, Galileo was shortly thereafter appointed mathematician and philosopher to the grand duke's court, as well as chair of mathematics at the University of Pisa. The latter position had no real duties associated with it, and Galileo promptly moved to Florence to continue his research. Galileo sent a copy of *The Starry Messenger* to German mathemetician and astronomer Johannes Kepler (1571–1630). After reading the book, Kepler was prompted to borrow a telescope and begin his own astronomical observations; he also answered Galileo with a most enthusiastic endorsement of his findings.

A devout Christian, Galileo retained his firm belief in God but doubted that the Bible, as it related to science, should be interpreted literally. In time, the church placed the writings of Copernicus and Kepler on the Index liborum prohibitorum (List of Forbidden Books) deeming their publications heretical and therefore dangerous to the church. Responding to the spreading word of Galileo's observations and his support of Copernicanism, which contradicted portions of the Bible, Pope Paul V instructed Cardinal Robert Bellarmine (1542–1621) to censure Galileo. After the cardinal met with Galileo in Rome in 1616, Bellarmine

20. Galilei, *Sidereus nuncius*, trans. van Helden, 84–85.

21. Ibid., 32.

informed the pope that Galileo had accepted the censure. Galileo did not abandon his theories, however, and returning to his villa in Arcetri, near Florence, he continued his work while maintaining a relatively low profile. After his friend Cardinal Maffeo Barbarini (1568–1644), who had opposed the church's condemnation of Galileo, became Pope Urban VIII, Galileo published *The Assayer* (*Il Saggiatore*), which dealt with his argument with the Jesuit astronomer Father Orazio Grassi (professor of mathematics at the Jesuit Collegio Romano) over the nature of comets, particularly three that appeared in 1618.

Galileo had several favorable meetings with the pope in 1623. The pope evidently claimed that he would not have placed a ban on Galileo for the book *The Assayer*, and he approved Galileo's plan to publish a work on Copernicus's heliocentric model if he would also cover the Ptolemaic (and Aristotelian) model impartially. Galileo published *Dialogue concerning the Two Chief World Systems* [*Ptolemaic and Copernican*] (*Dialogo dei due massimi sistemi del mondo*) in 1632. The book took the form of conversations between three friends, a Copernican (Salviati), an Aristotelian (Simplicio), and a convert to heliocentrism as championed by Copernicus (Sagredo). Needless to say, Galileo made a most persuasive argument for Copernicanism, even though some of the statements were not strictly correct. (In Italian, the word *simplicio* implies the sense of "simpleton," but Galileo claimed that the character of Simplicio was named for Simplicius, a famous Aristotelian philosopher.) Galileo chose to have Simplicio utter words used by Urban VIII in defending the Aristotelian view, thereby appearing to deride His Holiness. Taking great offense at what he considered a personal attack and public ridicule, Urban directed Galileo to appear before the Inquisition in Rome in 1633. The outcome is well known. Galileo had to withdraw his support of heliocentrism; he was imprisoned, and *Dialogue concerning the Two Chief World Systems* was banned, as were all future writings.

The imprisonment of Galileo was commuted, but he was placed under house arrest, first in a residence of the Medicis in Rome and then with the archbishop in Siena. Finally allowed to return to his villa at Arcetri in 1634,[22] he remained under house arrest for the remainder of his life. In Arcetri, Galileo continued his research on physics, with an emphasis on a mathematical formulation of the laws of motion and the strength of materials. Suffering from failing eyesight, Galileo dictated *Discourses and Mechanical Demonstrations concerning Two New Sciences, of Mechanics and of Motions* (*Discorsi e dimostrazioni matematiche, intorno a due nuove scienze*) to his disciple Vincenzo Viviani, who took it to The Netherlands where it was published in 1638, thereby avoiding the possibility of censorship by the Catholic Church. Blind at that point, Galileo died in 1642, just before turning seventy-eight years of age.

Grand Duke Ferdinando II wished to inter Galileo in the Francescan Basilica of Santa Croce in Florence, but because of the objections of Pope Urban VIII, he was instead buried in a small room below the bell tower.[23] In 1737 Galileo was reburied in the Basilica proper, a demonstration of Florence's independence from church oversight. A bizarre part of this story was the removal of three digits on Galileo's hand(s), a thumb, index finger, and middle

22. Settle, "Traces of Science and Engineering" (http://www.imss.fi.it/~tsettle/traces.html).

23. Galluzzi, "Sepulchers of Galileo."

finger, along with a tooth (molar), and a vertebra, by the historian and naturalist Giovanni Targioni Tozzetti, who retained them as relics. All of these relics are now in the Galileo Museum. After purchasing the Palazzo dei Cartelloni, Viviani, in defiance of the church, had the façade altered to include a bronze bust of Galileo, two marble scrolls flanking the main entrance, and images of the Jovian moons. The building with its tribute to Galileo remains.

Galileo never married, but he fathered three children with Marina Gamba, two daughters and a son. The daughters, considered unmarriageable in view of their illegitimate birth, were placed into the convent of San Matteo in Arcetri; the son was legitimized and later married. One of Galileo's daughters took the name of Maria Celeste upon entering the convent and had a lifelong correspondence with her father. Dava Sobel wrote an interesting book, *Galileo's Daughter*, based on their correspondence.[24]

Galileo's work was arguably one of the major catalytic factors in the start of the Scientific Revolution. Because of this, many honors have been bestowed on him in the intervening centuries, one of the most recent being the renaming of the Museum of the History of Science in Florence as the Galileo Museum.

5.5 Astronomical Instruments and Their Artistic Dimensions

Centuries before the telescope was developed for studying the moon and the planets, astronomers used armillary spheres and astrolabes. Armillary spheres were invented by the Greek Eratosthenes (276–194 BCE) to show the universe as it was understood based on the Aristotelian concept of the relative positions of Earth, the sun, and planets, and later on the Ptolemaic formulation. Another type of instrument, the astrolabe, was believed to have been first created in the second century BCE by the Greek astronomer Hipparchus. Like the armillary sphere, the astrolabe was used for astronomical determinations and was quite advanced in scope and design. Used by the Greeks, as well as in later centuries for determining location and position, both navigational and terrestrial, astrolabes could even serve as a rather crude form of time measurement. From Greece, the concept of the earliest astrolabe spread to Egypt (Alexandria) and then to the Islamic world, where modifications and new designs were made by various Islamic scholars, including Ali ibn Isa, astronomer in the court of Caliph al-Mansur in the eighth century. The astrolabe entered Europe by way of Toledo in Spain, where by the tenth century brass rather than wood became the preferred material for fabrication. Metal astrolabes, less prone to warping than their wooden counterparts, could be made larger and thus more accurate. They were in common use by the twelfth century, when one of the leading writers of his time, Adelard of Bath (ca. 1080–1152), penned a treatise on the astrolabe in which he detailed the mechanism and its applications. Astrolabes were extensively used by Europeans for nautical purposes during the era of exploration, but were eventually displaced by more convenient instruments such as the sextant. The increased emphasis on beauty and functionality of

24. Sobel, *Galileo's Daughter*.

telescopes, armillary spheres, and astrolabes during the Florentine Renaissance is a further example of the intimate relationship between art and science.

The Galileo Museum, whose website (http://www.museogalileo.it/en/explore/virtual-museum.html) includes a virtual tour, displays numerous astronomical instruments and models, including telescopes and lenses fabricated and used by Galileo in the early seventeenth century. In addition to focusing on improving scientific apparatus, Galileo strove to design aesthetically pleasing instruments. The museum contains two of Galileo's telescopes, the objective lens through which he discovered the moons of Jupiter, several of Galileo's compasses, and even several of his digits, as well as many scientific instruments from the collections of the Medici and the house of Lorraine. One of the telescopes on exhibit (inventory #2428), measuring just under a meter in length, is made of strips of wood covered with red leather, which with aging has become more brown in color, and is decorated with gold tooling. The extra effort required to make the telescope beautiful documents Galileo's desire to incorporate artistic quality into his scientific work.

Also on view are numerous astrolabes, including one believed to have been designed and constructed by Egnazio Danti in about 1570[25] (fig. 5.7). The astrolabe, with a diameter of eighty-four centimeters, was made with brass and wood. The magnificent wood and metal armillary sphere created between 1588 and 1593 by Antonio Santucci di Pomarance (d. 1613) (fig. 5.8), cosmographer to Grand Duke Ferdinando I de' Medici, is also displayed. The sphere, measuring nearly four meters high and two and a half meters wide, is elaborately painted and decorated, with carved wooden parts covered in fine gold leaf. Ferdinando I, inspired by the map room (Guardaroba) in the Palazzo Vecchio (see 3.2.2.1), designed a cosmographic room in the Uffizi that housed representations of the domain of the Medicis and, in the center of the room, grandly displayed Santucci's armillary sphere.

Objects such as these provided an opportunity to display not only the latest scientific knowledge and technical developments, but also to show, on a grand scale, the artistic talents of wood carvers, goldsmiths, painters, and others. Each and every one is intricate and artistic, and can be appreciated for its beauty even by those who do not understand their operation. Over the years the Medicis amassed a special collection of scientific apparatus[26] that was displayed in the Uffizi alongside their collections of more traditional artworks.

5.6 Astronomy Depicted in Art

5.6.1 Campanile

The Campanile (bell tower) of the Duomo (discussed in 4.5.1) is decorated with reliefs relating to both the arts and sciences; several dedicated to astronomy are on the first and second levels of the tower. The lower level series of hexagonal tiles was created by Andrea Pisano (1290–1345) and his workshop between 1334 and 1336. Incorporated into their

25. Settle, "Gnomons and Astronomical Instruments," 74–75.

26. Camerota and Miniati, *I Medici e le scienze.*

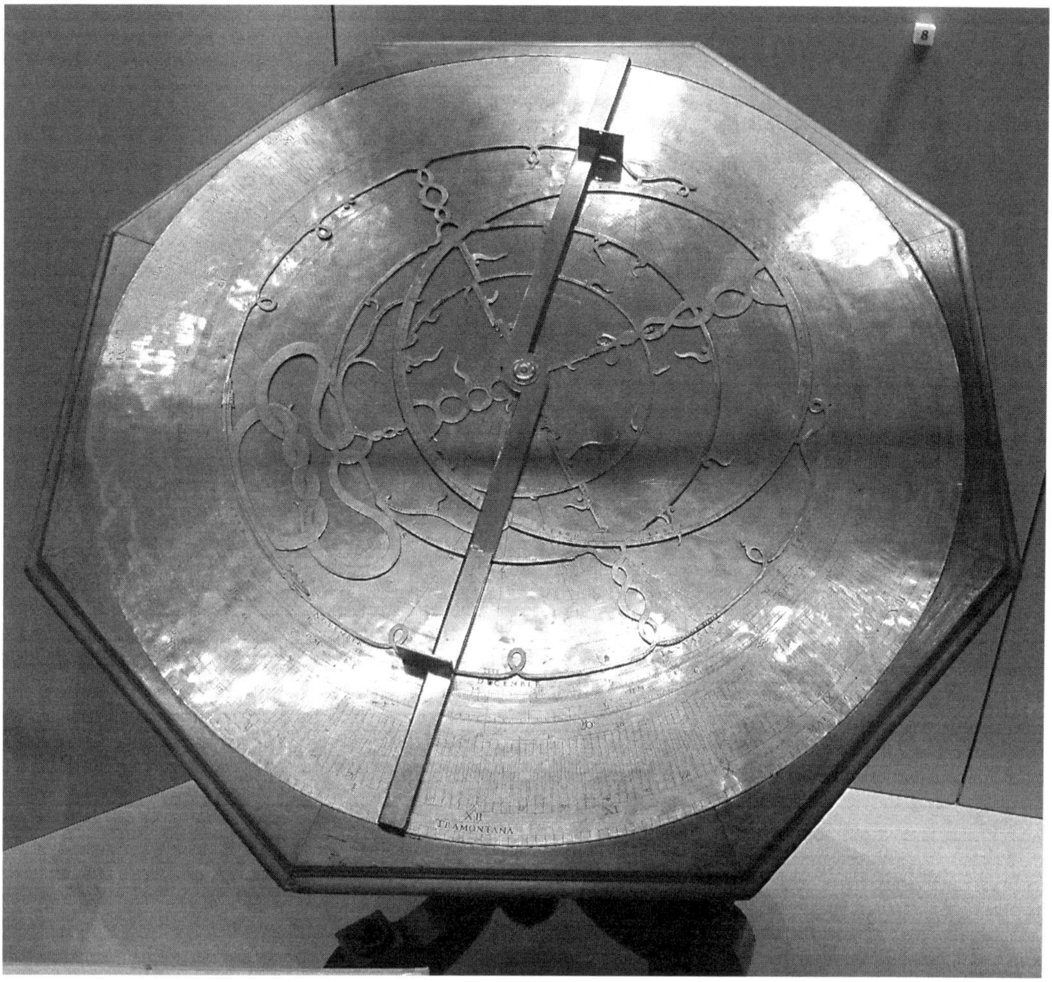

Figure 5.7: Egnazio Danti (attr.), Astrolabe (ca. 1570), Galileo Museum (photograph by Sailko, retrieved from Wikimedia Commons, Creative Commons Attribution 3.0 Imported License).

designs are astronomical instruments that were in common use at that time. Using such familiar instruments in the reliefs made it obvious to viewers which figures and scientific topics were being depicted. The tiles on the lower level represent man and his skills; those on the south side of the Campanile include a depiction of Gionitus, mythical inventor of astronomy, shown using a quadrant to determine the altitude from the horizon of the celestial bodies within his viewing area (fig. 5.9). On the second level, where lozenge (rhombus-shaped) tiles are dedicated primarily to moral, spiritual, and natural forces, those on the west side are devoted to the known celestial bodies. Saturn with the wheel of time, Jupiter with Christian symbols, Mars as an armed cavalier, Venus with two admirers, Mercury, the Sun with a scepter, and the Moon are all shown against blue backgrounds. Pisano and his assistants worked on the decorative tiles for the second, third, and fourth levels of the Campanile from 1337 to 1347, again using familiar symbols to depict celestial bodies and the qualities associated with each.

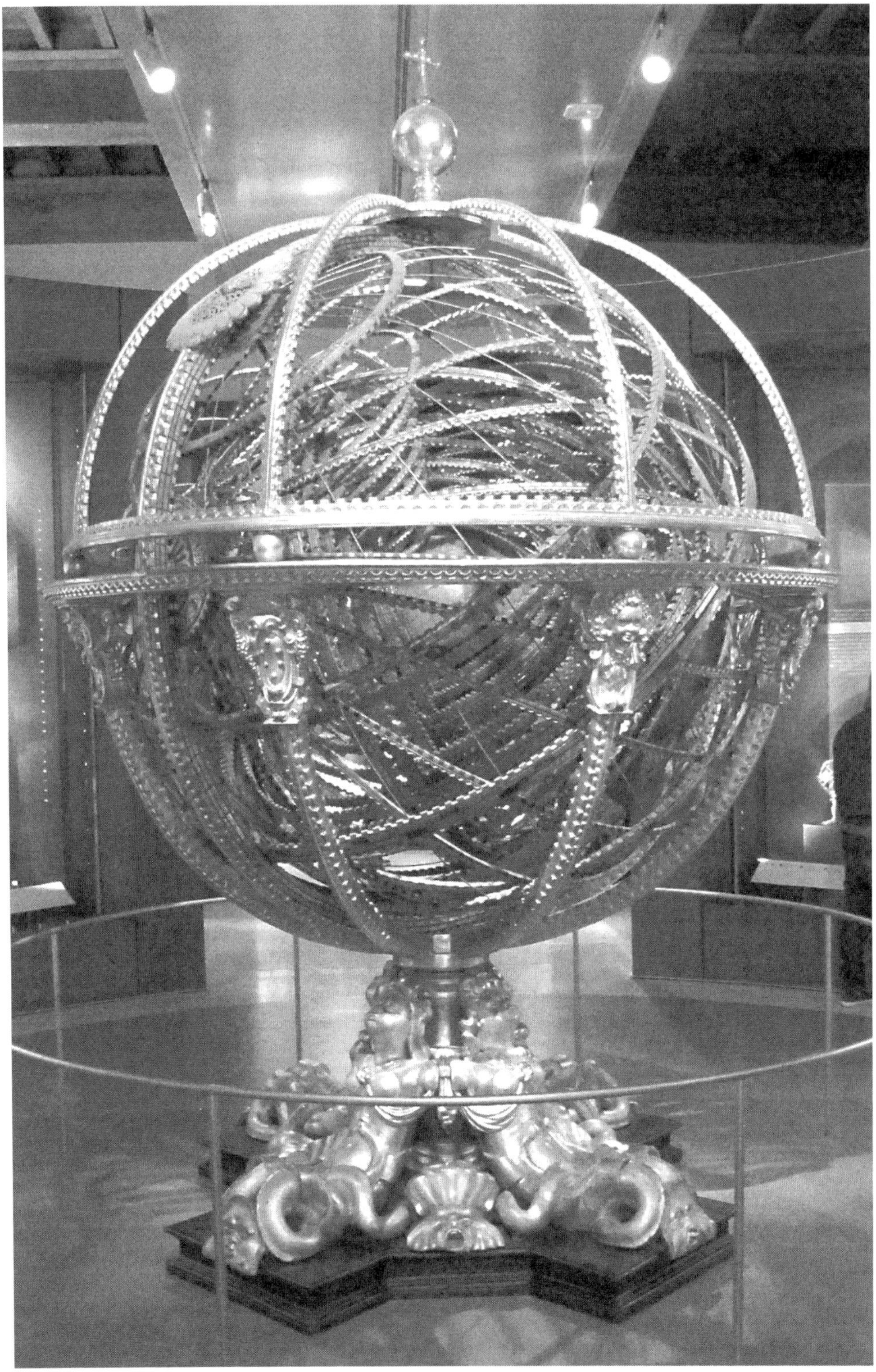

Figure 5.8: Antonio Santucci, Armillary sphere (1588–93), Galileo Museum (photograph by Sailko, retrieved from Wikimedia Commons, Creative Commons Attribution 3.0 Imported License).

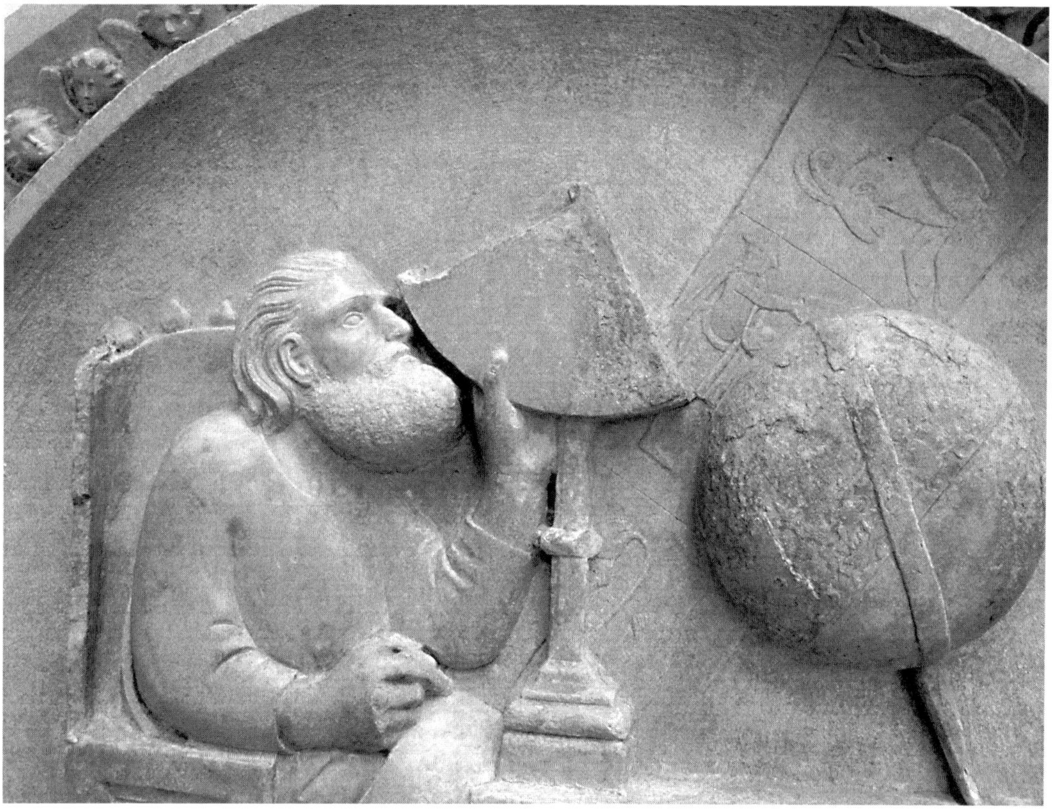

Figure 5.9: Andrea Pisano, *Gionitus* tile (ca. 1334–36) on the Campanile, Museo dell'Opera del Duomo (photograph by Sailko, retrieved from Wikimedia Commons, GNU Free Documentation License).

5.6.2 Santa Croce

A fresco in the Baroncelli Chapel at Santa Croce features an astronomical occurrence that would have been known to Renaissance Florentines. The *Annunciation to the Shepherds* was painted after 1338 by Taddeo Gaddi (d. 1366), a student of Giotto (ca. 1266/67–1337), and is positioned in the middle register to the left of the stained glass window. The fresco depicts two brightly illuminated shepherds, their flock of sheep, and the immediate landscape, set in a dark background. The light is emanating from an angel hovering overhead, and one shepherd is viewing the source of the light while the other is turned away with his head facing down and covered with a hood. This scene is believed to represent a solar eclipse and may refer to an experience of Gaddi in which he observed a partial eclipse directly and suffered some degree of blindness.[27] A letter to Fra Simone Fidati from a Taddeus of Florence describes such an experience, and some interpret the author to be

27. Olson and Pasachoff, "Comets, Meteors, and Eclipses," 1567–71.

Taddeo Gaddi,[28] although this interpretation has been questioned.[29] Unfortunately the date of the letter to Fidati is unknown, as is the year of the eclipse referred to by Taddeus in his correspondence.

A more dramatic example of a painting that depicts an astronomical phenomenon is a fresco by Andrea di Cione Arcangolo (known as Orcagna, d. 1368), executed in the 1340s, which shows the effects of a solar eclipse on those witnessing the event. In this badly faded fresco, under restoration for the Museo dell'Opera di Santa Croce, two men are observing the movement of the moon across the sun while a third man turns away. This painting is believed to refer to an eclipse in 1339 that Orcagna may have witnessed, but about which he was surely aware. Astronomical events such as eclipses and comets were often interpreted to indicate coming calamities such as earthquakes, plague, and the like, and so witnessing such an event might have been fascinating, but could also be frightening as indicated by the figures turning away and refusing to watch.

5.6.3 San Lorenzo

Astronomical topics and events continue to appear in several paintings in the quattrocento. One of the most impressive and stunning examples of the amalgamation of art and science is located in the magnificent Dominican church of San Lorenzo, designed by Brunelleschi. This church was lavishly supported by the Medicis, and the Laurentian Library and Medici tombs are also located on its grounds. Brunelleschi's Old Sacristy (Sagrestia Vecchia) contains artwork by Donatello and others, and the interior domed ceiling area has an unusual painting showing a representation of the sky. The fresco, on a beautiful azurite background, uses gilding effectively to show the sun, the moon, the five known planets, individual stars, and the equator, and includes a depiction of the constellations and many other aspects of the celestial vault visible at dawn in the Northern Hemisphere. Recent studies have determined that the fresco depicts the sky above Florence on July 4, 1442. The work has been attributed to Pesello (Giuliano d'Arrigo, d. 1446) and was probably done in 1442, but the precision and accuracy of the astronomical details suggest that Paolo Toscanelli may have served as a consultant in some capacity (fig. 5.10).

Isabella Lapi Ballerini directed the 1985 to 1986 restoration of the Old Sacristy fresco and has provided an excellent summary of her new findings from the detailed inspection:

> My studies have confirmed the accuracy, exceptional for its time, with which the painting depicts both the latitude of Florence and the relevant obliquity of the ecliptic, as well as the precise positioning of various stars in relation to the precession of the equinoxes. More importantly, when measuring the heavenly bodies on the ecliptic, we discovered that they include not only the sun and the moon but also all five of the planets known at the time.[30]

28. Smart, "Taddeo Gaddi."

29. Ladis, *Taddeo Gaddi*, 276.

30. Ballerini, "The 'Sky' of San Lorenzo," 31.

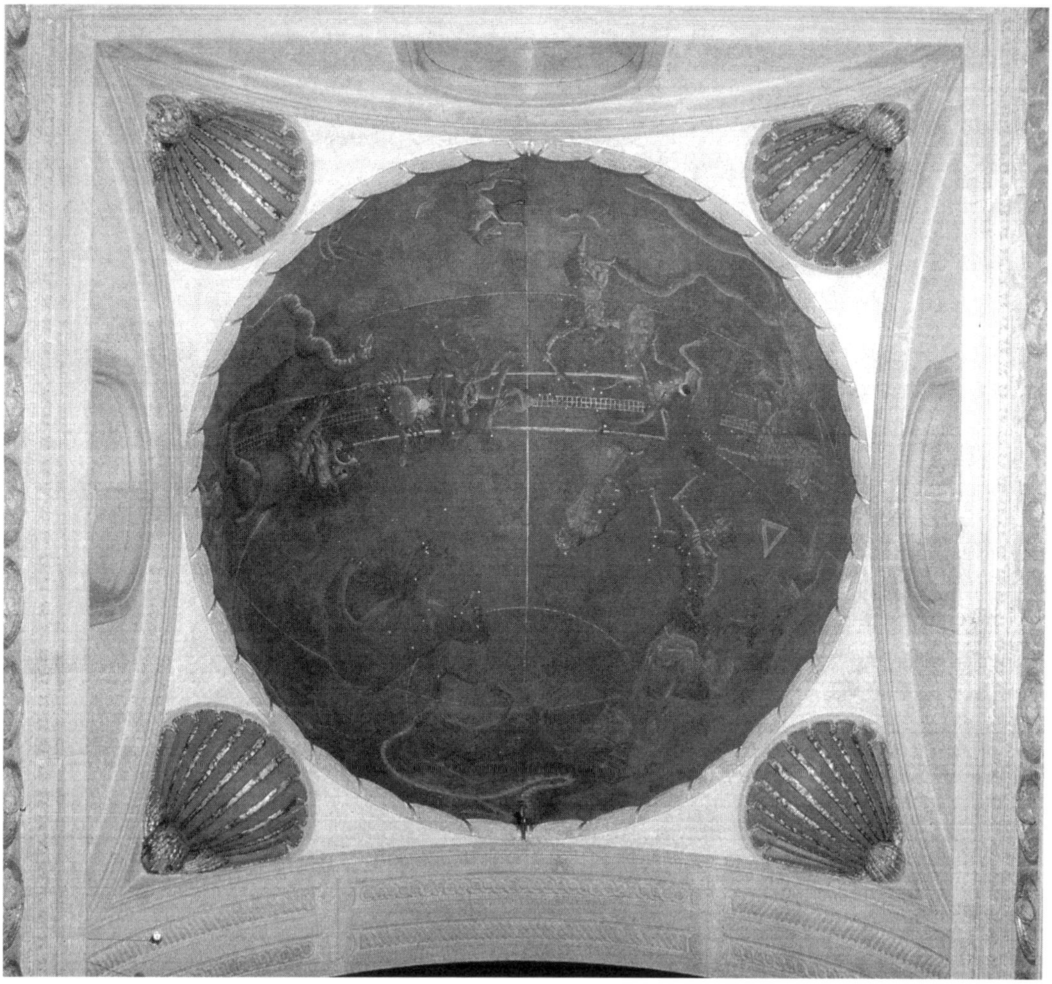

Figure 5.10: Interior of cupola depicting the sky over Florence, July 4, 1442, attributed to Pesello (1442–43), Old Sacristy, Church of San Lorenzo (Alinari Archives, Florence, photograph by George Tatge [CAL-F-001256–0000]).

An identical fresco is located in the Pazzi Chapel in Santa Croce, although the latter has faded considerably, obscuring the finer details that one can see in the San Lorenzo fresco. Discussing the two frescoes, Ballerini suggests (and it is generally accepted) that the date of July 4, 1442, was selected for depiction to mark the arrival of Renè of Anjou (1409–80) in Florence, which was apparently timed to coincide with the attendance of Pope Eugene IV (1383–1447) at a council in Florence. The frescoes at San Lorenzo and the Pazzi Chapel are among the finest examples in the Renaissance of the conjunction of art and astronomy.

5.6.4 Duomo, Uffizi, and Ognissanti

There are several other fifteenth-century paintings with astronomical connections that merit attention. The Duomo contains a painting dated to 1465 in which Domenico di Michelino (1417–91) depicts the celestial bodies in his famous composition of Dante holding a copy of the *Divine Comedy*. To the left of Dante, demons are leading the damned to hell; the mountain representing purgatory, with Adam and Eve shown at the top in the Garden of Paradise, is painted against a background of nine heavenly spheres.

Near the end of the fifteenth century, Filippino Lippi (ca. 1457–1504) executed a painting in tempera on wood entitled *Adoration of the Magi* (1496), now on display in the Uffizi. The man crouching in the lower left-hand portion of the painting is holding an astrolabe, one of the few depictions of the astrolabe in Florentine Renaissance art. Another exquisite example of scientific instrumentation featured in art is Botticelli's sixteenth-century fresco of *Saint Augustine in His Study*, which shows an astrolabe, an armillary sphere, and a clock (fig. 5.11). Commissioned by the family of explorer Amerigo Vespucci, the painting may show the instrumentation to pay homage to the importance of the astrolabe in navigational pursuits. In addition, Botticelli himself had a keen interest in humanistic philosophy and scientific inquiry. The fresco was originally located next to the choir in the church of Ognissanti, but during an eighteenth-century renovation it was detached and placed in its current location in the nave of the church.

Several later paintings depict astronomy, astronomical instruments, and notable Florentines. A seventeenth-century painting by Alessandro Gori (b. 1600) in the Uffizi Gallery (Corridoio di Ponente) shows Galileo, Toscanelli, and Torricelli with astronomical instruments and personifications of Saturn and the moon. Two paintings executed after the Renaissance are also worth mentioning. In the nineteenth century, Florentine artists Tito Lessi (1858–1917) and Annibale Gatti (1827–1909) made retrospective paintings, now displayed in the Galileo Museum. Lessi's painting shows Galileo conversing with his associate Vincenzo Viviani, and Gatti's highlights Galileo with (purportedly) John Milton in his villa at Arcetri. Evoking the glory of astronomical studies in the past, both artists included an armillary sphere in the scene.

5.7 Astronomy in Post-Renaissance Florence: The Sites and Scientists

Although installed during a later time period, two additional post-Renaissance observatories are mentioned because of their close connection to the material discussed in this chapter. The chapter closes with brief mention of a recent addition in Florence that deals with solar time measurements for hourly determinations and a calendar.

5.7.1 The Ximenes Observatory (Osservatorio Ximeniano)

In 1756, Leonardo Ximenes, the Jesuit priest who made careful measurements of and with Toscanelli's gnomon in Santa Maria del Fiore, as well as modifications and astronomical determinations (see 5.2.2), opened a small observatory on the top floor of the monastery of San Giovannino, located near the Dominican church of San Lorenzo (fig. 5.12). Today the former monastery/observatory serves as a museum that houses astronomical equipment, including a gnomon, as well as scientific items used for meteorological, seismographic, and environmental research. The museum's library contains about three thousand volumes, including Ptolemy's *Almagest* and Newton's *Optics*, making it a valuable resource for historical research.

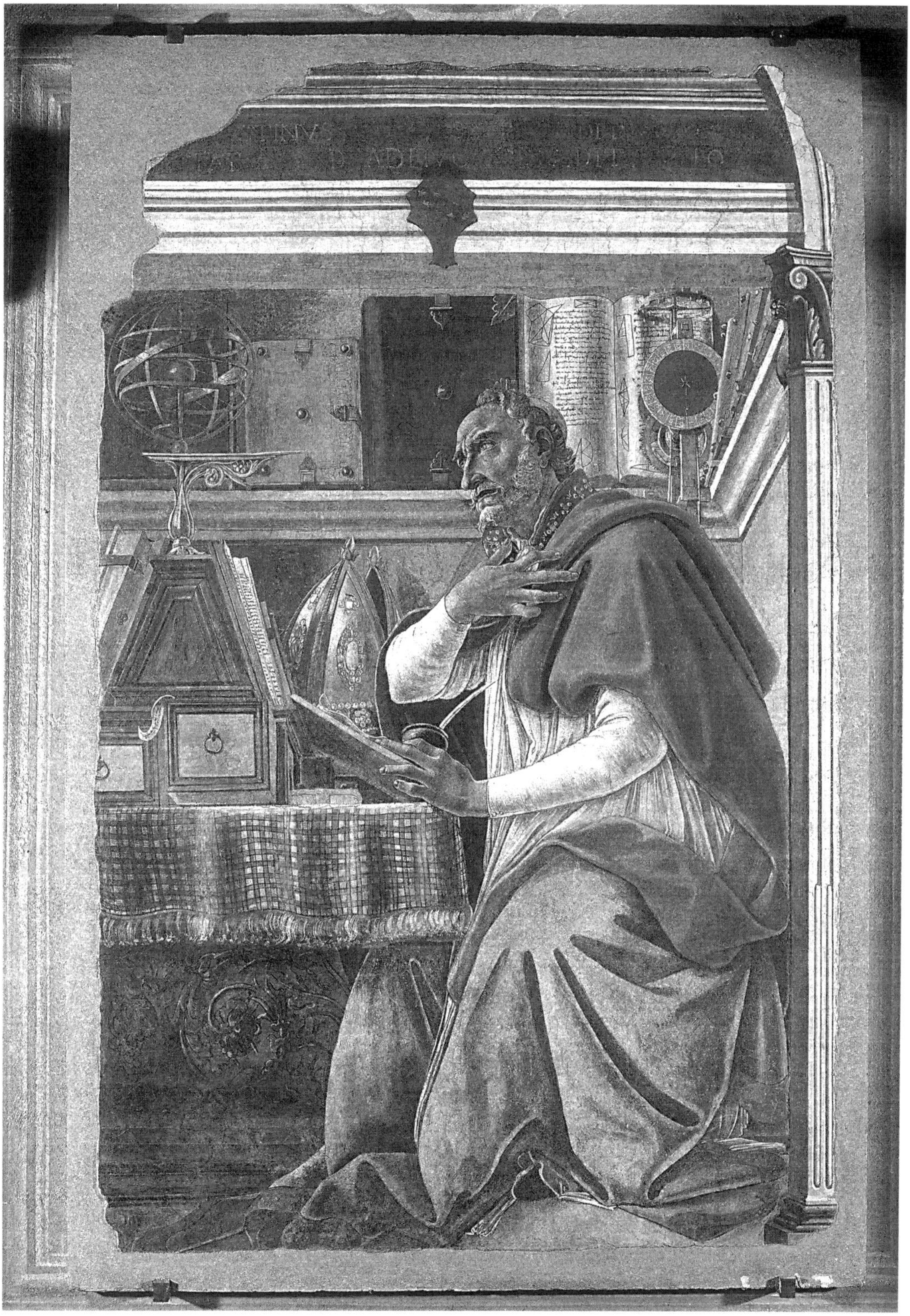

Figure 5.11: Sandro Botticelli, *Saint Augustine in His Study* (ca. 1480), fresco, Church of Ognissanti, Florence (Archive/Alinari Archives, photograph by George Tatge [SEA-S-F12000–0006]).

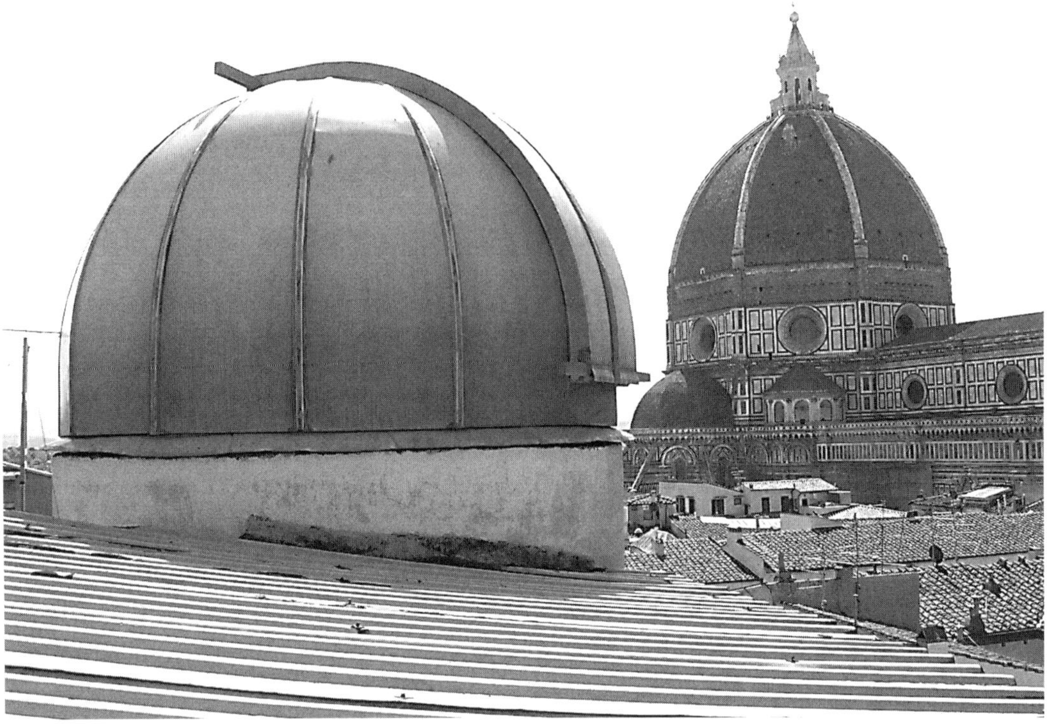

Figure 5.12: Juxtaposition of the Ximenes Observatory with the Duomo used by Paolo dal Pozzo Toscanelli for his astronomical measurements (photograph by authors).

5.7.2 La Specola

The astronomical observatory La Specola was established in the 1780s at the Imperial and Royal Museum of Physics and Natural History (Museo di Fisica e Storia Naturale). The building, originally known as the Palazzo Paoletti, is located near the Pitti Palace; it was purchased in 1771 by Peter Leopold I of Lorraine and renovated to accommodate a museum and observatory. After the installation of numerous astronomical instruments, including telescopes, the observatory became functional in 1808 with the arrival of the astronomer Domenico de Vecci. In about 1830 Giovanni Battista Amici (1786–1863) was named the first director of the observatory. A respected astronomer and instrument maker, Amici brought with him a group of talented technicians who aided him in the fabrication of many astronomical instruments. The apparatus was transferred to the newly constructed Arcetri Observatory in 1872, but some features of the original observatory remain, including the structure of the top floor observatory and a gnomon in the Stork Room (so named for the sculpted storks throughout the room).

The gnomon was originally constructed by Giuseppe Slop in 1784 and restored in 2005, a restoration that included a rebuilt shutter around the opening. This most unusual gnomon contained a string above the meridian (now a silver thread, but composed of knotted hairs in the eighteenth century) that was used for measuring the moment that the sun passed over the local meridian, thus permitting synchronization with the local time. The opening

(3 mm in diameter) is located at a height of about 3.3 meters, and the length of the meridian is some 7.8 meters from the vertical position to the ellipse denoting the winter solstice. A copper strip defines the center of the line, and the surrounding marble has embedded signs of the zodiac in blue scagliola (artificial marble). The summer and winter solstices are marked by copper ellipses that correspond to the image of the sun. The Latin inscription carved in the marble panels exterior to the meridian reads, "This meridian line was laid out in the Observatory of the Royal Science Museum of Florence while Pietro Leopoldo was ruler. 1784."[31] The gnomon in La Specola can be seen today, and follows a direct line from the earlier installations and measurements started by Toscanelli and Danti. As with those in the Renaissance, it represents a fine example of craftsmanship that is artistically appealing as well as scientifically useful. Today the museum contains an excellent exhibit focused on natural history and an unusual set of wax human anatomical models that were used for instruction in anatomy (see 4.2.2).

5.7.3 Piazza dei Giudici

The most recent addition to the solar-based astronomical instruments in Florence for time keeping and calendric purposes pays tribute to Florence's legacy of scientific and astronomical developments. Following a major renovation in 2010, the Museum of the History of Science, which occupies the former Palazzo dei Castellani, was renamed the Galileo Museum. In front of the building, the museum, in collaboration with the city of Florence, installed a sundial modeled after those of the past. It gives the local time and functions as a calendar with the signs of the zodiac flanking the meridian. The six-meter-high structure casts a shadow that, due to the limitations of space, can indicate time only from about 9:00 am to 3:00 pm.[32] This modern installation, situated in a public space along the Arno, recalls the sundial on the Ponte Vecchio used by thirteenth-century Florentines to determine time and serves as an interesting link to the richness of the astronomical work of the past.

31. Barbolini, "Gnomon in La Specola," 105.

32. Schnabel, "Sundial in Piazza dei Giudici."

Chapter 6

Art and Technology

> "There are many men whom nature has made small and insignificant, but who are so fiercely consumed by emotion and ambition that they know no peace unless they are grappling with difficult or indeed almost impossible tasks and achieving astonishing results."
>
> —Giorgio Vasari, *Lives of the Artists*

The observational approach so important to the Florentines, coupled with increased mechanical, mathematical, logistical, and technical inventiveness, resulted in the construction of iconic monuments, glorious art, and the production of dramatic performances. The goals of creating grand Florentine buildings and innovative sculpture required the efficient transport of marble from quarries in surrounding areas and the development of technology for construction and sculpting. Witness Brunelleschi's magnificent dome of the Florence cathedral, the dramatic interior of the Palazzo Vecchio by Vasari, Michelangelo's superb sculptures and ingenious fortifications, and the complex machinery used in sacred and secular theatrical presentations—all of these accomplishments represent a true integration of art and technology.

6.1 Rediscovery and Refinement of Ancient Technology

The sculptors, architects, and builders of the Florentine Renaissance created monumental masterpieces, and the necessity of handling huge blocks of marble and moving them safely became paramount. The Florentines designed and built palaces, churches, and the soaring domed cathedral of Santa Maria del Fiore (Duomo), requiring new sophisticated technology for construction and the machines necessary to raise and place the marble with great precision. As the Renaissance dawned, it was known that the early Egyptians had built massive pyramids and that they and the Greeks and Romans had constructed temples and carved epic statues. In addition, the Romans had laid road beds and arched bridges, many of which were still being used. Much of the knowledge of how these marvels were built had been temporarily lost at the dawn of the Renaissance.

Many ancient cultures designed and configured machines to perform specialized tasks, including the building of aqueducts, canals, dams, and reservoirs to move and collect water; cranes, pulleys, and levers to hoist large heavy objects; various types of gears and cranks to operate machinery; and weapons of war such as catapults and battering rams. Despite these accomplishments, innovations with broad applications were not typically part of the technological landscape of ancient cultures, at least as judged by surviving documents.

The earliest known writings on engineering and construction technology are those of Roman engineer and architect Marcus Vitruvius Pollio (ca. 80/70 BCE–after 15 BCE), whose illustrated manuscript *De architectura* dates from the first century BCE, and the writings of Greek mathematician and engineer Hero of Alexandria (ca. 10–ca. 70) from the first century CE. From the fourth century, we have Proclus Lycaeus's (412–85) commentary on Euclid's *Elements of Geometry* and Pappus of Alexandria's (ca. 290–ca. 350) *Mathematical Collection*. These fourth-century works describe ancient machines, including the components necessary to build cranes and hoists. The Romans made considerable advances on Greek, Egyptian, and Etruscan technology. Roman contributions were mainly in construction of buildings and public works, and comparable technological work was continued by the Byzantine empire during the millennium following the fall of Rome. The Arabic/Islamic focus was more oriented toward science, mathematics, astronomy, and medicine than toward technological advances, although many exceptions do exist. During the Middle Ages in Europe, technology was primarily concerned with improving agricultural productivity: for example, advances in metal working to fabricate heavier plows, refinement in the design of yokes to increase the load that could be pulled; and improved carts that could transport larger and heavier loads. Also during that same period, architects and builders in Europe were involved in the construction of large cathedrals, necessitating the invention of new methods of supporting the massive outward thrust that resulted from increasingly large roofs. The most notable contribution to architecture and building was the flying buttress, first introduced in France in the twelfth century. This period also witnessed improvements in mechanical clocks, military technology, efficiency of mills (waterwheels and looms), and the harnessing of wind power. These and other technological advances spread throughout Europe.

Writings such as those of Vitruvius survived the centuries, but many were buried in monastery libraries and remained unknown. In 1414 Gian Francesco Poggio Bracciolini (1380–1459), a Florentine humanist and papal secretary with a passion for uncovering old manuscripts, discovered a copy of Vitrivius's *De architectura* languishing in the Abbey of St. Gallen in Switzerland. However, it was not until about 1450 that Leon Battista Alberti (1404–72) translated Vitruvius's opus, which then became widely available. In 1423 the Sicilian Giovanni Aurispa brought a bundle of 238 manuscripts back to Italy from Constantinople. Among the manuscripts, which included works by Plutarch, Sophocles, Strabo, Aeschylus, and others, were copies of the Greek treatises by Proclus Lycaeus of Alexandria and Pappus of Alexandria (see above).[1] Florentines suddenly had access to explanations of fundamental early technological concepts and detailed drawings and diagrams that

1. King, *Brunelleschi's Dome*, 63.

expanded their understanding of engineering and construction. These seeds of knowledge from the Greeks and Romans augmented existing innovative Florentine technologies and inspired further development, providing an impetus for expanding on the basic principles and laying the groundwork for creative designs.

The Florentine inventive mind-set was applied to more efficient transportation of materials, such as marble and other types of stone needed for construction and for massive sculptures. During this period, engineers and inventors also made significant improvements to military machinery, to devices used for surveying, and to a wide range of trades and industries. Perhaps the most visible and spectacular advances were made in the public arena, where technological developments transformed not only the appearance of churches and other public buildings, but also, to some extent, the activities that took place in those buildings. For example, the long tradition of producing sacred plays (*sacre rappresentazioni*) and secular presentations (*intermezzi*) to entertain and educate the public blossomed with the availability of technology to create a variety of special effects, turning once simple performances into spectacular theatrical events.

The architectural masterpiece that most symbolizes the advances in technology is the dome of the Florence Cathedral (Duomo), Santa Maria del Fiore. Even before the early Greek and Roman treatises on construction were available in Florence, artists and architects were focused on new methods for advancing their disciplines. Although general plans for a dome on the cathedral had been proposed, it took the genius of Filippo Brunelleschi (1377–1446) to create a plan for a monumental dome and to imagine the intricate and innovative methods for the actual construction. For such a massive and ambitious project, Brunelleschi required new and specialized hoisting devices to lift the construction materials needed for the dome. Lacking knowledge of earlier Greek and Roman inventions, Brunelleschi had to fabricate unique apparatus for the required lifting tasks.

Although Brunelleschi left no notes or detailed drawings, some of his inventions survive and are on display in the Duomo Museum (Museo dell'Opera del Duomo) and in a small interior space located in an area of the dome itself. Fortunately for the historical record, others made drawings of Brunelleschi's hoisting devices, reversible gear systems, and the great crane (*colla grande*). Artist and engineer Mariano di Jacopo, known as Il Taccola (ca. 1382–ca. 1455), not only made sketches of Brunelleschi's devices, but he also designed his own machines for a number of practical purposes. He produced two manuscripts, *About Engines* (*De ingeneis*) in 1433 and *About Machines* (*De machinis*) in 1449, each filled with drawings and notes describing devices involved with hydraulic engineering, construction, and instruments of war. The idea of the artist-engineer who not only designed machines, but was also capable of graphically illustrating his inventions to enhance the understanding of the text, was reborn.

With many manuscripts on engineering and mechanics available to consult, and the desire of wealthy donors to commission projects, Florence became an important center for technological development in architectural endeavors. Giuliano da Sangallo (ca. 1443–1516), a Florentine sculptor as well as an architect and military engineer, had a strong interest in the study of ancient architecture, as documented in his many notebooks. He was frequently commissioned by Lorenzo de' Medici for projects in greater Florence, including the Medici villa at Poggio a Caiano, fortifications for Florence and other cities, and the church of Santa Maria delle Carceri in Prato. Giuliano and his brother, Antonio da Sangallo il Vecchio (ca. 1453–1534), also a Florentine architect and skillful engineer, were partners until Antonio's death. Following the example set by Il Taccola, the Sienese painter, sculptor, and architect Francesco di Giorgio Martini (1439–1501/2) composed a treatise on architecture in 1482. Although his work was not centered in Florence, his treatise was extremely influential to those pursuing architecture, including Leonardo da Vinci (1452–1519), the most prolific and innovative of all the Florentine polymaths.

In addition to being an artist, scientist, and architect, Leonardo created hundreds of pages of designs for machinery of many types and description, some to solve very specific problems and others to enhance the capabilities of ever-widening societal necessities: for example, plans for portable bridges, military armaments such as cannons and catapults, reinforced ships with the capability of resisting fire from cannons, machines designed for flight, machines that used water for power, and numerous other creative ideas. As stated by Marco Ceccarelli, Leonardo moved the discipline to a new level "to give rigorousness and scientific dignity to mechanism design by including experimental activity as additional necessary background within an early scientific perspective for interpretation and investigation of operations and tasks of mechanisms."[2] For Leonardo, there was no division between art and science: an understanding of one was not possible without an understanding of the other. Leonardo believed that knowledge of anatomy was essential to creating art and sculpture that portrayed the human form. To be an engineer, he believed, one also had to be an artist in order to depict technological innovations so that others could adequately conceptualize them.

At age fourteen Leonardo was apprenticed to Andrea di Cione, known as Verrocchio (1435–88), with whom he studied sculpture, painting, and drawing, and received training in metalworking, drafting, and mechanics. There is evidence in Leonardo's early notebooks of his interest in engineering, including (but certainly not limited to) a plan to redirect the Arno River to make it a navigable canal, designs for flour mills, and hydraulics such as pumps to elevate water that would create energy as it was released. The range of Leonardo's studies was so varied and eclectic during that Florentine period that it cannot be easily summarized. Because of his broad experience, in 1482 Leonardo was appointed by the Duke of Milan to serve as an artist/engineer to the court. His income during that time was derived from commissions awarded by wealthy patrons desirous of innovative military equipment, hydraulics, and engineered theatrical performances, to name but a few of the categories. His other more esoteric engineering studies involved explorations of general scientific principles underlying the technology. As stated by renowned historian of science and director of the Galileo Museum Paolo Galluzzi, "Leonardo now emerges as the culmination, as the most mature and original product of a collective development lasting several decades, to which many highly talented figures made sizable contributions."[3]

2. Ceccarelli, "Renaissance of Machines in Italy," 1533.

3. Galluzzi, *Art of Invention*, 11.

The range of interests of Leonardo and other engineers, architects, and inventors was manifest in a surge of technological developments that propelled Florence to the forefront. The evolution of technological innovation and its corollary artistic dimension set into motion in Florence can be seen through an examination of several specific areas: the transportation of massive amounts of marble needed for artistic and architectural projects; the design, construction, and facing of the dome, the crowning glory of the Florence Cathedral; the ingenuity of Vasari's Palazzo Vecchio renovations; military engineering; and the intriguing machinery fashioned for the Florentine religious festivals, plays, and court weddings.

6.2 Transport of Marble for Sculpting and Building

As mentioned in chapter 2, a discussion of marble would be incomplete without referencing the challenging technological problems involved in hauling large pieces of stone from the quarries to the artists' workshops and construction sites. For example, from the average density of marble (2.7 grams/cubic centimeter; the range for various types of marble is 2.6–2.8), one can calculate that a 1-cubic-meter block of marble weighs 2,700 kilograms (5,947 pounds). Therefore, a block sufficiently large to carve an oversize statue such as Michelangelo's *David* could easily require a block weighing up to 9,000 kilograms (nearly 10 tons or 20,000 pounds). During the construction boom in Renaissance Florence, massive amounts of marble were required for facing exterior walls of buildings, for columns, flooring, interior walls, pulpits, altars, and the like. Although other types of stone were utilized, marble has been prized since the classical period for its durability, sheen, beauty, and variety of coloration. Quarrying and transporting the marble presented a formidable technological challenge that was met admirably by the Florentines, building on the knowledge of the past.

6.2.1 Marble Quarries

The Florentines' primary sources for marble were the large quarries at Carrara, situated about 105 kilometers (65 miles) northwest of Florence. Carrara's white marble (*bianchi marmi*) was highly valued; it was hard, very white, and of excellent quality. Marble had been quarried at Carrara since Roman times, and the city was favorably located relative to water routes, had a well-organized workforce to quarry the stone and arrange for its sale and transport, and had a long history of providing high-quality stone.

Florentine builders and artists also used several other quarries, each with certain advantages, including marble coloration and pricing. In the early 1500s, when the towns of Seravezza and Pietrasanta, which had been briefly under the control of the Republic of Lucca, were returned to the territory of Florence by Pope Leo X, Seravezza granted Florence exclusive rights to quarry the marble in the area and to set marble prices and collect taxes on the sale of the marble. The city's disadvantage, and one that became problematic for artists, was its lack of infrastructure to support the quarrying operation. Seravezza did not have a group of quarry men as well trained as those in Carrara to coordinate and perform the work.

The Florence Cathedral needed a constant supply of marble for its commissioned interior and exterior sculptures, as well as for its massive exterior façade. The Board of Works

(Opera di Santa Maria del Fiore), which was charged with managing the operation of the Duomo, contracted with various quarries for three colors of marble for the cathedral: a blackish-green marble (*verde di prato* or *marmum nigrum*) found in the area around Monte Ferrato; a red marble (*marmum rubeum*) from multiple quarries including San Giusto a Monte Rantali, Montagnola Senese, and quarries near Valdinievole; and the valued white marble (*bianchi marmi* or *marmum album*) found in Carrara. In addition to Carrara, the Opera di Santa Maria del Fiore also considered quarries located near the western coast of Tuscany about forty miles south of Siena in Campiglia as sources for the white marble, but the contractors in that area were unable to reliably fill the required orders.[4]

6.2.2 Selection and Extraction

For large marble orders, contractors were hired to oversee the entire project from extraction of the marble to its transportation to the required sites in Florence. A typical agreement listed the area from which the marble was to be cut, the preliminary carving that was to be accomplished at the quarry according to specifications shown on templates, arrangements for transportation of the marble to Florence, and the cost of delivery. So adept was the Opera di Santa Maria del Fiore at arranging marble contracts that they became the organization charged with negotiating marble acquisition for most of the major construction in Florence. Frequently, artists would travel to the quarries with measurements and drawings of their commissions to personally choose the best marble and supervise the carving to be done on the premises. Some stayed for only a few weeks, while others remained at the quarries for many months. Michelangelo spent long periods of time in both Seravezza and Carrara selecting the desired marble for his projects in Rome and Florence, including the never-completed façade of the church of San Lorenzo. Brunelleschi, along with master stonecutters who would work on his project, inspected various quarries to select the best unblemished stone for the dome of the Santa Maria del Fiore.

After the artist or agent had selected the desired marble, the extraction process began. The quarrymen cut blocks of marble from the mountain using hammers, crowbars, picks, and wedges. More than strength was required for the task: the workers needed to possess an understanding of the seams (natural fissures) in order to accurately place the cuts, ensuring the stability of the selected block. It often took as long as two months to remove a marble block from the beds (*abbattimenti*). Once a block had been removed, it had to be lowered down the mountain, using essentially the same techniques that had been employed in Roman and Egyptian times. The blocks, extremely heavy and difficult to maneuver, were secured with ropes to a wooden sledge (*lizza*) and gradually moved downward along a prepared slope; their descent was controlled by pulleys and ropes attached to posts placed into the rock. The method known as *lizzatura* was very dangerous; workmen were occasionally killed and blocks lost when the sledge came loose from its ropes or the strength of the rope was compromised. It was extremely important that workers possessed

4 Saalman, *Filippo Brunelleschi: The Cupola of Santa Maria del Fiore*, 202–3.

the highest quality rope, pulleys, and iron rings. Those three types of equipment were studied and improved during the Renaissance.

Once the rough block had been extracted, an artisan was employed to cut the stone to the shape specified by the artist, using tempered iron tools. The amount of carving done on site varied. Some artists preferred to have stone delivered to Florence in a rough form while others provided cutout patterns to the quarry masters, models made from wood or plaster, and detailed drawings and measurements so that preliminary work could be accomplished before the marble was transported. Carving that could be executed at the quarry reduced the weight of the block and, therefore, lowered the cost of transport. Like many artists, Michelangelo preferred to do much of his marble carving in Florence, but he provided quarry managers with notebooks containing size specifications and drawings of the blocks he desired.[5]

6.2.3 Transportation

Several types of transportation were needed in order to move the extraordinarily heavy blocks of marble from the quarry to Florence without damaging the precious cargo. An important aspect in the movement of marble was the construction of new roads or the improvement of old Roman roads. Obviously these roads needed to be able to bear greater weight than those used for ordinary traffic of trade goods and travelers. Financing was required for the creation, upkeep, and rebuilding of adequate avenues of transport and of new or updated port facilities. In the case of the Florence Cathedral, the Opera expended a great deal of money and manpower on roads meant for marble conveyance. The journey from quarries to Florence sometimes required as much as a year, optimally somewhat less depending on circumstances.

A block of marble being transported from Carrara, for example, would first be loaded onto an oxcart using a ramp and pulleys. The oxcart traveled on reinforced roads that connected the quarry to the nearby town where export taxes were paid. From there, the cart would continue on to the wharf, where the marble was transferred from the cart to a system of rollers (*curri*)—a method employed since early times—that moved the marble into position alongside a boat. Various devices, including winches, were then used to lift the stone onto a large flat, reinforced boat; some boats were capable of carrying many blocks while others were considerably smaller. Once secured on the vessel, the marble was shipped via the Ligurian Sea to Pisa. When contractors were involved in the transaction, it was their responsibility to see that water transport was hired and available. Artists were often not involved in the transport of marble, but for the smaller jobs, if contractors did not provide their own designated vessel, the artist had to locate and hire someone willing to haul their materials. After the ship arrived at Pisa, the marble was transferred to a flat-bottomed river barge (*scafi)* that would carry it fifty miles up the Arno to the site of rapids and shallow water at Signa. There the marble had to again be moved on rollers to oxcarts for the final ten-mile portion of the journey to Florence. The entire process was hazardous and often resulted in damage to the marble, especially during the loading and offloading.

5. Wallace, *Michelangelo at San Lorenzo*, 41–46.

Because of the difficulties encountered with transportation and the potential for damage to an expensive block of stone, inventors and engineers directed their ingenuity toward identifying better techniques of conveyance. For centuries, marble had been moved using wooden rollers, and blocks were lifted by using ramps and winches. Michelozzo di Bartolomeo (1396–1472), a Florentine architect and sculptor who was much admired by Cosimo de' Medici, devised a different type of winch (*colla*) that made the lifting less perilous. Three-legged hoists, known since Roman times, were also utilized. Even these hoists were not without problems, however, and sometimes a block of marble would fall while being lifted, either because of the failure of the hoist, the rope, or the handler, and bury itself deep in the muddy harbor or river bottom.

For the offloading of some of Michelangelo's marble from the boats, a large machine, a *capocavallo*, was built in 1518 in Pisa. It took six months to complete construction of the *capocavallo*, which had about twelve pulleys that distributed the lifting power, thus making it easier to transfer heavy marble slabs to boats that were either pulled alongside or beneath the machine. Although the *capocavallo* malfunctioned during its first use, after repair it was used by not only Michelangelo, but also by the Opera di Santa Maria del Fiore. Between 1520 and 1525, the *capocavallo* in Pisa was used to offload about 3,000 blocks of marble, comprising about 250 shipments. Despite the need of frequent repairs during that period of use, the *capocavallo* proved to be a particularly valuable innovation.

Brunelleschi, who invented a variety of hoisting machines to lift materials during construction of the dome of Santa Maria del Fiore, also applied his ingenuity to the issue of transporting marble. Bothered by the myriad problems arising from the need to load and unload marble at multiple points during the journey from the quarries to Florence, he attempted to improve on the design of cranes. For example, he developed a crane capable of moving marble directly from the boats to the carts waiting at the waterfront in Signa rather than having to offload with rollers and then use a hoist to lift the blocks to carts.[6] In addition, Antonio Manetti di Ciaccheri, who worked closely with Brunelleschi on the design and construction of the lantern of the dome and who was appointed *capomaestro* to direct the building of the lantern following Brunelleschi's death in 1446, designed a much-improved cart in 1454 or 1455. His reinforced cart made it possible to transport marble from the quarry to the coast and from Signa to Florence with the least amount of stress on the stone.

In addition to the design of machinery to move blocks from the quarry to the ships, improvements were also needed to the ships themselves and to the coordination of water transport. Prior to the quattrocento, barge traffic on the fifty-mile stretch of the Arno from Pisa to Signa was controlled by the maritime Pisan authority, which was responsible for the scheduling and inspection of barges and the contracts and payment to barge operators. This coordination was necessary because of the amount of barge traffic on a notoriously difficult river to navigate. During that time, Florentine shipping companies provided many of their own boats that, along with those of private operators, included numerous types of navigational vessels usually propelled by oars or poles. After Florence acquired control of Pisa, the

6. Ibid., 56–57.

aim of the Medici was to make navigation on the Arno more profitable for Florence rather than the Pisans, in effect by creating a state monopoly for navigation.[7]

Frustrated at the length of time involved in moving the marble with existing barges, Brunelleschi turned his attention to the creation of a new type of river vessel capable of being powered by multiple means depending on the navigational conditions of the river. Although his innovative design proved to be unsuccessful, the story behind it sheds interesting light on Renaissance invention. Brunelleschi probably worked on the plans sometime between 1415 and 1421, presumably after visiting the pier in Pisa while awaiting a shipment of marble. Although most of Brunelleschi's focus was on hoists and machines needed for construction of the dome, the same principles applied to lifting materials during transportation. Reasoning that it would be more practical and economical to have the hoisting machines on the barge itself, he designed a flat-bottomed boat he dubbed *Il Badalone* (*The Monster*).

Convinced that his new design would prove more economical and efficient than the standard barges, and therefore was likely to be copied by others, Brunelleschi kept his plans from the public until 1421. At that time the Council of Florence, hoping the city would benefit from this new technology, granted the first known inventor's patent on his design. The final version of the patent specified that anyone who copied Brunelleschi's boat within a period of three years would have that boat condemned to flames.[8] Even after he was given the protection of the patent, Brunelleschi remained secretive about his invention. None of the plans for *Il Badalone* remain, nor does any information regarding its construction, but it has been speculated that the design involved something similar to paddle wheels attached to a flat boat that could also be towed by ropes attached to other small boats or to oxen walking alongside the river banks. The building of the boat was a lengthy process, and records indicate that it was not until 1427 that Brunelleschi was finally able to accept a contract to transport one hundred tons of marble from Pisa to Florence for use on the ribs of the dome. Evidence shows that *Il Badalone* failed to be the technological marvel envisioned by its creator. In fact, when launched in 1428, the barge was able to complete only a portion of the distance from Pisa to Florence (probably no farther than Empoli) before breaking down. The marble was either lost or, according to some accounts, offloaded to smaller boats to complete the journey. Failing to function and probably destroyed during its maiden voyage, *Il Badalone* proved an expensive and embarrassing episode in Brunelleschi's illustrious career. By the time the bulk of the required marble finally began arriving for the dome, Brunelleschi had turned his attention to other projects, effectively removing himself from involvement in the problems of transportation.

Even when machines and vessels functioned perfectly, inclement weather was often another enemy to marble transport. Heavy rains could make roads impassable, often leaving entire sections washed away in a downpour, or quickly transforming a river into a cascade of crashing water. Storms could cause boats to flounder or even capsize, resulting in marble being lost at the bottom of the sea or river. In 1421, one of the barges carrying *bianchi marmi*

7. Prager and Scaglia, *Brunelleschi: Studies*, 114–16.

8. King, *Brunelleschi's Dome*, 113.

chosen for the dome of Santa Maria del Fiore sank during such a storm. Too little rain could also be a problem. During dry periods the Arno, naturally filled with silt, became too shallow for the barges to negotiate the route. It was often impossible to predict what conditions to expect. The Arno and the difficulties it presented prompted Florentines to explore ways of making the river more navigable. Leonardo went so far as to propose bypassing the Arno altogether by building a canal of his design, but this was one of his many innovative ideas that was never realized.

The loss of significant amounts of marble at sea led to innovations related to salvaging operations. Mariano di Jacopo (called Il Taccola) and architect Francesco di Giorgio invented a type of apparatus for breathing underwater and an inflatable housing in which divers could be raised and lowered.[9] These innovations made it possible for divers, using special diving suits, to locate sunken caches of marble and then wrap the marble with heavy ropes, enabling salvagers to attempt to lift the blocks onto ships using hoists. Despite the effort expended, much of the marble was never recovered.

Inherent in the process of procuring and transporting marble were a range of problems that provided abundant technological challenges for early engineers. Spurred by the need to address these problems, Florentines began to not only develop devices that met current needs, but also created careful drawings and plans with annotated specifications that became blueprints for future generations to improve, modify, and generalize.

6.3 Technology Developed for Construction

6.3.1 The Florence Cathedral (Duomo), Santa Maria del Fiore

For most people, a mention of Florence conjures up an image of the dome of Santa Maria del Fiore, usually known as the Duomo (cathedral) (fig. 6.1). The story of the design and construction of the Duomo spans the trecento and quattrocento from its beginnings in 1296 to its completion a century and a half later and reveals much about Florentine methods of planning and implementing large-scale construction projects during this pivotal period in their history. The original structure was designed by Arnolfo di Cambio (ca. 1240–1300/10), but the Duomo is best known for its dome, creation of the innovative and brilliant architect and builder Filippo Brunelleschi. Before we delve into his life, however, it is instructive to develop the background preceding the vexing problem he was given and then cleverly, and most successfully, pursued.

The building of the dome provides insight into techniques, practices, and difficulties in understanding the myriad details of such a project. As with many complex topics, the planning and construction of a large structure of unusual design often have conflicting or at least varying interpretations of the exact sequence of events and the contributions of the major individuals involved. Exacerbating this problem was the reluctance of Brunelleschi, the master architect and builder, to reveal his thoughts and plans in writing and often in

9. Ibid., 117.

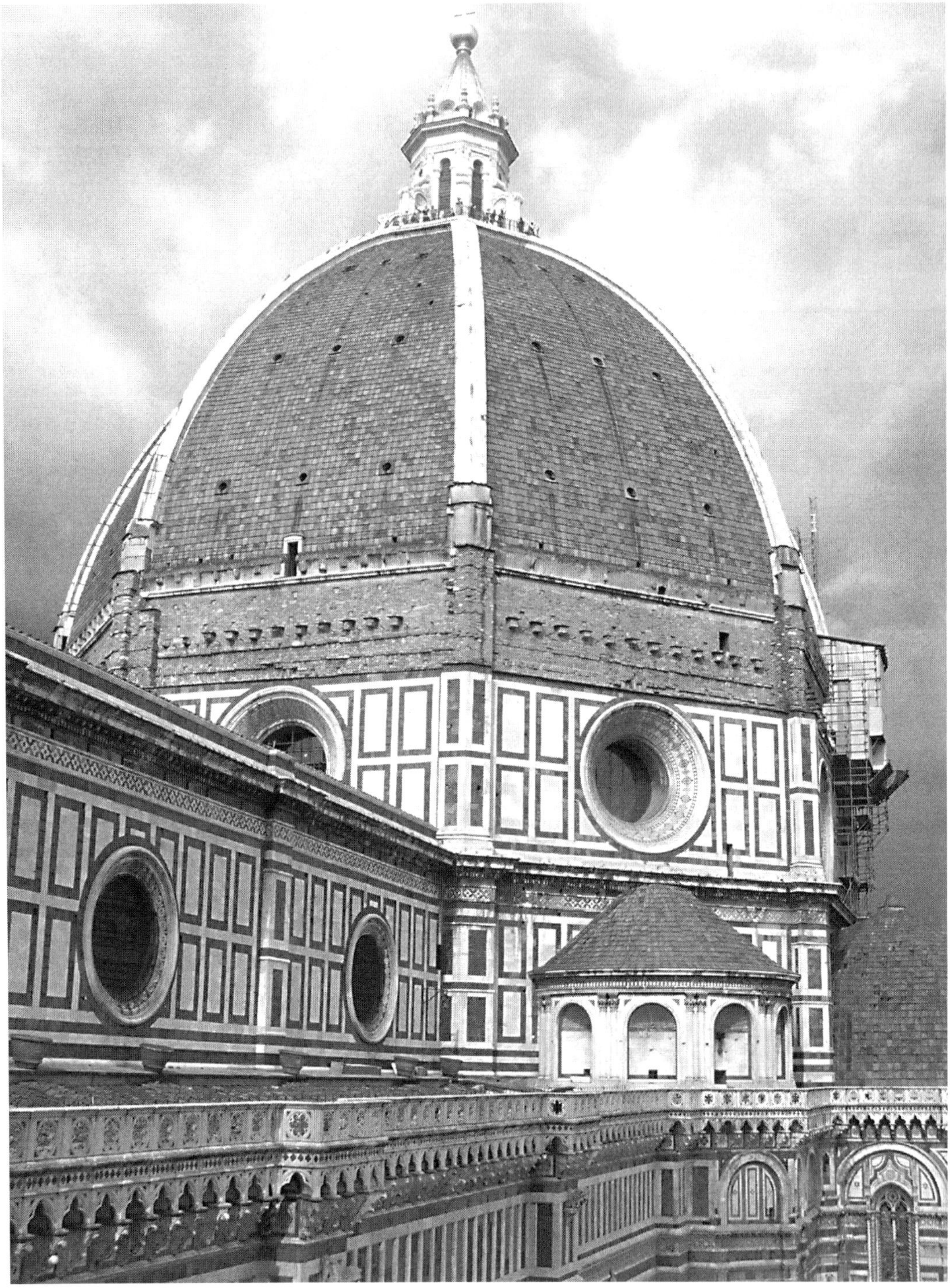

Figure 6.1: The dome (1420–36) of the Florence Cathedral (Duomo), Santa Maria del Fiore (photograph by Sailko, retrieved from Wikimedia Commons, GNU Free Documentation License).

dialogue. Although considerable documentation has survived,[10] there are major gaps in our knowledge, particularly about the building of the church in the century preceding the raising of the dome. The following précis is an attempt to integrate many of these reports into a cohesive framework that conveys the spirit of this undertaking, realizing that interpretations sometimes differ as to the details.

6.3.1.1 Building the Cathedral

Santa Maria del Fiore was constructed on the site of an early cathedral, Santa Reparata, which was built in the sixth or seventh century. By the thirteenth century, Santa Reparata was too small for Florence's population and, despite numerous repairs, was in poor condition. The increasing wealth of Florence in the thirteenth century, in large part from the wool industry but also from the various trades and businesses associated with the guilds, inspired the city's Comune to begin planning for a large and impressive cathedral to replace Santa Reparata. In 1292 the Tuscan architect Arnolfo di Cambio was hired to submit a design for the new structure. The Comune approved his plan in 1296 and the foundation stone was placed in the same year. While he most probably envisioned the construction of a dome, there are no surviving records to show whether he proposed a semicircular dome or one of a pointed Gothic design. In preparation for the eventual construction of the cathedral, the Comune purchased marble, requisitioned timber, and prepared the site. The new cathedral, with a footprint of 8,300 square meters, would be significantly larger than Santa Reparata. Workers cleared a large tract of land sufficient for the cathedral itself and the surrounding open plaza space. In the process they destroyed Santa Reparata and the nearby church of San Michele Visdomini, exhuming and reburying remains from a cemetery near the Baptistery, and demolished a neighborhood in the vicinity, forcing the inhabitants to relocate.

Such a major and, for Florence, unprecedented undertaking presented a number of challenges. First, the project required considerable funding to pay the designers, the construction crews, specialized craftsmen, and a range of administrative staff to organize the myriad tasks: hiring and supervising workers, and the timely purchase and procurement of materials. Experienced employees were needed to arrange methods for transporting marble, stone, bricks, tiles, timber, rope, and lead, often from many miles away, to the construction site in Florence and then for its storage until needed. Another hurdle to overcome was the vertical movement of the large stones, timbers, and other heavy items from the staging area to the craftsmen, tasks that increased in difficulty as they progressed upward in their building. Considerable ingenuity, tenacity, and talent were required to ensure completion of this magnificent structure that would represent the pride of Florence and make a statement to others about the importance of this strong city-state on the cusp of becoming much more powerful. The cathedral, and particularly the design and construction of the dome, represents a splendid intersection of science, technology, and the arts. As we examine in historical context this century-and-a-half-long challenging under-

10. Saalman, *Filippo Brunelleschi: The Cupola of Santa Maria del Fiore*. Interested readers are referred to Saalman, who relates much of the earlier work with clarity and also provides a comprehensive description of this interesting history and dome.

taking that inspired so many innovations, it can be considered one of the major catalysts of the emerging Renaissance.

In 1321 the Florentine Comune charged five of the major guilds—Wool (Lana), Silk (Seta), Merchants (Calimala), Bankers (Cambio), and Physicians and Apothecaries (Medici e Speziali), represented by their respective consuls—with overseeing the planning and building of the new cathedral. For the first ten years each guild assumed primary responsibility for a one-year period, but in 1331 the Wool Guild was assigned all duties. Within two years, the Opera di Santa Maria del Fiore (originally the Opera di Santa Reparata) adopted a constitution and appointed several officials to positions charged with conducting the overall business related to construction of the cathedral. With continued oversight by the appointees and consuls of the Wool Guild and the invited input of experts, work progressed, albeit slowly, and in about 1350 Francesco Talenti (1300–after 1369) was appointed master builder (*capomaestro*). While work on the cathedral complex was under way, construction had begun on the adjacent but free-standing Bell Tower (Campanile), which was designed by Giotto and built by three master craftsmen. Giotto worked on the square tower from 1334 until his death in 1337. Andrea Pisano (d. 1348), known for his bronze south doors on the Baptistery, succeeded Giotto and continued the planned design for reliefs on the first and second levels, but added niches for statues and false niches in the next two levels. Francesco Talenti, the final master to work on the Campanile (between 1350 and 1359), created a lighter structure above the two niche levels and added Gothic pointed windows.

During the construction of the Campanile, work on the Florence Cathedral moved slowly and mainly involved the procurement of materials and the building of the central nave and aisles. In 1357, after numerous meetings with architects and builders, the Opera di Santa Maria del Fiore adopted a resolution approving alterations to Arnolfo di Cambio's plan of 1296 by adding specificity such as the length of the nave and the diameter of the drum (the base that would eventually support a dome that was yet to be designed). According to the resolution, the overall length of the church would be 164 *braccia* (95.3 meters) and the diameter of a dome would be 62 *braccia* (36.0 meters). A Florentine *braccio* is equal to 22 and 7/8 inches or 0.581 meters; numbers have been rounded to avoid unwieldy fractions.[11] Francesco Talenti and another builder assigned to the project, Giovanni di Lapo Ghini, favored a Gothic-style dome using external buttresses to support the structure. The Opera requested plans and models for the dome from Ghini and a highly experienced mason, Neri di Fioravanti. The two models were quite different: Ghini's had external buttresses, while Neri's used chains around the dome to provide the necessary support and prevent buckling. The chains would be made with wood or stone, thus serving the same purpose as external buttresses—to prevent the lateral force of the weight of the structure from collapsing the dome. After much discussion and debate, the Opera chose the model submitted by Neri, although they decided in about 1366 to increase the supporting pillars, thus making the span of the dome even larger than was approved in 1357.

11. Fanelli and Fanelli, *Brunelleschi's Cupola*, 275.

In 1367 a group of eight architects and artists, including Orcagna (Andrea di Cione), Neri di Fioravanti, and Taddeo Gaddi, was appointed to work with the representatives of the guilds to refine the next phase of construction for the church, including the octagonal dome. Brunelleschi's father was one of the representatives on this special committee and he surely would have shared his insight into the deliberations and decisions being made with his son in later years. The committee's plan proposed that the dome was to be 144 *braccia* (83.7 meters) in height and 72 *braccia* (41.8 meters) in width (increased by 10 *braccia* over that proposed ten years earlier), making the width of the actual dome equal to the width of the nave and aisles. This proposal, along with other specifications, was approved by a referendum the same year, leading to a binding agreement in 1368 dictating adherence to these dimensions and the overall plan during construction. Some changes were later made as planning and construction progressed. The Opera asked Ghini to build a model of a dome using Neri's plans as modified by the larger group during their discussions in the 1360s. This model was publicly displayed, announcing the Opera's determination to complete the dome project. Despite the significant investment in time and money by many Florentines, there was no clear consensus as to how such a massive dome could be constructed. The audacity of this approach reflects the Florentines' supreme confidence in viewing the design as a challenge that could and would be met.

Thus, over seventy years had elapsed from the laying of the cornerstone in 1296 to agreement on a specified plan for the dome of Santa Maria del Fiore. The 1368 plan was further refined over the next years, but in the basic design it was decided that the total height of the drum plus the dome would equal 72 *braccia* (41.8 meters) and would be divided into four equal parts of 18 *braccia* (10.5 meters) each. The drum would be 18 *braccia* (10.5 meters) in height (although this was increased to 21 *braccia* (12.2 meters) in 1414 in order to add large round windows and a gallery). The remaining 54 *braccia* (31.4 meters) were to be devoted to the dome itself. Further, the total height of 144 *braccia* (83.7 meters) was interpreted to refer to the distance from the ground to the oculus ring (*serraglio*), not the hypothetical apex of the dome (some 4 *braccia*, 2.3 meters, above the oculus ring), resulting in an oculus diameter of about 12 *braccia* (7.0 meters). Thus, construction of the pointed octagonal dome would begin at about 93 *braccia* (54.0 meters) and then rise another 54 *braccia* (31.4 meters). Numerous other parameters were specified, and some of these will be covered as we discuss the actual construction of the dome. This plan and the models met with the approval of the Opera, the guilds, and many prominent Florentine citizens; thus, it was widely scrutinized by numerous individuals and organizations before finally being endorsed.

With this plan for the dome in mind, construction on the cathedral continued and by 1380 the nave and aisles had been completed and vaulted, the piers and a major arch were installed at the eastern end, and work would proceed on the tribunes, sacristy piers, and tambour. In about 1400, the architect Giovanni d'Ambrogio was hired and immediately began work on the eastern tribune (the one opposite the nave). The walls were already in place for the five chapels, and the vaulting was completed. D'Ambrogio decided that the windows should be raised, which he proceeded to do and at the same time extended the supports, leading to an interruption of the gallery. Serving on a nineteen-member advisory committee in 1404, Brunelleschi, Ghiberti, and the other committee members noted this change and recommended that the buttresses be lowered, a suggestion that was followed. A mathematician, Giovanni dell'Abbaco, was hired in 1417 to offer advice on the overall geometry and curvature of the dome and to ensure there were no serious flaws in the design. By the following year only the dome remained incomplete. Giovanni d'Ambrogio continued in his role until 1418, and in the same year Antonio di Battista was appointed as acting architect. With no viable plan of a method for constructing such a large dome, the Opera decided to hold a major competition in which models of a potential dome could be submitted. The interested reader is referred to Saalman for a thorough description of the events and available documentation for the period 1294 to 1421.[12]

At this point, the Florentines had been building the cathedral for more than a century. The commitment to continue planning and construction of the dome speaks well for the tenacity of the Florentines at the time and to their dedication to this major project. During the trecento they faced numerous hurdles of many different types, including the initial outbreak of the plague (Black Death) in 1348, followed by frequent recurring epidemics, reducing the population by more than half; political factions vying for power as Florence emerged from a feudal system toward a more republican type of government; famine; and military skirmishes that reduced the number of men available to work on the new church. Through all of these tribulations, the Florentines clung to their vision of a grand cathedral, one of the largest Christian churches ever built, that would demonstrate to the world the stature and prestige the city envisioned for itself. In a spirit of determination and with a confident optimism, the Opera di Santa Maria del Fiore announced its competition for building the dome.

6.3.1.2 The Dome Rises

a. Enter Brunelleschi

In 1418, Filippo Brunelleschi, already known in Florence as an artist and for his studies of architecture, was on the committee that advised the Opera on the cathedral project. Brunelleschi was born in Florence in 1377 and spent most of his life there; after his death in 1446, he was buried in Santa Maria del Fiore. His early biographer, Antonio di Tuccio Manetti (1423–97), informs us that Brunelleschi's father, Ser Brunellesco di Lippo Lapi, was a competent and well-traveled notary who participated in the 1367 referendum to adopt Neri di Fioravanti's provocative model for the dome, and his mother was of the Spini family. He had two brothers, one of whom became a priest. At an early age Brunelleschi learned to read, write, and use the abacus; he also gained some proficiency in Latin. From childhood he displayed a talent for drawing and painting, as well as a desire to achieve distinction in his endeavors.[13] Brunelleschi probably learned mathematics from his friend Paolo dal Pozzo Toscanelli (see 5.2.2), and later introduced a rigorous mathematically and geometrically based approach to perspective painting (see 3.1.2.1). Never married, Brunelleschi

12. Saalman, *Filippo Brunelleschi: The Cupola of Santa Maria del Fiore*, 32–57.

13. Manetti, *Life of Brunelleschi*, 36–39.

instead focused on his studies in art, architecture, mathematics, and engineering. By 1418, he was known and respected for his knowledge in a wide range of fields; to use a modern phrase, Brunelleschi was a true "Renaissance man." His biographer also tells us that he had a keen sense of humor that at times could border on maliciousness. This proud, capable, and extraordinarily talented man was destined to be one of the leaders during the pivotal transition of Florence from the late Middle Ages to the Renaissance.

Growing up a short distance from the Florence cathedral, the young Brunelleschi had ample opportunity to see the plans develop and change, as well as to view the progress being made as construction proceeded; he witnessed firsthand the logistical support required for such a massive undertaking. He probably saw Neri's model on display in a completed area of the church. Moreover, he must have observed and been intrigued with the myriad machines used to transport and lift the large stones and timbers. Given his interest in architecture and building, it is likely that young Brunelleschi learned from his father about the cathedral design and the challenges it posed. According to Manetti, Brunelleschi's father was disappointed that his son did not choose to become a notary, but nonetheless supported his desire to pursue a career that would use his prodigious creative talents. In 1392, Brunelleschi was apprenticed to the goldsmith Benincasa Lotti, where he quickly excelled, learning the established techniques, but also experimenting and developing mechanical devices, including clocks. As a testimony to his abilities, Brunelleschi became a master goldsmith in just six years.

In the late 1390s, Brunelleschi went to Pistoia for work and to escape the plague in Florence, but returned to participate in the 1401 competition for the bronze doors of the Baptistery. After the commission was awarded to Lorenzo Ghiberti (see 3.1.2.3b), Brunelleschi and his friend Donatello traveled to Rome to study Roman architecture and building methods. Brunelleschi lived there for over a decade, often traveling to Florence for visits. At the time, Rome was a squalid city with a population a fraction of the size that it had been during the Roman Empire. While in Rome, Brunelleschi and Donatello made careful measurements of many classical buildings, including the Pantheon and the Domus Aurea. In several cases, they had to excavate ruins to be able to take measurements. They were among the first to systematically measure and study Roman antiquities. By mapping the strict mathematical relationships of the three architectural orders—Ionic, Doric, and Corinthian—the Romans had copied from the Greeks, Brunelleschi gained a deep understanding of the role of mathematical proportions in yielding aesthetic appearances, a principle he would later apply in his use of hemispheres and squares/rectangles in his design of Florentine buildings, including San Lorenzo, Santo Spirito, and the Ospedale degli Innocenti.

On a return visit to Florence in 1404, Brunelleschi, along with Ghiberti and other accomplished masters of their specialties and respected Florentine citizens, were asked to serve on a committee of nineteen to advise the Opera on the ongoing construction of the cathedral; this is the group that intervened to correct Giovanni d'Ambrogio's modification to the eastern tribune. In 1417, Brunelleschi was again asked by the Opera to provide advice on the construction of the octagonal dome, for which he was paid 10 florins, a sizable sum. This act shows the respect in which Brunelleschi was held by the Opera, and by extension, by many Florentines. At that point, construction had progressed such that the 4.7-meter-thick drum had been installed, two (of three) tribunes were vaulted, and a crane for hoisting heavy loads was in place in the section that would become the presbytery. Despite this progress, the Opera still lacked definitive plans for construction of the dome itself.

In 1418, with the design for the dome finally settled, the Opera announced a competition for models that focused on the construction techniques needed to sustain such a massive load, using a pointed octagonal dome with no external buttresses. The Opera assumed that the dome would be made entirely of stone or brick held together by mortar, but expected the model to address the problem of the centering, the way to configure the temporary wooden structure that would hold the masonry intact while the mortar cured. The winner would receive 200 florins, to be paid at the discretion of the Opera, while the other competitors would be reimbursed for their costs of labor and materials. The competition attracted experienced Florentines, including Ghiberti, whose design for the Baptistery doors had been selected over Brunelleschi's, as well as masters from neighboring cities. Brunelleschi hired his friend Donatello and another sculptor, Nanni di Banco, to work with four masons and construct a large brick model according to his plans. Artist Stefano del Nero made a painting of the model, but sadly both the structure and the painting have been lost. The exact size of the model is unknown, but estimates of the inside diameter, based on the amount of masonry used, range from 3.4 to 3.7 meters, and we know that the model was tall enough to be entered by adults.[14] Brunelleschi's model showed how brickwork could be applied to construct the vault without the use of internal wooden centering, thereby avoiding the necessity of constructing a wooden device of such magnitude. There were undoubtedly skeptics who questioned the wisdom of extrapolating from the model to a full-size vault.

The models submitted by Brunelleschi and Ghiberti were both viewed favorably by the Opera, but rather than choosing a winner, they decided to schedule a second competition in 1420. Only Brunelleschi and Ghiberti were invited to participate, and each received 10 florins for their continuing efforts. Interestingly, the Opera never paid the 200-florin prize offered in the first competition, although years later Brunelleschi was awarded funds in the amount of 200 florins for certain innovative designs. For the second phase of the competition, Brunelleschi and Ghiberti were each instructed by a panel of four Officials of the Cupola to submit drawings and provide advice about the construction parameters associated with their respective models. Brunelleschi had brought innovative ideas to his model and was reluctant to share the details of his design and construction methods out of concern that others would take advantage of his plans. After a small committee appointed by the Opera reviewed the two models, they chose Brunelleschi's; however, lacking critical details, and much to Brunelleschi's displeasure, they appointed both him and his nemesis, Ghiberti, as superintendents for the dome. The committee also added Antonio di Battista, acting architect since 1418, as a third superintendent. This unusual threesome was to receive 3 florins each month for their work. In addition, the committee commissioned eight masons, each to assume responsibility for one portion of the octagon. In 1419 four members of the Wool Guild had been appointed to serve a term of six months to coordinate and oversee future

14. Prager and Scaglia, *Brunelleschi: Studies*, 30–31; Saalman, *Filippo Brunelleschi: The Cupola of Santa Maria del Fiore*, 61–69.

construction. With all of these appointments in place, the Opera was ready to proceed in building the much-anticipated dome. Brunelleschi's model was prominently placed in the Piazza del Duomo near the Campanile and protected by a fence. It remained in place until the early 1430s, when the Opera decided to dismantle it. Today the Opera di Santa Maria del Fiore Museum displays two wooden models, one of the dome situated on the drum with the attached chapels and the other of the lantern that some attribute to Brunelleschi, but this claim is not universally accepted.

In 1420, shortly after their appointment, the three superintendents (the title denoting a higher position than that of *capomaestro*) signed and submitted a detailed document outlining the overall construction plans for the dome. It was clearly written and scholars believe it was prepared primarily by Brunelleschi. The document contained twelve points (unnumbered in the original) for the construction; later amendments in 1422 and 1426 addressed important alterations.[15] The full document clearly addressed the parameters specified by the Opera and provided a solution to the central problem of constructing an octagonal pointed dome of specified curvature with an internal diameter of about 45 meters, a reasonable average of various dimensions reported. The plan further specified that the dome would rest on a drum 24.5 meters in height with a wall thickness of approximately 4.65 meters.[16] The octagonal drum was to sit atop four large piers rising 28 meters. Complicating logistical and building matters, construction of the drum had to begin well over 50 meters above ground. The document laid out a complex plan to construct the dome with an inner shell that supported the outer shell (see fig. 6.2), and it is clear from the myriad details in the plan that Brunelleschi was responsible for its creation. Even a summary of the scheme gives one an appreciation of the technical and engineering challenges presented by a project of this magnitude, and a sense of the extensive knowledge and the technical and creative skills needed to address those challenges.

b. Specifications for the Shells

Since the dome was to consist of an inner and outer shell, plans required complex calculations to maintain the correct curvature as the dome ascended. Unlike a spherical dome that would resemble a semicircle when viewed from the side, the dome of the Duomo was to be built as a pointed fifth (*quinto acuto*). In this technique the corners of the inner shell were to be vaulted in five sections. Although rather unfamiliar to us today, the pointed fifth was understood to mean that the radius of curvature of each of the eight segments of the dome was to be four-fifths of the interior base diameter of the dome. Thus, with a diameter of forty-five meters, the radius of curvature was dictated as thirty-six meters. Using this value to make a circle of this radius at each of the eight sections, their intersection would define the point of the dome. In order to maintain the correct curvature as specified by the Opera in 1367, one would take one-fifth of the diameter of the base of the dome, i.e., nine meters, and

15. Saalman, *Filippo Brunelleschi: The Cupola of Santa Maria del Fiore*, 70–77.

16. Fanelli and Fanelli, *Brunelleschi's Cupola*, 184.

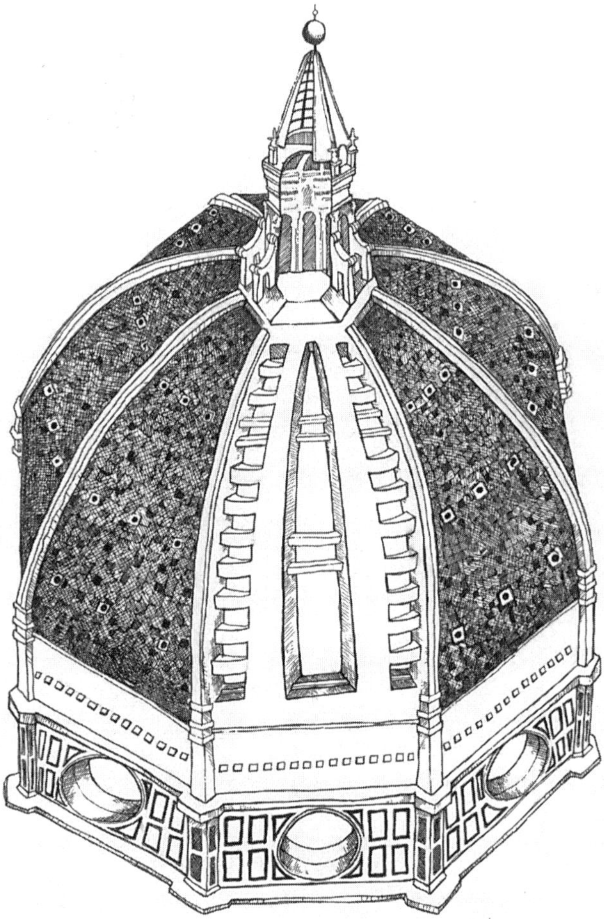

Figure 6.2: Sketch showing partial structural components of the dome, Santa Maria del Fiore (drawing by Justine E. Stevens [modified from illustration by George Rich], commissioned by the authors).

from each octagonal segment draw a circle with a radius of thirty-six meters. The mason in charge of each segment of the dome was to work with his own radius point within each octagonal segment.

The specifications for the pointed fifth were carefully calculated and resulted in an extremely accurate execution. Although modern architect Hans Samelson noted some apparent discrepancies between the accepted understanding of *quinto acuto* and the earlier interpretations and diagrams of the dome, remarkably the differences amounted to at most five millimeters (less than 0.2 inches), a value that is well within the accuracy of measurement for a building that size.[17] He reached his conclusions by analyzing publications from the eighteenth and nineteenth centuries by Nelli and Nardini, using figures that were reproduced by Howard Saalman.[18] It should be noted that because the structure of the dome has two shells, the inner and outer shells have different base diameters, and consequently the radii of curvature of the two shells must be slightly different.

Proceeding from the base of the dome to its top, the walls of the dome were to decrease in thickness. The thickness of the inner shell was specified to be 3.75 *braccia* (2.2 meters) at the base and 2.5 *braccia* (1.5 meters) at the top. The outer shell was to protect the inner shell from the elements and was designed to decrease from a thickness of 1.25 *braccia* (0.7 meters) at the base to 0.66 *braccia* (0.4 meters) at the top. The two shells comprising the dome would be separated by a narrow open stairway.

17. Samelson, "Letter to the Editor."

18. Saalman, *Filippo Brunelleschi: The Cupola of Santa Maria del Fiore*, plates 7 and 8.

Sixteen oculi of slightly less than fifty-eight centimeters in diameter were to be evenly spaced around the inner shell, two per segment, enabling one to observe a part of the interior of the cathedral from these vantage points. The outer shell was to contain seventy-two oculi, each also about fifty-eight centimeters in diameter, to allow ambient light into the narrow space. The lower part of the walkway between the shells was to provide access to an internal open gallery that was positioned to circle the drum at the base of the dome. The plan called for three walkways above the first section, with the uppermost located inside the oculus ring and the other two about 10.5 meters and 21.0 meters above the beginning of the curvature of the dome. With stairs and walkways providing access to the top of the dome and areas between the two shells, one can today view many aspects of the internal support, as well as the interior of the dome itself.

c. Structural Specifications

According to the plan, a total of twenty-four ribs or spurs (*sproni*) were to join the two shells and provide stability and rigidity. Of these, eight major ribs were to emanate upward from each corner of the octagonal base, and sixteen minor ribs were to be evenly distributed between the eight portions of the octagon, yielding two minor ribs per octagonal segment (see fig. 6.2). The major ribs were to be 7 *braccia* (4.1 meters) wide at the base, and the minor ribs were to be 4 *braccia* (2.3 meters) wide at the base. Every 12 *braccia* (7 meters) of elevation of the ribs, there would be a small barrel vault for passage; below this passage would be an oak beam chain affixing one rib to the next, with each oak beam having an iron chain above it. The ribs were to be built of sandstone (red and gray) up to a height of 24 *braccia* (14 meters), with the higher portions made of a brick or porous stone that was lighter in weight than sandstone.

Connecting these vertical ribs would be three massive sets of concentric or parallel stone rings (*cerchi*). The first set of sandstone rings was to rest on large sandstone blocks 3.2 meters in height that were placed above the drum such that both shells would be resting on the stones as they began their upward curvature. The second pair of stone rings was to be positioned at 10.5 meters above the beginning of the first set of rings, and the third pair was to be at 21 meters. (In 1426 a fourth pair of sandstone rings was added just under the base of the lantern.) The first, second, and third sets of parallel sandstone rings (from bottom to top) were to be 2, 1.33, and 1 *braccia* high (1.2, 0.8, and 0.6 meters, respectively) and 4.5 *braccia* (2.6 meters) in length. As construction proceeded, the cross-sections were modified to 0.76 *braccia* (0.44 meters) for all rings, presumably to reduce weight. The stone blocks in each ring were held together by lead-coated iron clamps and topped by a large iron chain. These were designed to aid in overcoming the hoop stress, i.e., the force pushing outward due to the heavy weight of the dome. The stone used was a particularly durable and hard sandstone, *macigno* (quartz-containing), of reasonably high tensile strength.

Shorter transverse sandstone beams of the same square area and approximately 1.4 to 2 *braccia* (0.8 to 1.2 meters) in length, also joined by iron clamps, were to be placed parallel to one another and under each set of rings, perpendicular to the sandstone blocks comprising the rings. In order to obtain a close fit one to the other, the transverse beams were designed to dovetail with the rings. The lowest transverse beams are visible today protruding from the dome (twelve in each octagon).

Four wooden chains were also planned, but only one was installed. This massive wooden chain contained twenty-four logs, each some 6.1 meters in length with a cross-sectional size of about 0.35 x 0.35 meters, connected by oak plates and iron bolts, with iron straps wrapped around the joints. The chain would be installed at a location some 13 *braccia* (7.6 meters) above the base of the dome.

Other planned components of the drum and dome included walkways and platforms. An outer walk for the workers was to be built above the lower eight round windows of the drum and two interior walkways, one over the other, with the upper one open. Brunelleschi's construction proposal further stated that the dome would be built to a height of 30 *braccia* (17.4 meters) using platforms for construction but without centering, truly an ambitious approach for the time. At higher elevations, Brunelleschi (or more probably the three superintendents) did not stipulate the method to use but rather left it up to the builders to decide when the time arose. Additional items specified in the document included eight permanent marble ridges on the dome to serve as gutters to transport rainwater to tan sandstone drain spouts.

Rather than viewing the two shells and supporting ribs separately, an unusual interpretation is to consider "the inner and outer shells as being one and the same solid structure, which is hollowed out between the ribs."[19] To implement this unique design concept for the dome, Brunelleschi used a variety of features and approaches, but, as mentioned earlier, he was always concerned about having his ideas copied, so did not leave detailed specifications. Much of what is known has been drawn from the available records and from detailed examination of the structure itself. Yet, since major portions of the structural elements of the dome are not visible from either the interior or the exterior, there is considerable room for speculation. Lastly, construction of a model,[20] as well as statics and structural analysis,[21] has provided valuable information.

d. Building Methods

From the time construction began in 1420 until the dome was completed and consecrated in 1436, there were numerous small alterations to the design and frequent adjustments to Brunelleschi's planned construction methods. One major change that was required was a reduction in the weight of the dome. Although it is impossible to know exactly what the weight would have been had the original specifications been followed, the final weights of the dome and lantern are estimated to be about 29,000 tons and 800 tons, respectively, yielding a total weight of just under 30,000 tons.[22] The need to reduce the overall weight led

19. Jones, Sereni, and Ricci, "Building Brunelleschi's Dome," 45.

20. Ibid., 39–61.

21. Fanelli and Fanelli, *Brunelleschi's Cupola*, 203–65.

22. Ibid., 184.

to a decision to begin using brickwork instead of heavier stone at a height of 12 *braccia* (7 meters) instead of 24 *braccia* (14 meters) as originally planned. It was also decided that the thickness of the minor ribs should be reduced, and to use chestnut instead of oak for the wooden chains. The thickness of the ribs was reduced to 3 *braccia* (1.7 meters instead of 2.4 meters as planned) at the base and 0.7 *braccia* (0.4 meters) at the top; the major ribs were roughly twice as thick. Another issue that arose was that the length of the octagonal sides of the drum, and later the dome, were not uniform, varying in length from 16.6 meters to 17.2 meters.[23] This construction error resulted in the inner octagonal angles deviating somewhat from the ideal 135°, thus requiring that adjustments be made to ensure that the shells would converge at the top. These are just a few of the many challenges that faced Brunelleschi as construction proceeded.

In the absence of the standard internal centering and ground-supported scaffolding, Brunelleschi cleverly devised a platform capable of holding the workers and their supplies and equipment. To support this platform, six square holes (called putlog holes), each 1 *braccia* (0.6 meters) square, were evenly spaced in the inner shell of each octagonal area at the lower part of the dome, where it rested on the drum. A large beam of chestnut was placed into each of the forty-eight putlog holes, projecting into the space between the two shells and protruding toward the center of the dome area to support a stable platform. This circular platform served the needs of the workmen and was also an important reference location during the construction process. Scaffolding was placed on the platform as construction proceeded, and the platform was probably given additional support by placing angled struts below.[24]

All evidence points to the eight segments' being constructed concurrently. As the dome rose, structural integrity was obtained through the use of different size bricks and a changing pattern of the herringbone brickwork as the elevation increased. In order to achieve the curvature required for the bricks and mortar beds as the building proceeded upward, workers used a technique denoted as slack line (*corda blanda*), which resulted in a concave arrangement. The records show that three wires (or ropes) were used to control the curvature in order to attain a pointed arch. Brunelleschi's exact methods are not known, although plausible interpretations have been offered.

Brunelleschi's innovative herringbone brickwork (*spinapesca*) involved laying bricks in both horizontal and vertical positions (see fig. 6.3). Bricks of varying sizes were used, but the two most common sizes were 17 x 34 x 5 centimeters and 22 x 44 x 5 centimeters.[25] While most scholars agree that Brunelleschi applied this method to provide localized support to adhere each new section of brick to those already in place and to prevent newly laid bricks from slipping as the mortar dried, another explanation has been advanced based on model building. This provocative hypothesis posits that workers used a radial method that dictated the positioning of each brick following a single principle, resulting in the herringbone

23. Ibid., 177–78.

24. Jones, Sereni, and Ricci, "Building Brunelleschi's Dome," 42.

25. Fanelli and Fanelli, *Brunelleschi's Cupola*, 187.

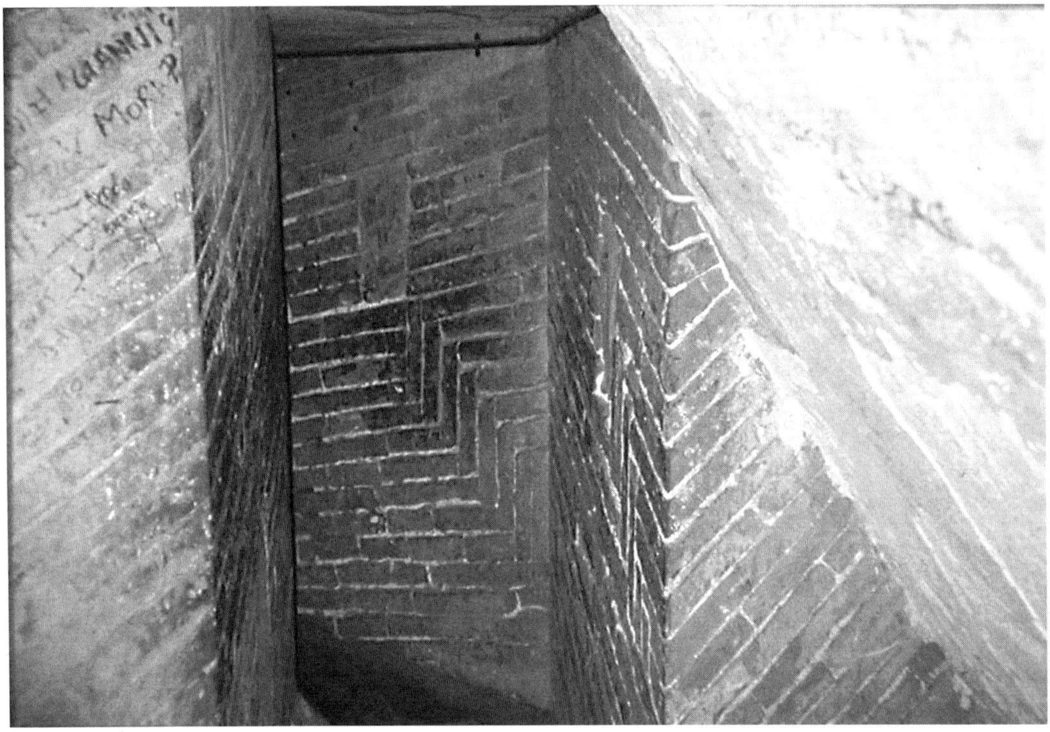

Figure 6.3: Herringbone brickwork designed by Brunelleschi for the dome of Santa Maria del Fiore, Florence (photograph by authors).

arrangement. The new interpretation is based on construction using fixed and mobile cords to determine location and curvature, thus demonstrating that segments within octagons could be built independent of others, unlike that previously surmised.[26] Ample discussion of this possibility will surely occur as this new suggestion is critically examined by others. Rather than using the more common semicircular tiles for the roof, Brunelleschi designed tiles measuring about 37 x 50 centimeters with lateral raised ribs. Some 30,000 were secured using lime mortar and nails, overlapping at the top and bottom.

e. The Timeframe of Construction

Although planning had progressed in various stages, construction by Brunelleschi and his team began in earnest in 1420 and continued without major interruption until the lantern was completed in 1471, twenty-five years after Brunelleschi's death. Late fourteenth-century and early fifteenth-century machines remained on site throughout the construction process, and building materials were continually being made available with supplies arriving regularly. Brunelleschi and Ghiberti celebrated the official beginning of the dome construction with breakfast on August 7, 1420, and received payment in October for the quarter ending

26. Jones, Sereni, and Ricci, "Building Brunelleschi's Dome," 50–54.

on September 30. Before construction started, Brunelleschi had already begun designing a hoist that was completed in early 1421 and put to use within the year. He also designed a new crane, with technological improvements, that was in place and operational in the summer of 1423. Building progressed steadily, with numerous meetings to monitor progress; some sessions, particularly those in 1422 and 1425, led to amendments to and alterations of the original plan. Construction was temporarily halted during portions of 1429 and 1430 due to the war with Lucca, when funding was reduced for a short period. The last section of the dome was constructed in the 1430s, and on August 30, 1436, it was consecrated by Bishop Benozo Federighi of Fiesole.[27] At that point the Opera turned its attention to the final building phase—construction of the lantern. Throughout the building of the dome, Brunelleschi continued to use innovative approaches to solve the myriad problems that arose, and there were many.

During the early years of design, adjustments, and construction of the dome, Brunelleschi was annoyed that Ghiberti was receiving the same wage as he and enjoying the prestige of the Florentines; in Brunelleschi's opinion, Ghiberti was contributing little to either the planning or the progress. As the large beams (6.1 meters long x 0.3 meters wide) began arriving early in the project, Brunelleschi is believed to have feigned an illness to force Ghiberti to assume responsibility for joining the beams together. Following an apparent miraculous recovery, Brunelleschi returned to the construction site and convincingly discredited Ghiberti's approach. Shortly after this, Brunelleschi's pay was increased to 100 florins per year. Ghiberti, on the other hand, was temporarily suspended from the project for six months in 1425, but was later reinstated at his earlier salary of 36 florin per year. By that time, it was clear to all that Brunelleschi was in charge of implementing his design for the dome.

6.3.1.3 The Lantern Completes the Construction

When Brunelleschi's workers placed the last brick in the dome, only one more section was needed to complete the project that had been started over a century before. A lantern would be the crowning jewel on the majestic dome. While the lantern had the practical purpose of admitting light to the interior, the desire was to create a structure both aesthetically superior and technically sound. Although the dome had been in the hands of Brunelleschi for sixteen years, the Opera decided to announce a competition for design of the lantern. As much as this must have been an insult to Brunelleschi, in 1436 he submitted a model for the competition. Brunelleschi drew plans and presented them to Antonio Manetti Ciaccheri, the carpenter who had assisted him in various phases of the dome construction. It was Antonio's task to build a model from those sketches. Brunelleschi's well-known fear of having his designs copied and his ideas stolen appears to have been well founded in this instance. One of the competitors submitting a model was none other than Antonio who, along with the model he built for Brunelleschi, submitted one of his own that was very similar in detail to Brunelleschi's. Lorenzo Ghiberti also entered

27. Saalman, *Filippo Brunelleschi: The Cupola of Santa Maria del Fiore*, 112–34.

a model for consideration, as did two others: Bruno di Ser Lapo Mazzei and someone identified only as Domenico.

Testifying to the importance of both art and technology in satisfying the criteria for selection, the committee included painters, theology masters, masons, goldsmiths, a mathematician, architects, and prominent citizens, including Cosimo de' Medici. Brunelleschi was awarded the commission based on his design, which was deemed strongest, lightest, most weatherproof, and most aesthetically pleasing. The lantern project was under his direction for the last ten years of his life. After Brunelleschi's death in 1446, work on the lantern was continued first by Michelozzo, then by Antonio Manetti Ciaccheri, and in 1460 by Bernardo Rossellino. There were minor modifications made to the original design, but the exact numbers or types of variations are not known. This is due, in part, to the fact that Brunelleschi, according to biographer Antonio Manetti, created models that focused on the central elements such as walls, showing the location of the physical features without addition of artistic components.[28] Since Brunelleschi had expected to personally oversee every detail himself, the lack of specificity in his models did not seem problematic. The project, however, was not completed until 1471, having been slowed considerably by the difficulty of supplying materials and working at the top of the dome; there was also a problem locating suitable marble.

A shared sentiment among art historians is that the lantern epitomizes the "close relationship between architecture and the applied arts."[29] Sitting on the marble keystone of the vault (*serraglio*) and supported by a sandstone chain, the completed lantern is twenty meters high with eight radial buttresses that are aligned with the ribs of the dome. The sides of the buttresses have hollowed niches, with half shells depicted in the arched top of each niche. Curling volutes top the buttresses and rest against half columns that frame the eight elongated arched windows. The Corinthian half columns, with capitals depicting acanthus leaves, give the illusion of supporting a base upon which the octagonal entablature rests. Above the columns, the corners of the entablature have pilasters capped by cornices and turned pinnacles bearing round balls; the columns and pinnacles are also in alignment with the dome ribs. That continuation of the ribs serves to lift the eyes of the observer upward toward the cross at the zenith. Filling the spaces between the cornice-topped pilasters are arched niches aligned with the windows.

The spire rises over the pilasters and is crowned by a gilded cross resting on a brass ball (*palla*) that is affixed by a copper bracket to a knob on the spire. The original ball, measuring about 2.4 meters in diameter and designed by Andrea del Verrocchio, was cast in eight pieces, soldered with silver, and gilded. Verrocchio's ball was damaged by lightning in 1601 and replaced by the one presently resting atop the spire. Hidden from external view is a stairway located within one of the buttresses, allowing access to the interior of the spire. Inside the spire, there are ladders leading to the ball, which was fitted with a small window covered by a flap. The aesthetic quality of the finished lantern was deemed superior, and the design was often emulated.

28. Manetti, *Life of Brunelleschi*, 116.

29. Di Cagno, *Cathedral, the Baptistery and the Campanile*, 72.

6.3.1.4 Materials for the Dome and the Lantern

Such a massive undertaking as construction of the dome required considerable funds and logistical support. The Opera was well organized to provide such support, particularly with the funds available through the guilds. The effort employed many artisans and laborers from Florence and surrounding regions to ensure availability and delivery of the materials required for construction. Fortunately, the greater Florentine territory was blessed with suitable quarries and forests, and much of the required iron ore could be imported from the nearby island of Elba.

The sandstone (*pietra forte*, or "strong stone") used by Brunelleschi and other Florentine Renaissance artists, and before them the Etruscans, was noted for its high tensile strength and durability. Although sandstone was obtained from sites near Florence, e.g., the Trassinaia quarry, the exact location of the material used by Brunelleschi was unknown until recently. Using records from the Opera di Santa Maria del Fiore, maps, and a geological and stratigraphic analysis, scholars have concluded that the sandstone used in construction of the dome was obtained from a nearby quarry in the Fiesole hills on a high ridge (400 meters) that formed part of the Oligocene-era Macigno Formation of central Italy.[30] The quartz- and mica-containing sandstone abundant in the hills above Florence was valued for its strength and beauty.

Both the dome and the lantern also used marble extensively; as noted above (6.2.1), the Opera called for marble of three colors—white, black-green, and red—that were quarried from a variety of sites. The finest white marble was from Carrara, although quarries in Campiglia and Gallena were also considered and minimally used by the Opera. Black-green and red marble were obtained from several quarries in regions near Florence. Timbering of the forests in the Casentino and Campigna was placed under the control of the Opera and those forests supplied the major needs for the types of timber required: chestnut, oak, beech, pine, elm, linden, olive, and walnut. The majority of rope was purchased from Pisa, although Genoa and Bologna supplied a small amount. Most of the iron was from Elba, although it was also mined in the hills of the Campiglia. Bricks were obtained from several kilns in and near Florence, as was mortar, made from sand and limestone. Brunelleschi was personally involved in overseeing production of the brick-making and mortar. In his biography of Brunelleschi, Saalman provides a much more complete description of the materials used, as well as their sites of origin.[31]

6.3.1.5 Machines Used in the Construction of the Dome

Ratchets, treadmills, winches, pulleys, gears, cranes, and the like used by the early Greeks and Romans were still the basic machines in use at the dawn of the Renaissance. For Florentines in the 1300s, the field of engineering was essentially limited to adapting these tools

30. Coli et al., "'Pietra Serena' Stones."

31. Saalman, *Filippo Brunelleschi: The Cupola of Santa Maria del Fiore*, 195–205.

to a specific purpose. When construction of the dome began in 1420, the combination of a large number of heavy stones and the extreme height to which materials had to be raised presented an entirely new situation to Brunelleschi and his workers. It was evident that the traditional method of lifting heavy materials—a winch controlled by a large horizontal wheel turned by men or animals—was completely inadequate for such an undertaking.

a. The Great Hoist

To address the need for new technology, Brunelleschi designed a hoist in the spring of 1420 (fig. 6.4), and work began on its construction as soon as the materials arrived in Florence. These included oak, elm, beech, and chestnut for the poles, and extremely long and heavy rope from Pisa. By the summer of the following year, the hoist had been completed, a platform had been built as a staging area, and the new device was ready for use. As with Brunelleschi's other designs, his written plans do not remain, but the interest generated by the hoist ensured that it was studied and drawn by several Renaissance engineers, including Buonaccorso Ghiberti, Giuliano da Sangallo, Francesco di Giorgio, Il Taccola, and Leonardo da Vinci. In addition, the Opera kept extensive records detailing the various artisans involved in building the components, the materials used, the specifications, and the amounts paid by the Opera for the work. From these facts and the drawings, we are able to discern with a degree of certainty the particulars of the Brunelleschi hoist.

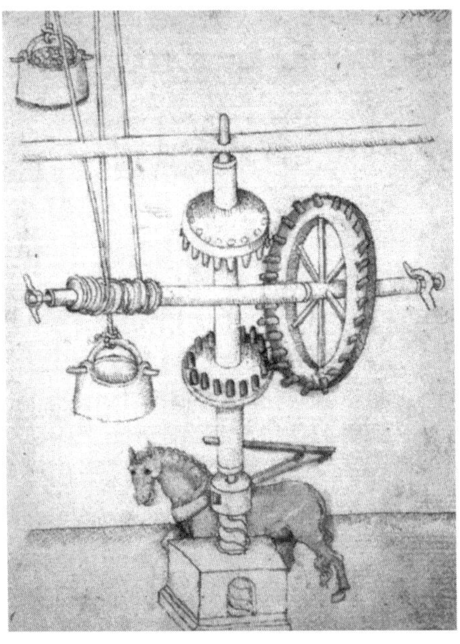

Figure 6.4: Il Taccola, sketch of great hoist (1420) designed by Brunelleschi (retrieved from Wikimedia Commons, public domain).

In addition to being exceptionally large and powerful, the hoist was unique in its complexity. The basic machine was a wooden frame with a number of vertical and horizontal poles turned by cogged wheels. Saalman calculated that the height of the hoist, including the axle shaft, was 12.5 *braccia* (7.3 meters).[32] One or two animals (oxen or horses) were yoked to a tiller that was attached with two pinions to the central vertical shaft. The central shaft, turned by the animal(s), was fitted with two horizontal cogged wheels that were separated by a larger vertical cogged wheel with twenty-four teeth mounted on a pole horizontal to the main shaft. The bottom of the central shaft ended as a large screw that could be turned to raise or lower the shaft and its attached cog wheels. When the horizontal cogged wheels were raised or lowered a few inches, one or the other of them would engage the large vertical wheel on the

32. Ibid., 156.

perpendicular pole. One of the horizontal wheels turned the pole so that it wound the rope and lifted the hoist; the other turned the pole in the opposite direction, unwinding the rope and lowering the hoist. In simple terms, when the animal(s) rotated the central shaft, a series of cogged wheels operated a hoist that could either lift or lower the load, depending on how the screw was set. This reversible gear system was the hallmark of Brunelleschi's engineering ingenuity, enabling a change of direction without having to remove animals from their yokes and physically turn them around.

With the gears set into motion, the workers could move materials by using one of three rope drums. Each drum was attached to a horizontal shaft of a different diameter, enabling the rope to be turned at different rates of speed, the largest (*subbio grosso*) capable of raising a load more rapidly than the one of smaller diameter. Because the rope was about 6.4 centimeters thick, it had to be frequently wetted to prevent combustion caused by the friction of the rope passing through the pulley wheels. Loads were attached to the hoist rope with special hangers designed by Brunelleschi that used iron bars inserted in a hole in the load created for that purpose.[33]

The hoist was extremely successful and although repairs were necessary over the twelve-year period of use, it performed exceedingly well. The hoist lifted many thousands of tons of building materials, as well as raising and lowering workers.[34]

b. The Great Crane

When construction of the dome began, workers used an existing crane to position the stones; however, as the height of the dome increased, it became obvious that the relatively small crane was not up to the task. Again, Brunelleschi's ingenuity and technological expertise were tested in his design for the great crane. The Opera accepted his design, assembly began in April of 1423, and it was completed that summer. Records are confusing as to the exact features and details of the crane and whether it was used to the exclusion of the other, smaller crane. The longer working arm of the new crane afforded more flexibility in positioning the stones. Brunelleschi's crane worked in tandem with the hoist and workers received the raised loads, moving them into place on the load positioner. Whether Brunelleschi invented or improved the design of the load positioner is unclear, but the precision with which the stones could be manipulated horizontally and vertically is remarkable. It was particularly effective in the positioning of the wooden and stone chains that reinforced the structural elements of the dome.

As greater heights were reached in the dome, the crane, because of its long positioning arm, became less useful. As the dome curved inward toward its peak, the horizontal distances became shorter, but the distance from the loading platform increased. Thus, in 1432 a new crane, probably designed by Brunelleschi and employing the technology of the great crane, was fabricated using proportions more suitable to the new parameters.

33. King, *Brunelleschi's Dome*, 60–62.

34. Prager and Scaglia, *Brunelleschi: Studies*, 89.

6.3.1.6 Additional Machines Used in the Construction of the Lantern

After the design of the lantern had been accepted, it again became obvious that new technology was required both to raise the tons of required marble to the top of the dome and to execute the intricate plan intended to create an artistic lantern befitting the massive structure. A hoist was required that was sufficiently small to be operated from a platform at the top of the dome and at the same time capable of lifting two-ton marble blocks from that loading platform. Again, the Opera held a competition for the submission of a model suitable for the task, and again Brunelleschi won. Construction of his smaller, innovative hoist was begun by Antonio Manetti Ciaccheri in 1442 and completed the following year. A sketch made by Buonaccorso Ghiberti survives that gives one a fairly accurate representation of the device.[35]

Although not as complex as the great hoist, the lantern hoist was equally as inventive, meeting the unique specifications required. It was activated by pushing two poles inserted into the base gear that engaged, by means of its teeth, a vertical gear. That gear, in turn, drove the pulleys and wound the rope. The hoist operated with a system of double pulleys, enabling a counterweight to be used to balance the load. Brunelleschi also designed a braking system that allowed the workers to stop the progress, when necessary, once the pulleys had been activated. The braking system was operated by a ratchet wheel that could be locked to hold the counterweight steady in an above-ground position, thus allowing the marble to be attached from below. After the load was secure, the brake could be released, enabling the weight and counterweight to ascend and descend simultaneously in a controlled manner.

Before any work could commence, a platform had to be built at the top of the dome to provide a space for assembling the crane and for working on the construction. That platform was completed in the summer of 1447 and by December the crane that would be used to position the stone was assembled. Again, although Antonio Manetti Ciaccheri directed construction of the crane, it was built using Brunelleschi's ingenious design.

In order to construct the lantern, workers used the great hoist to lift the marble to the lantern platform, then used the lantern hoist to raise the blocks to the working level and the crane to position them into place. Other smaller cranes and machines were also in use for hauling mortar and needed materials. As construction of the lantern progressed, the operation became more intricate. Clearly, the technology involved was advanced and cleverly designed. Although there was potential for adapting Brunelleschi's innovative technologies to future endeavors, his reluctance to make public his designs thwarted subsequent use. At the time of his death on April 15, 1446, shortly after the first stone of the lantern was consecrated by Cardinal Antoninus, Brunelleschi had become known as one of the first in a line of technological geniuses who advanced fields that would become known as mechanical and civil engineering.

The construction of the dome and lantern, with all of the complex engineering and technology required for completion, represents a splendid example of art driving technology and, in turn, technology driving art.

35. Saalman, *Filippo Brunelleschi: The Cupola of Santa Maria del Fiore*, plate 123.

6.3.1.7 Art and the Cupola

The impact of the dome of Santa Maria del Fiore, when viewed from Florentine streets or the surrounding hills, never fails to amaze. Shifting rays of the sun, varying angles, and distances all create an ever-changing yet predictably permanent statement of the glory of the Renaissance. The art in the interior of the dome only adds to the drama of the Florence Cathedral.

Structurally, the dome's interior space is segmented, with two walkways giving access to the area and to steps leading upward between the two shells. Sixteen round openings about one *braccio* (58 cm) in diameter are located along the centerline of each of the levels. These were created to provide places for supports to be used by those who would later decorate the dome. From these openings it is possible to see the interior of the Duomo from the internal walkways.

The vaulted space lights the lower portion of the Duomo with its eight round windows and central lantern. Because of the specialized skills required to create stained glass windows, the Opera identified glass craftsmen for the project even before they selected artists, a common practice during this period. Glassmaker Bernardo di Francesco was given the commission to do the work on five of the dome's eight roundels, Angelo Lippi was commissioned to do one alone and another in collaboration with Domenico di Pietro da Pisa, and the craftsman who worked on the eighth is not known.

Once the craftsmen had been selected, the Opera turned to the artists. In 1433, the Opera first hired Ghiberti, the designer of windows in other sections of the cathedral, to prepare a cartoon for the first round window; they also asked Donatello to submit a design. Donatello's cartoon was ultimately the one selected. He, along with craftsmen Domenico di Pietro and Angelo Lippi, executed *Coronation of the Virgin* from 1434 to 1437, with the two single figures being characterized by their simplicity and nobility. Possessing but a few dominant colors bisected by the lead framework, the window projects a commanding presence. Although Donatello's cartoon was to set the style for all eight of the windows, this was the only one he actually designed.

Paolo Uccello is credited with three of the eight windows. *Nativity* (1434–37), crafted by Angelo Lippi, depicts Mary, Joseph, the Christ child, and the stable scene figures without regard to any realism in terms of size or perspective, but with a brilliance of color resulting in a forceful presentation. In *Resurrection* (1443–44), according to Giuseppe Marchini, "the extreme foreshortening lends a significant realism to the rigidity of the body and the sarcophagus."[36] Both *Resurrection* and *Annunciation* were executed by Bernardo di Francesco. The *Annunciation* scene was removed in 1828 and was replaced by opaque glass.

Ghiberti laid the plans for another three windows: *Ascension* (1443–44), *Sermon in the Garden* (1443–44), and *Presentation in the Temple* (1443–45), all of which were crafted by glassmaker Bernardo di Francesco. The spatial relationships between the figures, the trees, and city portrayed in the background in *Sermon in the Garden* are not calculated perspectiv-

36. Marchini, *Italian Stained Glass Windows*, 51.

ally, and in regard to the brilliance of color, all elements are equally bold. In contrast, in the *Presentation in the Temple*, the elements coordinate in a more spatially symmetrical arrangement.

Andrea del Castagno was the fourth artist involved in the windows of the dome. His *Deposition*, completed in 1444 (the craftsman is not known), is devoid of background elements, rather focusing on Christ and the seven figures in their grief, each clearly accentuated by their vibrant colors.[37]

Frescoes (see 2.2.1.1) adorn the interior surface of the dome, which Brunelleschi originally left white, anticipating the installation of mosaics similar to those in the Baptistery. The decision to install mosaics had been made as early as 1426, with Francesco di Domenico Livi da Gambassi being given the commission to make the colored glass for both the roundels and the mosaics. In 1436 the interior of the dome was plastered in such a way as to create a bed for the mosaics. Nothing, however, was done in terms of finishing the inner dome. At one point, Lorenzo the Magnificent urged the installation of mosaics, but for approximately the next century, the white surface remained unadorned. It fell to Cosimo I de' Medici, Grand Duke of Tuscany (1519–74), to implement the project. Having made the decision to fresco the dome walls rather than install mosaics, Cosimo awarded the commission for the planning stage to Don Vincenzo Borghini, a Florentine architect favored by Cosimo I. Like in the Baptistery, the chosen theme was *Last* [or *Universal*] *Judgment*, but was executed in a version reflective of the later era (fig. 6.5). From 1571 until his death in 1574, the noted artist, architect, and biographer Giorgio Vasari frescoed a portion with the help of his assistants. Although he was not a Florentine and was working in England at the time, Federico Zuccari agreed to complete the project after Vasari's death. Vasari had designed the frescoes, including clever portrayals of objects that disguised the holes left by Brunelleschi for the scaffolding, and for the two-year period of 1578 to 1579, Zuccari, with an associate named Passignano and other assistants, painted the remaining sections. While remaining faithful to the overall theme, Zuccari did not use all of Vasari's illustrations, many of which are now housed in the Louvre in Paris. One example of Zuccari's alterations was his introduction of portraits of contemporary Florentines in the unfinished scenes, including Cosimo I, Francesco de' Medici, and a self-portrait that he signed and dated.[38]

The completed conceptualization is arranged in five levels, with the highest level, executed by Vasari, painted to give the impression of niches containing *The Twenty-Four Elders* and the *Apocalypse.* Angels, blue seraphims, and lesser ranked yellow-ochre cherubims occupy the next level as does, on the east side, the fresco of *Christ in Judgment* with the Madonna, Saint John, Adam, Eve, and other saints. Figures of the blessed are also depicted on this level. Beneath the image of Christ are *The Virtues*: Faith, Hope, and Charity. Still lower, scenes show both the passing and stopping of time with a *Winged Elder with an Hourglass* and *The Seasons Asleep.* Depictions of hell were begun by Vasari and completed by Zuccari, with the latter's illustrations including skeletal and nearly skeletal human figures. For much of the dome, Zuccari used

37. Di Cagno, *Cathedral, the Baptistery, and the Campanile*, 97–99.

38. Ibid., 94–96.

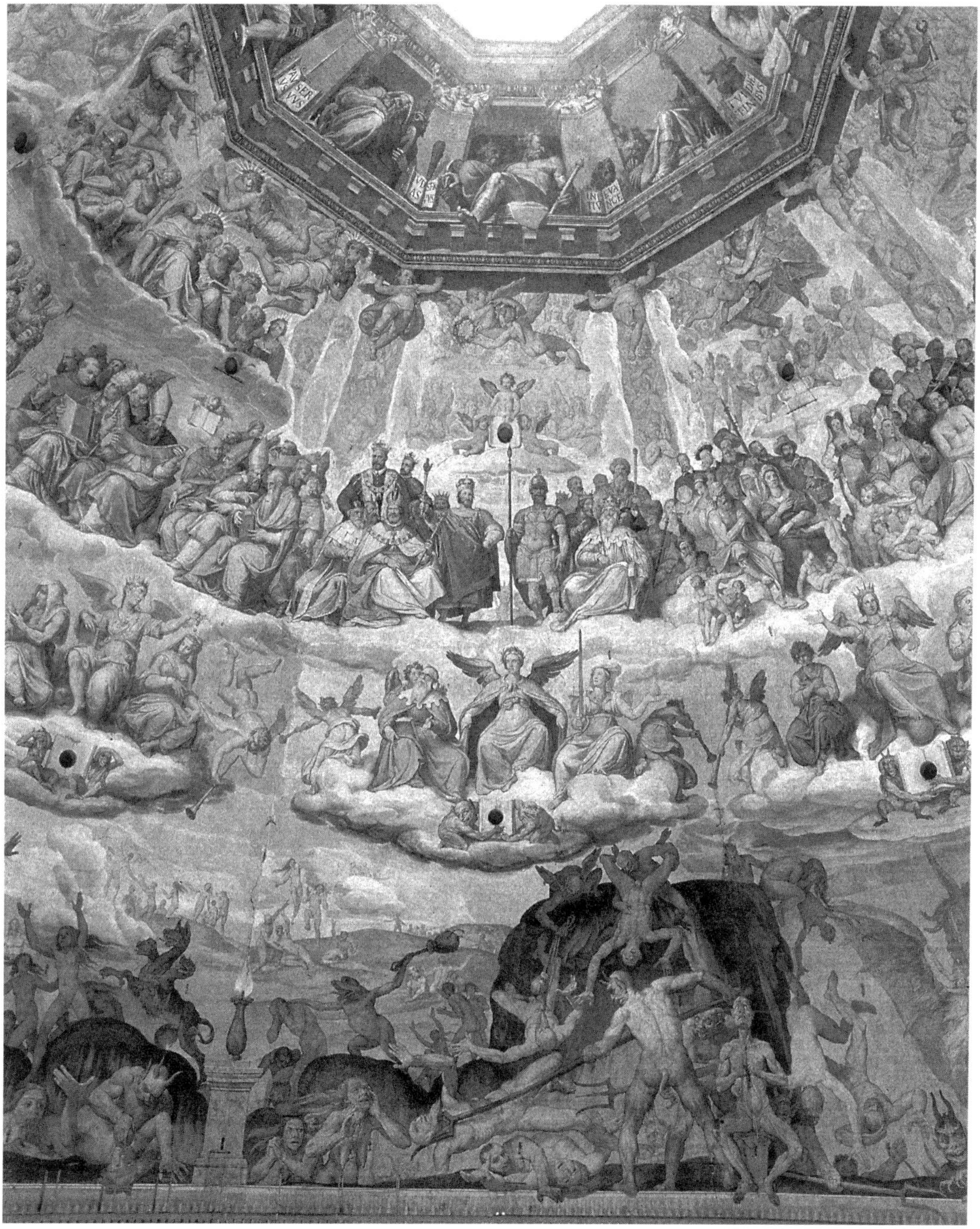

Figure 6.5: Giorgio Vasari and Federico Zuccari, *Last Judgment* (1571–79), fresco, interior of the dome of Santa Maria del Fiore, Florence (Nicolò Orsi Battaglini/Alinari Archives, Florence, photograph by Nicolò Orsi Battaglini, reproduced with the permission of Ministero per i beni e le Attivita Culturali [OBN-F-000043–0000]).

the technique of painting on dry plaster (*fresco secco*). Although this technique was not nearly as durable as *fresco,* its location in the dome prevented the deterioration that might otherwise be expected. While Vasari used regular fresco painting on the majority of his portion of the dome, he did, however, also utilize *fresco secco* for the area directly beneath the lantern. That section, subject to light and dampness, has unfortunately suffered the most damage.

6.3.1.8 Intersection of Art and Contemporary Science

Beginning in the 1990s, researchers employed modern analytical techniques to examine the frescoes, using Mössbauer spectroscopy to determine the nature of the pigments in the frescoes. Results of one study identified the pigments selected by Vasari and Zuccari for the interior of the dome. Red earth (red ochre, hematite) was detected, although the presence of a small amount of titanium, substituting for iron in the crystal lattice of the pigment, was found in the samples; the green earths, glauconite and celadonite, were also present. Importantly, the study also clarified the potential role of organic components in the pigments that led to deterioration of the frescoes.[39] Another investigation took advantage of fiber optics reflectance spectroscopy in the visible and near infrared region of the spectrum to determine the blue pigment in two samples taken from Zuccari's frescoes, identifying smalt, a widely used cobalt-containing pigment for blue.[40] In such a large space as the interior of the dome, the use of a less costly blue pigment, rather than the more brilliant lapis lazuli, is not surprising. Lastly, direct pyrolysis in a mass spectrometer enabled determination of several amino acids (alanine, aspartic acid, glutamic acid, glycine, hydroxyproline, leucine, phenylalanine, and proline) in the frescoes from Vasari and Zuccari. What was found, although not surprising, was the presence of animal glue, differentiated from other protein-containing binders such as egg and milk by the presence of hydroxyproline, an integral component of collagen and a major protein in animal skin.[41]

For all their beauty and the careful craftsmanship involved in creating the frescoes and stained glass images in the interior space, the height of the dome and the difficulty of viewing the frescoes from any suitable angle is problematic. It is only from the two walkways in the dome that the full impact of the commanding space is felt. Just as noted when looking at the design, the construction techniques and machines, and the materials used in other aspects of the dome, once again we see the science of technology and art working in concert to create the memorable symbol of Florence.

6.3.2 Renovation of the Palazzo Vecchio

The Republic of Florence made the decision at the end of the thirteenth century to construct an imposing building to house the governmental counterpart to the Florence Cathedral,

39. Casellato et al., "Mössbauer Approach."

40. Bacci and Picollo, "Non-Destructive Spectroscopic Detection," 136–44.

41. Musumarra and Fichera, "Chemometrics and Cultural Heritage."

with Arnolfo di Cambio, the architect for the cathedral, hired as the principal designer. Completed in 1322, it was originally named the Palazzo dei Priori, and in the fifteenth century became known as the Palazzo della Signoria as the civic organization underwent transitional changes. In 1540, Cosimo I de' Medici felt the need to consolidate the Medici power and moved his offices and the family residence to the Palazzo della Signoria, thus making a strong statement of his control of the government. Cosimo commissioned extensive renovations in order to create a magnificent palace worthy of the Medici name and the glory of Florence. The building later became known as the Palazzo Vecchio (Old Palace) when Cosimo relocated his family residence across the Arno to the Pitti Palace in the 1560s.

6.3.2.1 Salone dei Cinquecento

The hall in the Palazzo Vecchio, originally constructed between 1495 to 1496, was the space in which 500 members of the 1,500-member assembly of the Republic of Florence met to deliberate the affairs of state. When Cosimo decided to convert the Palazzo Vecchio to include a ducal residence, he first commissioned Baccio Bandinelli and Giuliano di Baccio d'Agnolo, in about 1542/43, to transform the Hall of Five Hundred (Salone dei Cinquecento) into an audience chamber. To emphasize his importance and power, Cosimo directed that a raised platform be built at the end of the room on which his ducal throne could be placed. The dais (*udienza*) stretched across the north end of the hall and was adorned with a triumphal arch theme with freestanding columns at each end honoring the Medici power, strength, and authority. While the hall was under renovation, Giovanni Battista del Tasso was at work redecorating and remodeling the upper floors to create the ducal apartments. In 1554, before much of the process could be completed, Tasso died and Cosimo tapped Giorgio Vasari (architect and artist, as well as biographer) to complete the renovation.

a. The Technological Challenge

For the next ten years, Vasari coordinated efforts to complete one ambitious project after another in order to satisfy the wishes of Cosimo. After the *udienza* was created in the hall, Cosimo was still not pleased with the overall appearance. The suggestion was made to raise the height of the ceiling by some 13 *braccia* (7.6 meters), thereby adding to the grandeur of the space, and to install high windows along the sides and ends of the hall. Raising the height of the ceiling posed an enormous technological challenge for a room that measured about 1,200 square meters. Under the direction of Vasari, this task was accomplished between 1559 and 1563, with the hydraulic lifting of the forty-two ceiling panels taking twenty-two months. In order to adequately stabilize the ceiling structure, Vasari designed an ingenious system of supports in the attic space over the large expanse (fig. 6.6). Iron was used to buttress beams of wood that formed a system of catwalks and twelve trusses. The trusses were engineered to pivot and pull with sufficient flexibility to withstand any earth tremors that might compromise the strength of the ceiling, a truly innovative technological accomplishment. Later, thirteen pine trusses were added to increase stability. In addition, galleries were created across the short ends of the hall, suitable for viewing

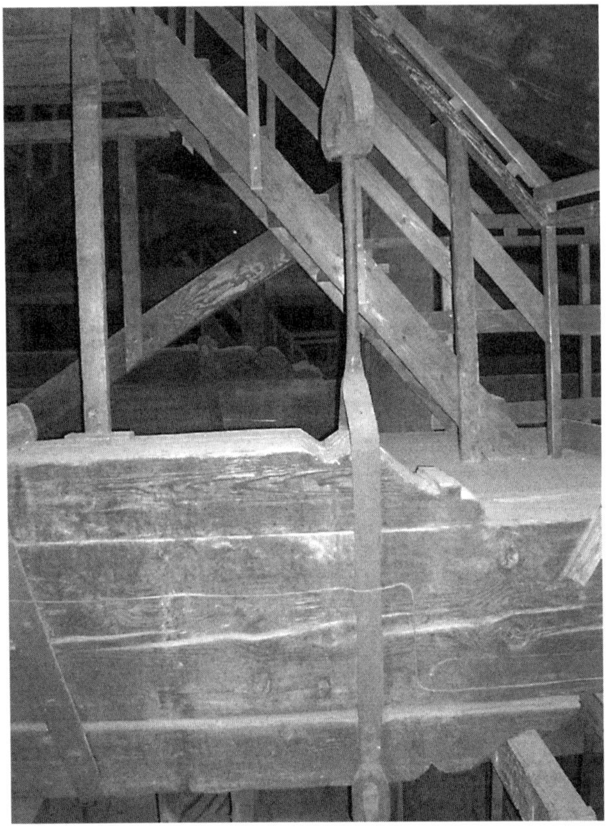

Figure 6.6: View of ceiling structural support (1559–63) designed by Vasari in the Palazzo Vecchio (photograph by authors).

the room below and as connectors for the ducal apartments located on either side of the building.[42]

b. Enhancement with Art

Once the ceiling had been raised, it was left to Vasari to augment the architectural improvements with decorations on the walls and ceiling that would fulfill the grandiose scheme envisioned by Cosimo, showing his rule as a continuation of the past and culminating in a triumphant and glorious present. Vasari and his assistants decorated the coffered ceiling with thirty-nine painted scenes honoring Cosimo and illustrating a brief history of Florence. In the center coffer, *Apotheosis of Cosimo I*, Cosimo is depicted as a god, surrounded by cherubs and coats of arms from the various guilds in Florence. Allegories portraying the glory of Florence encircle the center coffer with paintings honoring the accomplishments of the Florentine state. At the two ends of the ceiling, the four districts of Florence (*quartieri*) are represented, as are their neighborhoods. On either side of those depictions, the cities and regions of Tuscany under Cosimo's control are symbolized.

The two long walls are also adorned, one with scenes of the Florentine war with Pisa (1494–1509) and the other, facing the entrance, containing scenes depicting the war with Siena (1554–55). These two wars contrast one another, with the Sienese War, carried out under Cosimo's direction, being more efficiently organized and of shorter duration. The frescoes have the appearance of large tapestries, with one battle scene barely distinguishable from the other, and large numbers of people and horses dominating the space. The overall impact of the frescoes is the evocation of tremendous power and force, a metaphor for the Medici rule. A sculpture by Michelangelo, originally created for the tomb of Pope Julius II, was given to Cosimo by the nephew of Michelangelo. That statue, *Genius of Victory*, also

42. Salemi, *Palazzo Vecchio a Firenze*, 90.

graces the space. The ceiling height, the walls, and the artistic references all give the room a feeling of magnificence, engineered expertly by Vasari.

For many years scholars have pondered the question of an unfinished mural by none other than Leonardo da Vinci in the Hall of Five Hundred that may have been covered by Vasari's renovation and paintings. It is known that in 1503 Leonardo was commissioned to commemorate Florence's 1440 defeat of Milan in the Battle of Anghiari. Many art historians believed that Vasari would never have destroyed a work by Leonardo, even one that was unfinished. Rather, some scholars suspect that Vasari simply added a brick wall in the Hall for his paintings, thus leaving Leonardo's undisturbed behind the bricks. Unfortunately no documentation exists, although some have interpreted a writing on one of Vasari's paintings, "Seek and ye shall find" as a tantalizing clue that indeed something may be preserved. For a number of years Maurizio Seracini and collaborators have used a variety of techniques to determine the overall architectural design of the hall before Vasari's renovations. Using radar scanning, they found that an air gap existed on the east wall,[43] and this discovery led to an examination of methods capable of penetrating the wall for preliminary mapping and identification purposes. One such method attempted was nanosecond neutron analysis, which could detect the existence of any pigments behind Vasari's wall.[44] On March 12, 2012, an exciting announcement was made: researchers reported the presence of several pigments for red, blue, and beige of the same type known to have been used by Leonardo in other paintings. The outcome of this recent development will surely catalyze further efforts to learn as much as possible about the purported lost work.

6.3.2.2 Studiolo

Francesco I succeeded his father as grand duke in 1574, maintaining his residence in the Palazzo Vecchio. The contrast between the two grand dukes cannot be better illustrated than with the projects Vasari undertook for each of them. For Cosimo I, he renovated the grandiose Salone dei Cinquecento into an audience chamber in which Cosimo conducted the affairs of state. Francesco's interests were directed toward science, discovery, alchemy, and the pursuit of experimentation, for which he sought places of solitude to indulge those passions. For him, Vasari built the small Studiolo (fig. 6.7), which served as both a study and a place in which to house Francesco's scientific collections. Although the renovations for Francesco did not present technological challenges, the Studiolo provides an interesting example of the mixing of art and science on a very basic level.

Technically, Francesco commissioned Vasari to design an artistically meaningful space for study and contemplation that could be totally hidden from the reality of his position as heir apparent grand duke. From 1570 to 1573, Vasari, with input from humanist Vincenzo Borghini (1515–80), used his ingenuity to create a private world into which Francesco could retreat. To the casual observer, the barrel-vaulted small room adjacent to the massive Salone

43. Pieraccini et al., "Non-Contact Intrawall Penetrating Radar."

44. Kuznetsov et al., "Nanosecond Neutron Analysis."

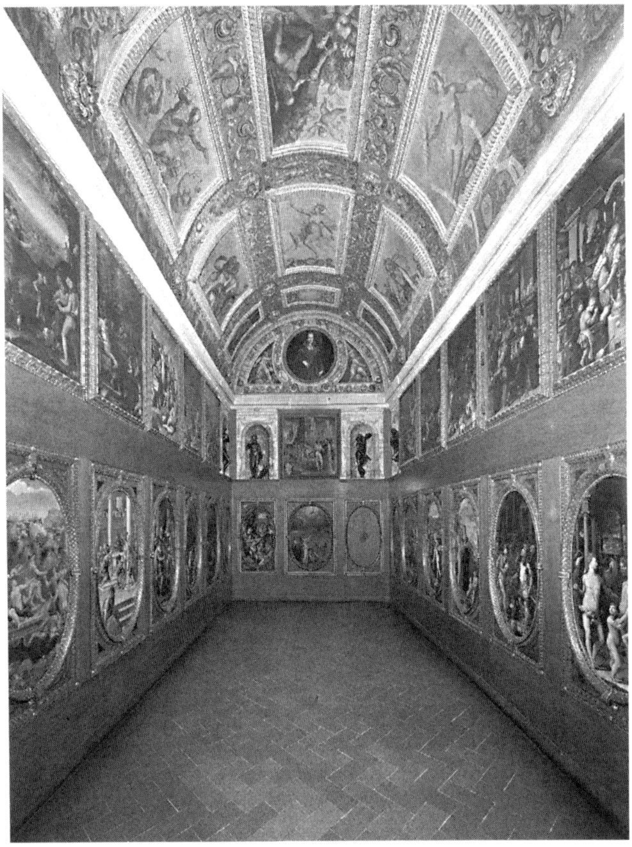

Figure 6.7: Vasari, Studiolo (1570–73) in the Palazzo Vecchio (Web Gallery of Art, retrieved from Wikimedia Commons).

dei Cinquecento appears to be little more than an interesting, decorated alcove. In Francesco's day, however, Vasari's design assured the secret nature of the area. Originally, the Studiolo was closed off from public view, entered not from the hall as it is today, but directly from Francesco's bedroom. Vasari also constructed a private doorway behind which a staircase led to an area providing direct access to the street adjacent to the Palazzo Vecchio. The doors were cleverly concealed as painted panels.

The cycle of paintings in the Studiolo reflects the work of twenty-one artists. Those on the ceiling are representations of the four basic elements—earth, fire, air, and water—a notion inherited from the early Greeks. The popular cosmological view of the time, relating the elements to human existence, is evident in those paintings. Of interest, there is a suite of rooms on the second floor of the Palazzo Vecchio, also completed during Vasari's renovation, known today as the Elements Chamber, in which the four elements represented in the Studiolo were also allegorically illustrated by Vasari. Earth is depicted in a scene in which Saturn is receiving fruit, fire shows Vulcan's forge, the birth of Venus represents water, and, on the ceiling, Apollo's chariot exemplifies air. Thus Cosimo, also with an interest in science that surely influenced Francesco, integrated scenes of the basic elements into the decorative renovation of his palazzo.

In order for Francesco to transform his innocuous-appearing alcove into a suitable area for scientific study and contemplation, Vasari hinged the lower register of painted panels in such a way as to hide the drawers and cabinets containing collections and instruments that Francesco used in his scientific investigations. With no natural light, the room was extremely dark, and thus candles were used for illumination, casting a glow on the works of art. Drawers were not labeled as to the contents, but rather, were allegorically identified by the paintings on the panels behind which they were hidden. With drawers containing natural gems, crystals, shells, dried insects, and material for medicinals such as tusks, Francesco's collection provided bountiful subject matter for the paintings in an unselfconscious blend of art and science.

One painting adorning the room—*The Alchemist's Studio*, completed in 1571 by Jan van der Stract (Giovanni Standano)—includes in the lower right corner a portrait of Francesco being instructed by a man who appears to be a learned alchemist (fig. 6.8). Alchemy, referring both to the transmutation of metals into gold and to the search for a potion that was both curative and an avenue to eternal life, had become an area of interest to members of the Medici family long before the birth of Francesco. When Hermetic manuscripts from the Hellenistic period were brought to Florence in about 1460, Cosimo the Elder commissioned the Platonic scholar Marsilio Ficino to translate and interpret works relating the narrative of a man who possessed the ability to divine aspects of himself and equate those revelations to mankind in general. After Cosimo's death in 1464, his grandson Lorenzo the Magnificent became intrigued by explorations into areas that would become the seeds of further experimental observation.[45] Francesco was especially intrigued by early chemical studies undertaken to better understand nature and manipulate existing formulations, many of which were developed centuries and even millennia before.

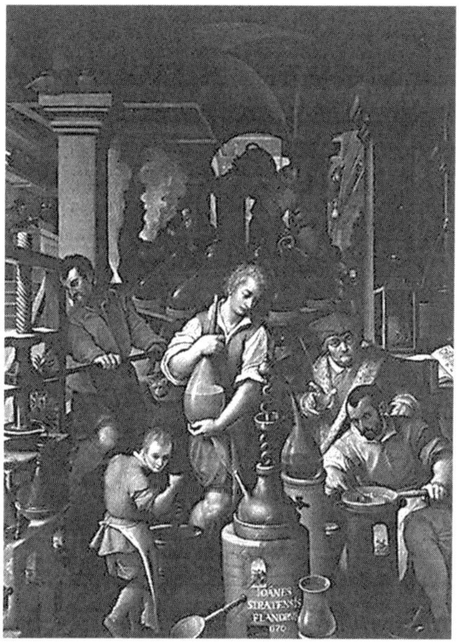

Figure 6.8: Giovanni Standano, *The Alchemist's Studio* (1570), painting, Studiolo, Palazzo Vecchio (paleopatologia.it, retrieved from Wikimedia Commons).

Corollary to his interest in having a Studiolo in which to work, Francesco ordered construction of a building in a portion of the Medici Gardens on Via Cavour. Bernardo Buontalenti was commissioned to build a scientific laboratory that became known as the Casino Mediceo di San Marco. Here Francesco established a laboratory for his chemical experimentation. After Francesco's death in 1587, his son Antonio de' Medici continued Francesco's interest, turning the Casino into a center for chemical studies that attracted scholars interested in alchemy. Today, Antonio's extensive library is located in the Biblioteca Nazionale Centrale di Firenze. Interestingly, the Casino is located in a part of the Medici Gardens where artists designated by Lorenzo the Magnificent, including Michelangelo, were invited to work and perfect their sculpting techniques.

6.4 Military Engineering and Fortification

Florence, like other Italian city-states of the period, relied heavily on fortifications to protect both the citizenry and the urban center. The military engineers who designed and oversaw the construction of these defenses were by necessity familiar with siege warfare and techniques of attack. Two of Florence's eminent artists, Leonardo da Vinci and Michelangelo,

45. Ronan, *Science: Its History*, 274–75.

also had experience in military engineering, and Florence turned to them to assist in periods of offensive and defensive actions.

6.4.1 Leonardo da Vinci

As documented in his notebooks, Leonardo was keenly interested in fortifications and in designing ships and machines for military use, including catapults, mangonels, massive crossbows, transportable cannons, equipment designed for maritime warfare, and much more.[46] While his many designs of military apparatus were most likely never constructed, his persuasive letter to Ludovico Sforza, ruler of Milan, describing his prowess as a military engineer earned him a good position at the court during the late quattrocento. Later in Florence, he formed an unlikely alliance with Niccolò Machiavelli to develop a plan to deprive Pisa of water. This plan arose from the continuing problem between Florence and Pisa concerning the Arno and access to the sea for commercial enterprise, which was most important for the Florentines. Initially Leonardo developed plans to breach the walls of Pisa, but then designed a scheme to divert the Arno. Leonardo's plans were detailed to the point of determining how much earth a single worker could move in a day. Unfortunately for Leonardo, and of course for Florence, the hydraulic engineer in charge of the operation, Colombino, modified Leonardo's original design and the effort failed.[47]

6.4.2 Michelangelo

The Medici's power was interrupted in 1527 when they were exiled and the Republic of Florence reemerged, headed by the Council of Ten. In 1529 Pope Clement VII decided to send the imperial army, led by the Prince of Orange, to overthrow the Republic of Florence and restore the Medici to power. Michelangelo was appointed general governor and procurator of the fortifications of Florence. In this capacity, he designed bastions and gates, and added a wall on the south side of the Arno. His clever use of polygonal defensive walls made the bastions less vulnerable to cannon balls, while still permitting firepower from the inside. It appears that although the fortifications were constructed, none were actually built to Michelangelo's specifications; moreover, none of the bastions remain today, having been removed by Cosimo I. Michelangelo's drawings for the fortifications are, however, located in the Casa Buonarroti.[48] Interestingly, the drawings reflect an intersection of the use of an artist's skills and the development of a technology. His unique ability to conceptualize forms led him to create plans for military uses that also reflected an artistic aesthetic.

46. Suh, *Leonardo's Notebooks*, 264–73.

47. Masters, *Fortune Is a River*, 122–33.

48. de Tolnay, "Michelangelo Studies."

6.5 Performing Arts and Technology

6.5.1 Background

The best-known art of the Florentine Renaissance is fixed or stationary, but another form emerged, one that encompassed motion and elaborate scenery and stage props. That interesting combination of theatrical art and technology was centered in sacred enactments (*sacre rappresentazioni*), secular performances (*intermezzi*), and operatic productions. Florence had a classical model—Greek and Roman drama—for these types of technical enhancements of dramatic presentations. Vitruvius described theatrical technology developed by the Greeks and Romans that was varied and extensive. Another classical source of information was Hero of Alexandria, an engineer and mathematician who probably lived around the first century AD. Vitruvius's volumes were available in Renaissance Florence, and Hero's writings were translated and published in Latin in 1501 and in Italian in 1589. In Florence, the blending of technology and performance art was considered to be so ingeniously developed and applied that when Vasari used a term literally meaning genius (*ingegno*) to describe the discipline, he was echoing the Florentine sentiment that the mechanical apparatus and the resulting theatrics deserved such an appellation.

6.5.2 Sacred Enactments

Making an appearance in Florence in the quattrocento, the *sacre rappresentazioni* were religious spectacles offered in various churches in celebration of important occasions and festivals. With his expertise in creating machines for the Duomo, Brunelleschi was the obvious person to whom the church leaders turned for developing the complex technical apparatus necessary for the task. Later, Francesco d'Angelo, known as Il Cecca (1446/47–88), combined his talents as a sculptor with his engineering skills to make mechanical sculptures that were carried during religious processions and also theatrical machinery, some of which replaced Brunelleschi's initial apparatus. Contemporary documents point to four Florentine churches in which *sacre rappresentazioni* were staged: the church of the Santissima Annunziata, Santa Maria del Carmine, San Felice in Piazza, and Santo Spirito. Machinery for the performances was left in place within the churches, and altered and improved upon as the need arose and new technologies were invented.

From diaries and notes, we have firsthand information concerning some of the staged events. The Russian bishop Abraham of Souzdal, while in Florence for the Council of Florence in 1439, attended *The Annunciation* at Annunziata. A narrative of the production, most probably penned by him, contains some interesting insights as to the drama created by the mechanized apparatus. His description mentions the actor portraying God being located above the tribune and surrounded by at least five hundred burning lamps that revolved while moving up and down. Angels were represented by boys in white, playing flutes, cymbals, and harps, and moving mechanically above the ground level. To depict the Annunciation itself, a singing angel descended on ropes sufficiently thin that the boy acting the angel, garbed in white and gold, appeared to be floating downward. Fire then descended along three ropes accompanied by the sound of thunder, and the flames caused sparks to flutter across the church as the angel ascended with wings flapping. The flames continued until the angel completed his ascent and the curtains were closed.[49] To effect the moving fire, Brunelleschi designed a type of rocket that traveled along the corded rope and could light lamps placed in a series along its route. According to written accounts, the rocket used a cold flame made from Greek pitch or gunpowder that did not endanger the scenic paper hangings in the immediate vicinity.[50] When one imagines the spectacle, with every facet in continuous movement, it is easy to understand its artistic impact on the attendees.

The apparatus Brunelleschi designed for *The Annunciation* in Santissima Annunziata was also constructed in San Felice di Piazza. Giorgio Vasari presented a vivid account of that particular *sacre rappresentazioni* in his *Lives of the Most Eminent Painters, Sculptors, and Architects*. Although Vasari's description appeared a century after the death of Brunelleschi and his statements cannot always be relied upon, corroborating analyses and the possibility that Vasari actually saw the machinery used in San Felice di Piazza give credence to his drawing and narrative.[51] Vasari related that one section of the apparatus depicted heaven with a place for God surrounded by angels, and a more complicated section comprised of movable parts was shaped like a half globe with a star located at the top that served as an iron fastener and enabled the half globe to rotate. An iron ring at the base contained twelve platforms on which children garbed as angels were secured. Lanterns above the children's heads appeared to spectators to be stars in the heavens. Clouds were simulated by covering the supporting beams with cotton.

Below and suspended from the half globe with ropes was a frame containing eight spokes. At the end of each spoke was a small platform on which a child dressed as an angel was secured. Those angels were raised and lowered using a windlass (cylinder around which rope was attached and turned by a crank). From the eight-spoked section, a small carriage (*mandorla*) was suspended that held small lanterns controlled by a spring that could, when depressed, make the light disappear from view. A person portraying the Angel Gabriel rode inside the *mandorla* wearing an iron brace that could be released by a machine operator so that when a second windlass allowed the *mandorla* to fully descend, Gabriel could emerge and walk toward the Virgin Mary.

Large movable doors, reported by Vasari to be ten feet wide with metal rollers, were attached to beams supporting the roof. When the doors were opened by means of a cord, the rollers on the grooves created a sound similar to thunder. With all of the above mechanized, from the thunderous opening of the doors revealing God and heaven, to the revolving lighted components with choir angels, the lowering of Gabriel, his appearance before Mary, ascension back up to the angels, and the final thunderous closing of the doors, the spectacle, known as *Paradiso*, was complete.[52] The basis for the systems devised for lifting and lowering

49. Battisti, *Filippo Brunelleschi*, 300.

50. Pallen, *Vasari on Theatre*, 16.

51. Larson, "Vasari's Descriptions," 293.

52. Vasari, *Lives of the Artists*, trans. Bull, 166–68.

heavy platforms and the *mandorla* were the complex hoists Brunelleschi had developed to such a remarkable degree during his work on the Duomo.

The Annunciation was also performed in the much larger church, Santo Spirito, presumably also using Brunelleschi's mechanisms. The platform representing heaven was located in the dome and was engineered to open, thus allowing the angels to move downward in a circular motion. Gabriel descended to appear before Mary, whose chamber was suspended above the floor; after the Annunciation, Gabriel was hoisted back toward heaven.[53]

A second popular *sacre rappresentazioni* was *The Ascension*. When it was presented in 1439, Bishop Abraham of Souzdal was in attendance and again he left a firsthand account of the spectacle. He reported that when Jesus appeared at the top of a mountain, God emerged encased in a brilliant light to the accompaniment of thunder. Angels appeared to float around him and music could be heard in the background. A cloud, suspended by seven ropes, descended from heaven. According to the bishop, the cloud surrounded Jesus, and with an angel on either side, all ascended to heaven. Lighted lamps created an all-encompassing radiance that embraced both of the angels and Jesus, who arose seemingly on his own with no obvious sign of ropes or pulleys.[54]

Vasari credited Il Cecca with adding to and improving the machinery for *The Ascension* for the church of Santa Maria del Carmine, but it is believed by many that the majority of the equipment attributed to Il Cecca by Vasari had actually been in use earlier.[55] Vasari's account of *The Ascension* at Santa Maria del Carmine is similar to that of the Bishop of Souzdal, but has additional technical details. He commented that the mountain on which Jesus appeared was constructed of wood, and because the sanctuary of the church of Santa Maria del Carmine is larger than that of San Felice, heaven was portrayed with more grandeur. One section was constructed to receive Jesus after his ascent, while another area of heaven, located above the central apse, was built in ten parts made of large wheels that could be turned in order to reversibly move the sections. Lamps representing stars were located in lanterns made of copper that ingeniously remained in upright positions even as the wheels turned. From this locale, two angels descended by means of bronze pulleys along two large cables on a platform held by an iron beam. Angels in various colors were slowly dropped by ropes to announce Jesus and the later ascension. Because the angels were bound at the waist to an iron rod attached to the platform, they could turn and move, seeming to float, by means of lead counterweights placed beneath their feet, thereby facing upward as the ascension progressed.[56]

Other *sacre rappresentazioni* were performed throughout Florence during the successive years, most notably those staged in 1471 on the occasion of a visit by Galeazzo Maria Sforza and in 1494 upon the arrival of King Charles VIII of France. Interesting technical apparatus was also used wherever possible during religious processions.

53. Nagler, *Theatre Festivals*, 34.

54. Battisti, *Filippo Brunelleschi*, 301.

55. Larson, "Vasari's Descriptions," 297.

56. Pallen, *Vasari on Theatre*, 53–54.

6.5.3 Secular Theatrical Presentations

6.5.3.1 Intermezzi

Following somewhat later were the *intermezzi*, spectacular dramatizations that were presented between acts of plays or court festivals and weddings. The comedy *La Pellegrina* was particularly notable. Giovanni de'Bardi was commissioned in 1586 to invent devices for six *intermezzi* to be performed during the comedy. According to contemporary sketches, account books, and descriptions, most notably that of Bastiano de'Rossi, who wrote the official narrative explaining the *La Pellegrina* presentation, each of the six *intermezzi* was a true amalgamation of technical and artistic creativity. The first *intermezzo* depicted *The Harmony of the Spheres*, the second was *The Contest of Muses and Pierides*, the third featured *Apollo and the Python*, followed by *Prophecy of the Golden Age*, *Arion and the Dolphin*, and *The Gods Send Rhythm and Harmony to Earth*. The *intermezzi* incorporated elaborate mechanical effects, including heavens with stars, rising clouds on which actors were perched, intricate ascents and descents, a machine capable of effecting rapid scene changes, mountains that rose seamlessly, lighting that gave the appearance of flying machines, a python emitting flames, a fiery inferno, and simulated waves, to name but a few of the complicated stage sets.[57] The Medici court adopted the practice of staging *intermezzi*, and the celebration of the marriage of Grand Duke Ferdinando to Christine of Lorraine in 1589 occasioned a *spectacolo* of previously unseen proportions. Lasting for three weeks and including multiple theatrical productions, the celebration was held at the Uffizi Theatre.

Specific details of the design of the machines used are lacking, but information gleaned from a theater in Parma that was modeled on the Uffizi space, combined with staging manuals, allows for some fairly accurate deductions as to the general structure, sizes, numbers of supporting technical personnel, and the system of operation. To give some idea of the complexity and the number of machines that needed to be functional in a limited space, there were about one hundred men positioned to work the apparatus. The area at the top of the stage was crisscrossed with beams in a grid pattern from which winches and pulleys operated rigging, allowing for the lifting and lowering of platforms, chariots, clouds, sunbursts, choirs, and the like. Each of the larger clouds, for example, required ten or twelve operators to release coiled ropes (*auguille*), a winch (*argano*), a windlass (*verricello*), and a folding device (*ripiegarsi*).

The areas both on and below the floor of the stage were also mechanized spaces. The stage contained grooved scene-changing tracks, trapdoors that required six men to open and close, and a wave machine that allowed for simulated dolphins, ships, and other devices to slide between the wave segments. Below stage level, numerous winches and platforms of various sizes were stowed and available for raising and lowering the special effects at specified times. An example of one type of scene enhancement was the appearance of large mountains. Each mountain was engineered to expand and contract, allowing it to appear at the desired height on stage while also making it possible to store the prop before its elevation. Eight-man teams

57. Saslow, *Medici Wedding of 1589*, 19–33.

operated the large winch and smaller winches, and a windlass was also utilized. After the mountains were lifted to their maximum heights, wooden supports were unobtrusively inserted to give them added stability. All of this is especially impressive when one considers that the vast amount of equipment and manpower required to create a choreographed operation had to be able to seamlessly perform the myriad functions in a limited space without interfering with the equal number of cast members.[58] The stages on which the *intermezzi* were performed with scenery perspectivally painted and with complex technological effects, including ascending, descending, and changing elements, became a pinnacle of theatrics.[59]

Extravagant celebratory events, like those at the Uffizi, were also performed at other venues in Florence. One example was a mock sea battle and performance in the courtyard of the Palazzo Pitti. Enhanced by a technologically superb *intermezzi* with intricate scenic devices, the artistic production featured fire-eating dragons, a volcano (Mt. Aetna) that exploded, an entire garden that was maneuvered smoothly onto the set, and finally an apparent ocean on which a battle featuring eighteen galleys was enacted.[60]

The tradition of performances of spectacles was carried on by Ferdinando's son Cosimo II. Upon his marriage in 1608 to Archduchess Maria Maddalena, he had a *spectacolo* staged on the Arno River as part of the nuptial festivities. The audience seated along the riverbank was entertained by a production of *Argonautica* featuring huge mechanical dolphins, lobsters, and hydras spitting fire as Jason negotiated his way around a man-made island in order to capture the Golden Fleece.[61] Jason's fleet consisted of sixteen ships, each an individual work of art as well as being technically adept at maneuvering in the space on the river. One of the ships was pulled by two mechanical whales carrying actors who sang and played instruments. After Jason fought fire-breathing bulls, hissing dragons, and warriors on the island, the battle returned to the sea and the island moved downstream to provide room for the naval conflict. As darkness fell, the production ended with the victorious Argonauts sailing to the bride's location along the riverbank, lighted by torches, and handing her the Golden Fleece.[62]

Elaborate *intermezzi* continued to be part of the Medici court festivals, nuptial celebrations, and theater productions in the ensuing years. The Uffizi, Palazzo Pitti, Boboli Gardens, the Arno River, and Piazza Santa Croce were all locations for *spectacoli* featuring dramatic scene changes and technically complicated apparatus employed to enhance the artistic images and advance the story lines.

6.5.3.2 Opera

Florentine interest in reviving Greek drama combined with music was stimulated when a professor from the University of Florence, Piero Vettori (1499–1585), and his student

58. Ibid., 84–86.

59. Strong, *Art and Power*, 38.

60. Nagler, *Theatre Festivals*, 91–92.

61. Hibbert, *House of Medici*, 280–82.

62. Nagler, *Theatre Festivals*, 111–15.

Girolamo Mei (1519–94) researched and found texts of works by Euripides and Aeschylus. Mei's contribution included reintroducing the Greek monodic style of a single melodic line. Vincenzo Galilei (ca. 1522–91), father of Galileo and an accomplished lute player and composer, was associated with the Camerata, an informal gathering of intellectuals founded by Giovanni Bardi, whose discussions often turned to Greek musical genres.[63] Vincenzo's writings included work on dissonance and the human voice as well as acoustical studies using mathematical ratios involving the perfect fifth. Spurred in part by information about Greek music learned from Mei, Vincenzo published his treatise, *Dialogue on Ancient and Modern Music*, in 1581, making a significant contribution to music theory. Along with others, Vincenzo promoted the idea of creating a union of plainly understood words and melody in solo performance, laying a foundation for opera. Another important contribution of Vincenzo was his use of experimental observations, rather than abstract deduction, to explain the physics of columns of air and vibrating strings, such as the roles of string length, dimension, and tension in the production of particular sounds. It is tempting to speculate that his insistence on properly developing and using experimental paradigms to explain natural phenomena may have stimulated his son Galileo to adopt a similar approach.

Giulio Caccini (1551–1618), an instrumentalist, composer, and singer, refined the technique of monody in his work, *Le nuovo musiche* (1602), and he and Jacopo Peri (1561–1633) each scored works in Florence that can be considered as the first operatic productions. Peri, born in Rome but educated in Florence, was appointed as a singer, organist, and composer in the Medici court. Sadly, the score of his first opera, *Dafne*, written about 1597 with libretto by Ottavio Rinuccini and staged in Florence, has not survived. However, his subsequent opera, *Euridice,* libretto also by Ottavio Rinuccini, had the honor of being performed in the Palazzo Pitti on October 6, 1600, as a part of the festivities celebrating the marriage of Maria de' Medici to King Henry IV of France. Caccini also scored music for an opera entitled *Euridice*, but it was not performed until 1602, two years after Peri's.

Like the *sacre rappresentazioni* and the *intermezzi*, operatic presentations used complicated technical apparatus to enhance scenes. For the production of *Euridice,* the actors, known as "instrumentalists" in reference to their vocal performances, sang their parts amid and/or on special devices, including clouds, mountains, a sea, a forest, a plain, the openings of crevices, movement of the earth, a whale, a chariot, and a winged cupid, all of which appeared to emerge, disappear, or mechanically fly in a seemingly effortless fashion at various times throughout the five acts and afterpiece.[64]

Performance art in Renaissance Florence reached new levels of sophistication and dramatic presentation with the addition of technological expertise not previously used in that arena. Florentines excelled in the mutual enhancement of disciplines across many areas of art and science, broadly interpreted to include mathematics, experimental observations, technology, and engineering, thus providing the infrastructure to the fabric of their Renaissance with its interwoven amalgamation that we appreciate today.

63. Grout and Williams, *Short History of Opera*, 34–36.

64. Nagler, *Theatre Festivals*, 93–100.

Concluding Remarks

Connections Between Science and Art

While numerous renaissances have appeared in various cultures throughout history, the renaissance arising in Florence, here broadly defined as the period from the early fifteenth century to the seventeenth century, was in many ways unique. Boundaries were extended politically, culturally, and intellectually. Upheavals in conventional thinking were actively promoted in the areas of science, technology, art, literature, and music: a massive dome that came to symbolize the power and glory of Florence was successfully designed and constructed by an architect/engineer; artists engaged in mathematically and geometrically correct perspectival painting; the first medicinal compendium was published, standardizing for the first time in Europe the many recipes used by Florentine physicians and apothecaries; astronomers inverted the Earth-centric view of the universe; opera, with elaborate mechanical devices designed for productions, emerged as a distinct musical form; and authors writing in the vernacular popularized quality literature. Having examined all of these areas, there can be no doubt that art and science were inextricably intertwined during the period of the Florentine Renaissance. However, even a clear understanding of that essential fact leaves some questions about the nature of the interaction between art and science.

Did the burgeoning emphasis on science and technology drive the Florentine Renaissance?

It was the environment of observational exploration, coupled with a desire among the artists to extend the boundaries of their discipline, as much as it was the relearning, reinventing, and expansion of the sciences that made the broadening and innovation that characterized the Florentine Renaissance possible. As the city expanded and the wealth of many households increased, the rising elite class was motivated by civic pride and competition for family honor to provide funding for exceptional artwork to adorn the escalating number of private and public construction projects. The turn to a more humanist outlook led to a moneyed segment of Florentine society enrolling in universities that valued classical literature, fostered inquiry less structured than Christian doctrine had previously allowed, and encouraged scholarly analysis. In time these were the individuals who commissioned the proliferation of new artwork. Humanists opened young minds to alternative possibilities and ways of viewing Man in relation to the Creator. An excellent example is Lorenzo the Magnificent's appointment of noted humanist poet and scholar Angelo Ambrogini (1454–94), known as Poliziano, as the tutor for his children. An erudite philologist, Poliziano moved in a circle of like-minded thinkers of the period, both while educating the Medici children and later while teaching university students. In addition, the early secular writings of Euclid, Ptolemy, Vitruvius, and Galen, preserved and expanded by Arabic scholars, were increasingly available in translation to the learned.

Thus while the increased interest in science and technology played an important role in the Florentine Renaissance, it was not the driving force; rather the impetus for the Renaissance was multifaceted, encompassing political expansion and civic pride, rediscovery of classical writings, creative explorations in the arts, a move toward a more humanist view of the world, and innovative developments in science and technology.

Did all of the notable artistic innovations result from collaborations between artists and individuals trained in mathematics, geometry, anatomy, astronomy, or other scientific disciplines?

Many artists of the Florentine Renaissance individually used the same kind of experimental techniques developed in the sciences to explore methods of attaining their desired outcomes. Also, the prevailing ambience of Florentine culture—with its tradition of cooperation between artists and scientists—helped ensure artistic successes. Certainly, the friendship shared by Brunelleschi and Toscanelli was key to the development of the mathematical formulations that explain Brunelleschi's experiments with perspective, experiments that were elegantly utilized by the artist Masaccio. In addition, by working with guild members trained in technical disciplines, artists and architects had a pool of experienced individuals from whom to draw. Perhaps the best example is the design and construction of the dome of Santa Maria del Fiore, the Duomo, which required a large number of skilled builders and technicians to implement the innovative design and clever construction methods developed by Brunelleschi. Artistic training gave scientists a way to depict their discoveries, the most notable example being Galileo's excellent drawings of Earth's moon and the moons of Jupiter. Studies of anatomy, plants, and maps also benefited greatly from artistic renderings.

We conclude that while many artistic innovations were the result of a direct collaboration between artists and scientists, some of the most spectacular advances came from individuals who studied both science and art, and who apparently saw no clear separation between what we today view as two separate disciplines.

In what way did Brunelleschi's creative mind embody a close connection between artistic achievement and scientific/technical developments during the Florentine Renaissance?

Brunelleschi combined his art and architectural expertise with his experiments in mathematical perspective, studies that were natural extensions of his inquisitive mind-set, to produce a paradigm that drastically altered the direction of Florentine painting. In addition, he integrated his mechanical knowledge and his artistic vision to create and implement his design for the dome of the cathedral. Brunelleschi's extraordinary grasp of the holistic approach stands as a remarkable testament to that which could be accomplished.

• • •

The interplay between art and science/technology, manifested so vividly by Brunelleschi's fertile imagination, became autocatalytic, so that each development in one area facilitated and stimulated another area to even greater achievements. Some of the other notable examples include Luca della Robbia and his experiments to develop a novel protocol for a durable terracotta; Cellini's adaption of methods for bronze casting that made it possible to create complicated, large, and detailed sculpture; Toscanelli's and Danti's pursuit of cartographic and astronomical innovations; artists' renderings that were directed to potential healing; apothecaries' use of herbal and other ingredients for both the artist and the physician; Leonardo da Vinci's creative drawings of technical machines and anatomically correct human forms based on his dissections; Ghiberti's and Donatello's use of perspective in sculpture; Michelangelo's incorporation of his knowledge of human anatomy in his art; the idea of tempering stone-carving instruments to render them sufficiently strong to sculpt porphyry; and the seemingly unrealistic Florentine conviction that they could eventually complete construction of an enduring cathedral begun with an opening for a massive and imposing dome, the design and construction of which had never before been attempted.

The Renaissance Florentines were emboldened by a turn toward humanism, a return to the classical, and a desire to attain artistic and scientific heights that, for the first time, seemed within the realm of possibility. This single-mindedness, manifesting itself in countless directions, would not have been possible without the synergistic interaction of art and science that nurtured their accomplishments. That integration of science and art was the driving force; the individuals involved and the prevailing intensely observational culture provided the momentum and the means.

Site Links and Topics Cross-Referenced by Location

Accademia

www.accademia.org
Marble: chapter 2.3.1.2c

Baptistery

www.museumsinflorence.com/musei/Baptistery_of_florence.html
see also
www.imss.fi.it/~tsettle/traces.html
(indexed as "Baptistery-San Giovanni Battista")
Bronze: chapter 2.3.5.2a
Mosaics: chapter 2.4.3.1
Perspective: chapter 3.1.2.3b
Solar Time Measurement: chapter 5.2.1

Bargello Museum

www.museumsinflorence.com/musei/museum_of_bargello.html
Bronze: chapter 2.3.5.2d
Medicine in Art: chapter 4.5.2
Porcelain: chapter 2.6.2.1
Pottery: chapter 2.6.1.1
Terracotta: chapter 2.3.4.1b
Wood: chapter 2.3.3.1c

Campanile (Duomo)

www.imss.fi.it/~tsettle/traces.html (indexed as "Campanile di Giotto")
see also
en.wikipedia.org/wiki/Giotto's_Campanile
Astronomy in Art: chapter 5.6.1
Medicine in Art: chapter 4.5.1

Duomo Museum (Museo dell'Opera del Duomo)

www.museumsinflorence.com/musei/cathedral_florence_museum.html
Marble: chapter 2.3.1.2b
Wood: chapter 2.3.3.1b

Florence Cathedral (Duomo), Santa Maria del Fiore

www.museumsinflorence.com/musei/cathedral_of_florence.html
see also
www.imss.fi.it/~tsettle/traces.html
(indexed as "Duomo-Cathedral-Santa Maria del Fiore")
Art and the Dome: chapter 6.3.1.7
Astronomy in Art: chapter 5.6.4
Fresco: chapter 2.2.1.2b
Hourly Time Measurement: chapter 5.3.3
Solar Time Measurement: chapter 5.2.2
Stained Glass: chapter 2.4.4.2a
Technology for Construction: chapter 6.3.1.1; 6.3.1.2a–6.3.1.2e; 6.3.1.3; 6.3.1.4; 6.3.1.5; 6.3.1.6
Terracotta: chapter 2.3.4.1a

Galileo Museum

www.museogalileo.it
Astronomical Instruments: chapter 5.5
Hourly Time Measurement: chapter 5.3.4

Garden of the Simples

www.museumsinflorence.com/musei/Botanical_garden.html
Medicinal Ingredients and Recipes: chapter 4.4.2

La Specola

www.museumsinflorence.com/musei/museum_of_natural_history.html
Solar Time Measurement: chapter 5.7.2

Loggia dei Lanzia (Loggia della Piazza Signoria)

www.turismo.intoscana.it/allthingstuscany/tuscanyarts/loggia-dei-lanzi-open-air-museum-in-florence/
Bronze: chapter 2.3.5.2e

Medici Tombs (Chapel of the Princes, San Lorenzo)

www.museumsinflorence.com/musei/Medici_chapel.html
Pietre Dure: chapter 2.5.1.2

Officina Profumo-Farmaceutica di Santa Maria Novella

www.museumsinflorence.com/musei/Pharmacy_S_Maria_Novella.html
Apothecaries: chapter 4.4.1

Ognissanti

www.museumsinflorence.com/musei/ognassanti_last_supper.html
Astronomy in Art: chapter 5.6.4
Fresco: chapter 2.2.1.2c
Perspective: chapter 3.1.2.3f

Opificio delle Pietre Dure

www.museumsinflorence.com/musei/opificio_delle_pietre_dure.html
Pietre Dure: chapter 2.5.1.1

Oratorio Buonomini di San Martino

en.wikipedia.org/wiki/San_Martino_del_Vescovo
Medicine in Art: chapter 4.5.4

Orsanmichele

www.museumsinflorence.com/musei/orsanmichele.html
Bronze: chapter 2.3.5.2c
Marble: chapter 2.3.1.2a
Perspective: chapter 3.1.2.3c

Ospedale degli Innocenti

www.museumsinflorence.com/musei/ospedale_degli_innocenti.html
History and Mission: chapter 4.3.2.1
Importance of Art: chapter 4.3.2.2

Palazzo Vecchio

www.imss.fi.it/~tsettle/traces.html
Cartography: chapter 3.2.2.1
Hourly Time Measurement: chapter 5.3.2
Medicine in Art: chapter 4.5.3
Porphyry: chapter 2.3.2.2c
Technology and Art in Renovation: chapter 6.3.2.1a–6.3.2.1b; 6.3.2.2

Piazza dei Giudici

www.museogalileo.it/en/aboutus/epalazzocastellani/sundial.html
Solar Time Measurement: chapter 5.7.3

Piazza Santa Trinita

www.florence-on-line.com/piazza-santa-trinita
Porphyry: chapter 2.3.2.2d

Pitti Palace

www.museumsinflorence.com/musei/Pitti_palace.html
Solar Time Measurement: chapter 5.2.5

Ponte Vecchio

brunelleschi.imss.fi.it/itineraries/place/PonteVecchio.html
see also
www.bluffton.edu/~sullivanm/pontevecchio/ponte.html
Hourly Time Measurement: chapter 5.3.1

San Lorenzo

www.museumsinflorence.com/musei/Medici_chapel.html
see also
www.imss.fi.it/~tsettle/traces.html
(indexed as "San Lorenzo: Sagrestia Vecchia")
Astronomy in Art: chapter 5.6.3
Bronze: chapter 2.3.5.2b
Perspective: chapter 3.1.2.3c
Porphyry: chapter 2.3.2.2a

San Marco

www.museumsinflorence.com/musei/museum_of_san_marco.html
Fresco: chapter 2.2.1.2g
Perspective: chapter 3.1.2.3e
Tempera: chapter 2.2.2.3c

Sant'Apollonia

www.museumsinflorence.com/musei/santa_apollonia_.html
Fresco: chapter 2.2.1.2e
Perspective: chapter 3.1.2.3f

Santa Croce

www.museumsinflorence.com/musei/museum_of_opera_s_croce.html
Astronomy in Art: chapter 5.6.2
Fresco: chapter 2.2.1.2h
Wood: chapter 2.3.3.1a

Santa Felicitá

en.wikipedia.org/wiki/Capponi_chapel
Medicine in Art: chapter 4.5.3

Santa Maria del Carmine (Brancacci Chapel)

www.museumsinflorence.com/musei/Brancacci_chapel.html
Fresco: chapter 2.2.1.2f
Perspective: chapter 3.1.2.3a

Santa Maria Novella

www.museumsinflorence.com/musei/santa_maria_novella.html
Fresco: chapter 2.2.1.2a
Perspective: chapter 3.1.2.3a; chapter 3.1.2.3d
Porphyry: chapter 2.3.2.2b
Solar Time Measurement: chapter 5.2.3
Stained Glass: chapter 2.4.4.2b
Tempera: chapter 2.2.2.3a
Wood: chapter 2.3.3.1a

Santa Maria Nuova and Sant'Egidio

en.wikipedia.org/wiki/Hospital_of_Santa_Maria_Nuova
see also
brunelleschi.imss.fi.it/itineraries/place/OspedaleSantaMariaNuova.html
History and Mission: chapter 4.3.1.1
Healing through Art: chapter 4.3.1.2

Santa Trinita

www.museumsinflorence.com/musei/santa-trinita.html
Fresco: chapter 2.2.1.2d

Uffizi Gallery

www.uffizi.org
see also
www.imss.fi.it/~tsettle/traces.html (indexed as "The Uffizi")
Astronomy in Art: chapter 5.6.4
Oils: chapter 2.2.3.3a
Perspective: chapter 3.1.2.3d; 3.1.2.3f
Printmaking and Engraving: chapter 2.8.1
Solar Time Measurement: chapter 5.2.4
Tempera: chapter 2.2.2.3b

Ximenes Observatory

brunelleschi.imss.fi.it/itineraries/image/img9.html
Astronomy: chapter 5.7.1

Works Cited

Adams, Laurie Schneider. *Italian Renaissance Art.* Boulder, CO: Westview Press, 2001.

Ahl, Diane Cole, ed. *The Cambridge Companion to Masaccio.* New York: Cambridge University Press, 2002.

Ajò, David, Umberto Casellato, E. Fiorin, and Pietro Alessandro Vigato. "Ciro Ferri's Frescoes: A Study of Painting Materials and Technique by SEM-EDS Microscopy, X-Ray Diffraction, MicroFT-IR, and Photoluminescence Spectroscopy." *Journal of Cultural Heritage* 5 (2004): 333–48.

Alberti, Leon Battista. *On Painting.* Translated by John R. Spencer. New Haven, CT: Yale University Press, 1967.

Andersen, Kirsti. *The Geometry of an Art: The History of the Mathematical Theory of Perspective from Albert to Monge.* New York: Springer, 2007.

Andres, Glenn M., John M. Hunisak, and A. Richard Turner. *The Art of Florence.* 2 vols. New York: Abbeville Press, 1988.

Arletti, Rossella, Sonia Conte, Mariangela Vandini, Cesare Fiori, Susanna Bracci, Mauro Bacci, and Simone Porcinai. "Florence Baptistery: Chemical and Mineralogical Investigation of Glass Mosaic Tesserae." *Journal of Archaeological Science* 38 (2011): 79–88.

Bacci, Mauro, Francesco Baldini, Roberto Carlà, and Roberto Linari. "A Color Analysis of the Brancacci Chapel Frescoes." *Applied Spectroscopy* 45 (1991): 26–31.

Bacci, Mauro, and Marcello Picollo. "Non-Destructive Spectroscopic Detection of Cobalt (II) in Paintings and Glass." *Studies in Conservation* 41 (1996): 136–44.

Ballerini, Isabella Lapi. "The 'Sky' of San Lorenzo." In *The Line of the Sun: Great Sundials in Florence,* edited by Filippo Camerota, 28–43. Translated by Catherine Frost. Florence, Italy: Edizioni della Meridiana, 2007.

Barbolini, Stefano. "The Gnomon in La Specola." In *The Line of the Sun: Great Sundials in Florence,* edited by Filippo Camerota, 101–7. Translated by Catherine Frost. Florence, Italy: Edizioni della Meridiana, 2007.

Barkoudah, Youssef I., and Julian Henderson. "Plant Ashes from Syria and the Manufacture of Ancient Glass: Ethnographic and Scientific Aspects." *Journal of Glass Studies* 48 (2006): 297–321.

Bartolini, Simone. *I Fori Gnomonici di Egnazio Danti in Santa Maria Novella.* Florence, Italy: Edizioni Polistampa, 2006.

Battisti, Eugenio. *Filippo Brunelleschi: The Complete Work.* Translated by Robert Erich Wolf. New York: Rizzoli International Publications, Inc., 1981.

Bedini, Silvio A. "A Renaissance Lapidary Lathe." *Technology and Culture* 6 (1965): 407–15.

Bellucci, Roberto, Pierluigi Carcagnì, Andrea Della Patria, Raffaella Fontana, Cecilia Frosinini, Maria Chiara Gambino, Michael Greco, Maria Mastroianni, Marzia Materazzi, Enrico Pampaloni, Luca Pezzati, Roberta Piccolo, and Pasquale Poggi. "Integration of Image Data from 2D and 3D Optical Techniques for Painting Conservation Applications." *Imaging Science Journal* 55 (2007): 80–89.

Berggren, John Lennart, and Alexander Jones. *Ptolemy's* Geography*: An Annotated Translation of the Theoretical Chapters.* Princeton, NJ: Princeton University Press, 2000.

Bianchin, Sara, Nicola Brianese, Umberto Casellato, Frederica Fenzi, Paolo Guerriero, Pietro Alessandro Vigato, Luca Nodari, Umberto Russo, Massimo Galgani, and Marja Mendera. "Medieval and Renaissance Glass Technology in Valdesa (Florence), Part 2: Vitreous Finds and Sands." *Journal of Cultural Heritage* 6 (2005): 39–54.

Bianchin, Sara, Nicola Brianese, Umberto Casellato, Frederica Fenzi, Paolo Guerriero, Pietro Alessandro Vigato, Marino Battagliarin, Luca Nodari, Umberto Russo, Massimo Galgani, and Marja Mendera. "Medieval and Renaissance Glass Technology in Valdelsa (Florence), Part 3: Vitreous Finds and Crucibles." *Journal of Cultural Heritage* 6 (2005): 165–82.

Boccaccio, Giovanni. *The Decameron.* Translated by Mark Musa and Peter E. Bondanella. New York: W. W. Norton, 1977.

Bomford, David, Jill Dunkerton, Dillian Gordon, and Ashok Roy. *Art in the Making: Italian Painting before 1400.* London: National Gallery Publication Ltd., 1990.

Bouquillon, Anne, Ildiko Katona, Angelina D'Alessandro, and Alessandro Zucchiatti. "PIXE Analysis of Artistic Blue Glazes and the Use of Cobalt Pigments in Ceramics during the Italian Renaissance." *International Journal of PIXE* 15 (2005): 317–22.

Brianese, Nicola, Umberto Casellato, Frederica Fenzi, Marja Mendera, Bruno Messiga, Luca Nodari, Maria Pia Riccardi, and Pietro Alessandro Vigato. "Incised Slipware Production at Castelfiorentino (Florence): The Via Bertini Site (15th–17th Centuries)." *Archaeometry* 50 (2008): 775–96.

Brianese, Nicola, Umberto Casellato, Frederica Fenzi, Sergio Sitran, Pietro Alessandro Vigato, and Marja Mendera. "Medieval and Renaissance Glass Technology in Tuscany, Part 4: The XIVth Sites of Santa Cristina (Gambassi-Firenze) and Poggio Imperiale (Siena)." *Journal of Cultural Heritage* 6 (2005): 213–25.

Brucker, Gene A. *Renaissance Florence.* Malabar, FL: Robert E. Krieger Publishing Company, 1975.

Bruijn, Rolf H. C., Karsten Kunze, David Mainprice, and Luigi Burlini. "Mechanical and Microstructural Development of Carrara Marble with Pre-Existing Strain Variation." *Tectonophysics* 503 (2011): 75–91.

Burckhardt, Jacob. *The Civilization of the Renaissance in Italy.* Translated by S. G. C. Middlemore (London, 1878). Reprint, New York: Modern Library, 1995. Originally published as *Die cultur der Renaissance in Italien* (1860).

Burke, Jill. *Changing Patrons: Social Identity and the Visual Arts in Renaissance Florence.* University Park: Pennsylvania State University Press, 2004.

Butters, Suzanne B. *The Triumph of Vulcan: Sculptors' Tools, Porphyry, and the Prince in Ducal Florence.* Vol. 1. Florence, Italy: Leo S. Olschki, 1996.

Cagno, Simone, Koen Janssens, and Marja Mendera. "Compositional Analysis of Tuscan Glass Samples: In Search of Raw Material Fingerprints." *Analytical and Bioanalytical Chemistry* 391 (2008): 1389–95.

Cagno, Simone, Marja Mendera, Teresa Jeffries, and Koen Janssens. "Raw Materials for Medieval to Post-Medieval Tuscan Glassmaking: New Insight from LA-ICP-MS Analyses." *Journal of Archaeological Science* 37 (2010): 3030–36.

Camerota, Filippo, ed. *The Line of the Sun: Great Sundials in Florence.* Translated by Catherine Frost. Florence, Italy: Edizioni della Meridiana, 2007.

———, and Mara Miniati, eds. *I Medici e le scienze: Strumenti e macchine nelle Collezioni Granducali.* Florence, Italy: Giunti, 2008.

Campbell, Thomas P. *Tapestry in the Renaissance: Art and Magnificence.* New York: Metropolitan Museum of Art, 2002.

Casellato, Umberto, Frederica Fenzi, Paolo Guerriero, Sergio Sitran, Pietro Alessandro Vigato, Umberto Russo, Massimo Galgani, Marja Mendera, and Andrea Manasse. "Medieval and Renaissance Glass Technology in Valdesa (Florence), Part 1: Raw Materials, Sands and Non-Vitreous Finds." *Journal of Cultural Heritage* 4 (2003): 337–53.

Casellato, Umberto, Pietro Alessandro Vigato, Umberto Russo, and Mauro Matteini. "A Mössbauer Approach to the Physico-Chemical Characterization of Iron-Containing Pigments for Historical Wall Paintings." *Journal of Cultural Heritage* 1 (2000): 217–32.

Ceccarelli, Marco. "Renaissance of Machines in Italy: From Brunelleschi to Galilei through Francesco to Giorgio and Leonardo." *Mechanism and Machine Theory* 43 (2008): 1530–42.

Cecci, Alessandro, and Paola Pacetti. "La Sala delle Carte Geografiche di Palazzo Vecchio." In *I Medici e le scienze: Strumenti e macchine nelle Collezioni Granducali*, edited by Filippo Camerota and Mara Miniati, 141–46. Florence, Italy: Giunti Editore, 2008.

Cellini, Benvenuto. *The Autobiography of Benvenuto Cellini.* Translated by George Bull. Harmondsworth, Middlesex, UK: Penguin Books, 1956.

Cennini, Cennino. *The Book of Art of Cennino Cennini: A Contemporary Practical Treatise on Quattrocento Painting.* Translated by Christiana J. P. Herringham. London: George Allen and Unwin, 1922.

Chiarelli, Caterina, Stefano Barbolini, and Giovanni Garofalo. "The Gnomon in Palazzo Pitti." In *The Line of the Sun: Great Sundials in Florence*, edited by Filippo Camerota, app. 1. Translated by Catherine Frost. Florence, Italy: Edizioni della Meridiana, 2007.

Cobb, Matthew. *Generation: The Seventeenth-Century Scientists Who Unraveled the Secrets of Sex, Life, and Growth.* New York: Bloomsbury Publishing, 2006.

Cole, Michael, and Mary Pardo. "Origins of the Studio." In *Inventions of the Studio, Renaissance to Romanticism*, edited by Michael Cole and Mary Pardo, 1–35. Chapel Hill: University of North Carolina Press, 2005.

Coli, Massimo, Chiara Tanini, Margaret Haines, Enrico Pandeli, Gabriele Pini, and Federica Bencini. "The 'Pietra Serena' Stones of Brunelleschi's Cupola." *Journal of Cultural Heritage* 9 (2008): 214–21.

Colomban, Philippe, Véronique Milande, and Hérve Lucas. "On-Site Raman Analysis of Medici Porcelain." *Journal of Raman Spectroscopy* 35 (2004): 68–72.

Costagliola, Pilario, Giovanni Baldi, Curzio Cipriani, Elena Pecchioni, and Antonella Buccianti. "Mineralogical and Chemical Characterisation of the Medicean Glass Mosaic Tesserae and Mortars of the Grotta del Buontalenti, Giardino di Boboli, Florence, Italy." *Journal of Cultural Heritage* 1 (2000): 287–99.

De Santis, Alberto, Elisabetta Mattei, Illaria Montini, and Claudia Pelosi. "A Micro-Raman and Internal Microstratigraphic Study of Ceramic Sherds from the Kilns of the Medici Castle at Cafaggiolo." *Archaeometry* 54 (2012): 114–28.

de Tolnay, Charles. "Michelangelo Studies." *Art Bulletin* 22 (1940): 127–37.

Degano, Ilaria, Jeanette Lucejko, and Maria Perla Colombini. "The Unprecedented Identification of Safflower Dyestuff in a 16th Century Tapestry through the Application of a New Reliable Diagnostic Procedure." *Journal of Cultural Heritage* 12 (2011): 295–99.

Di Cagno, Gabriella. *The Cathedral, the Baptistery, and the Campanile*. Florence, Italy: Mandragora, 2001.

Donati, Silvia. "Precious Renaissance Tapestries Returned to Palazzo Vecchio in Florence." *Italy Magazine*, February 7, 2014. http://www.italymagazine.com (accessed July 26, 2014).

Dunkerton, Jill, Susan Foister, Dillian Gordon, and Nicholas Penny. *Giotto to Durer: Early Renaissance Painting in the National Gallery*. London: Yale University Press, National Gallery Publication, 1991.

Edgerton, Samuel Y. *The Mirror, the Window, and the Telescope*. Ithaca, NY: Cornell University Press, 2009.

———. *The Renaissance Rediscovery of Linear Perspective*. New York: Basic Books, 1975.

Edson, Evelyn. *The World Map, 1300–1492: The Persistence of Tradition and Transformation*. Baltimore: Johns Hopkins University Press, 2007.

Eknoyan, Garabed. "Michelangelo: Art, Anatomy, and the Kidney." *Kidney International* 57 (2000): 1190–1201.

Fanelli, Giovanni, and Michele Fanelli. *Brunelleschi's Cupola: Past and Present of an Architectural Masterpiece*. Translated by Jeremy Carden, Michele Fanelli, Andrea Paoletti, and Mark Roberts. Florence, Italy: Mandragora, 2004.

Festa, Gaetano, Carla Andreani, M. P. De Pascale, Roberto Senesi, G. Vitali, Simone Porcinai, A. M. Giusti, R. Schulze, L. Canella, P. Kudejova, Marcus Mühlbauer, B. Schillinger, and the Ancient Charm Collaboration. "A Nondestructive Stratigraphic and Radiographic Neutron Study of Lorenzo Ghiberti's Reliefs from Paradise and North Doors of Florence Baptistery." *Journal of Applied Physics* 106 (2009): 1–5.

Field, Judith Veronica. "Piero della Francesca's Mathematics." In *The Cambridge Companion Guide to Piero della Francesca*, edited by Jeryldene M. Wood, 152–70. Cambridge: Cambridge University Press, 2002.

Field, Judith Veronica, Roberto Lunardi, and Thomas B. Settle. "The Perspective Scheme of Masaccio's *Trinity* Fresco." *Nuncius* 4 (1989): 28–81 (appendices pp. 82–117).

Fiorani, Francesca. *The Marvel of Maps, Art, Cartography and Politics in Renaissance Italy*. New Haven, CT: Yale University Press, 2005.

Freely, John. *Aladdin's Lamp: How Greek Science Came to Europe through the Islamic World*. New York: Alfred A. Knopf, 2009.

Frick, Carole Collier. *Dressing Renaissance Florence: Families, Fortunes, and Fine Clothing*. Baltimore: Johns Hopkins University Press, 2005.

Galilei, Galileo. *Sidereus Nuncius, or, The Sidereal Messenger*. Translated by Albert van Helden. Chicago: University of Chicago Press, 1989.

Galluzzi, Paolo. *The Art of Invention: Leonardo and Renaissance Engineers*. Institute and Museum of the History of Science. Florence, Italy: Giunti, 1999.

———. *Renaissance Engineers from Brunelleschi to Leonardo da Vinci*. Florence, Italy: Giunti, 1996.

———. "The Sepulchers of Galileo: The 'Living' Remains of a Hero of Science." In *Cambridge Companion to Galileo,* edited by Peter Machamer, 417–47. Cambridge: Cambridge University Press, 1998.

Gill, Anton. *Il Gigante: Michelangelo, Florence, and the David 1492–1504*. New York: St. Martin's Press, 2004.

Ginn, Sheryl R., and Lorenzo Lorusso. "Brain, Mind, and Body: Interactions with Art in Renaissance Italy." *Journal of the History of the Neurosciences* 17 (2008): 295–313.

Giusti, Annamaria. *Museum of the Opificio delle Pietre Dure*. Livorno, Italy: Sillabe, 2007.

Goldthwaite, Richard. *The Economy of Renaissance Florence*. Baltimore: Johns Hopkins University Press, 2009.

Green, Judy, and Paul S. Green. "Alberti's Perspective: A Mathematical Comment." *Art Bulletin* 69 (1987): 641–45.

Grout, Donald J., and Hermine Weigel Williams. *A Short History of Opera*. 4th ed. New York: Columbia University Press, 2003.

Hall, Marcia B. *Color and Meaning: Practice and Theory in Renaissance Painting*. Cambridge: Cambridge University Press, 1992.

Heilbron, J. L. *The Sun in the Cathedral: Cathedrals as Solar Observatories*. Cambridge, MA: Harvard University Press, 1999.

Henderson, John. *The Renaissance Hospital: Healing the Body and Healing the Soul*. New Haven, CT: Yale University Press, 2006.

Hibbert, Christopher. *The House of Medici: Its Rise and Fall*. New York: Morrow Quill Paperbacks, 1980.

Hoffmann, Volker. "Giotto and Renaissance Perspective." *Nexus Network Journal* 12 (2010): 5–32.

Huguet-Termes, Teresa. "Islamic Pharmacology and Pharmacy in the Latin West: An Approach to Early Pharmacopoeias." *European Review* 16 (2008): 229–39.

Innocenti, Piero degli. "The Solstitial Sundial in the Baptistery of San Giovanni." In *The Line of the Sun: Great Sundials in Florence,* edited by Filippo Camerota, 13–19. Translated by Catherine Frost. Florence, Italy: Edizioni della Meridiana, 2007.

Jones, Barry, Andrea Sereni, and Massimo Ricci. "Building Brunelleschi's Dome: A Practical Methodology Verified by Experiment." *Journal of the Society of Architectural Historians* 69 (2010): 39–61.

Kemp, Martin. *Behind the Picture: Art and Evidence in the Italian Renaissance.* New Haven, CT: Yale University Press, 1997.

———. *The Science of Art.* New Haven, CT: Yale University Press, 1990.

King, Ross. *Brunelleschi's Dome: How a Renaissance Genius Reinvented Architecture.* New York: Penguin Books, 2001.

Kuksis, Aris. "Yolk Lipids." *Biochimica et Biophysica Acta* 1124 (1992) 205–22.

Kuznetsov, Andrey V., Igor Yu Gorshkov, Andrey V. Evsenin, Oleg I. Osetrov, Dmitry N. Vakhtin, Antonio Cosentino, and Maurizio Seracini. "Nanosecond Neutron Analysis for the Search of the Lost Leonardo's Masterpiece, *The Battle of Anghiari.*" *Nuclear Instruments and Methods in Physics Research B* 267 (2009): 3694–97.

Ladis, Andrew. *Taddeo Gaddi: Critical Reappraisal and Catalogue Raisonne.* Columbia: University of Missouri Press, 1982.

Larson, Orville K. "Vasari's Descriptions of Stage Machinery." *Educational Theatre Journal* 9 (1957): 287–99.

Levey, Michael. *Florence, A Portrait.* Cambridge, MA: Harvard University Press, 1998.

Lippi, Donatella. "The Diseases of the Medici Family and the Use of Phytotherapy." *Evidence-Based Complementary and Alternative Medicine* 4 (2007): 9–11.

———. "The Strange Destiny of Caterina de' Medici and a Case of Puerperal Fever in the Sixteenth Century." *Journal of Maternal-Fetal and Neonatal Medicine* 24 (2011): 650–51.

———, and Carlo M. Rotella. "Bronzino or Pontormo? Look at the Neck!" *Journal of Endocrinological Investigation* 33 (2010): 72–73.

———, and Laura Vannucci. "Maria Petrocini Ferretti, the First Female Surgeon in Florence, Italy." *Archives of Surgery* 146 (2011): 1231–32.

Magrini, Graziano. "Ponte Vecchio." Translated by Catherine Frost. http://Brunelleschi.imss.fi.it/itineraries/place/PonteVecchio.html (accessed June 25, 2016).

Manetti, Antonio di Tuccio. *The Life of Brunelleschi.* Introduction and Critical Text Edition by Howard Saalman, translated by Catherine Enggass. University Park: Pennsylvania State University Press, 1970.

Marchini, Giuseppe. *Italian Stained Glass Windows.* New York: H. N. Abrams, 1957.

Masters, Roger D. *Fortune Is a River: Leonardo da Vinci and Niccolò Machiavelli's Magnificent Dream to Change the Course of Florentine History.* New York: Free Press, 1998.

Mazzucconi, Fabrizio, Piero Ranfagni, and Alberto Righini. "The Gnomon of Santa Maria del Fiore." In *The Line of the Sun: Great Sundials in Florence,* edited by Filippo Camerota, 45–62. Translated by Catherine Frost. Florence, Italy: Edizioni della Meridiana, 2007.

McClellan, James E., III, and Harold Dorn. *Science and Technology in World History: An Introduction.* Baltimore: Johns Hopkins University Press, 1999.

Morelli, Giovanna, Yajun Song, Camilia J. Mazzoni, Mark Eppinger, Philippe Roumagnac, David M. Wagner, Mirjam Feldkamp, Barica Kusecek, Amy J. Vogler, Yanjun Li, Yujun Cui, Nicholas R. Thomson, Thibaut Jombart, Raphael Leblois, Peter Lichtner, Lila Rahalison, Jeannine M. Petersen, Francois Balloux, Paul Keim, Thierry Wirth, Jacques Ravel, Ruifu Yang, Elisabeth Carniel, and Market Achtman. "*Yersinia Pestis* Genome Sequencing Identifies Patterns of Global Phylogenetic Diversity." *Nature Genetics* 42 (2010): 1140–43.

Morrison, Alan S., Julius Kirshner, and Anthony Molho. "Epidemics in Renaissance Florence." *American Journal of Public Health* 75 (1985): 528–35.

Musumarra, Giuseppe, and Maria Fichera. "Chemometrics and Cultural Heritage." *Chemometrics and Intelligent Laboratory Systems* 44 (1998): 363–72.

Nagler, Alois Maria. *Theatre Festivals of the Medici, 1539–1637.* Translated by George Hickenlooper. New Haven, CT: Yale University Press, 1964.

Najemy, John M. *A History of Florence, 1200–1575.* Malden, MA: Blackwell Publishing, 2008.

Neugebauer, Otto. *A History of Ancient Mathematical Astronomy.* New York: Springer-Verlag, 1975.

Norton, Stata. "The Pharmacology of Mithridatum: A 2000-Year-Old Remedy." *Molecular Interventions* 6 (2006): 60–66.

Olson, Roberta J. M., and Jay M. Pasachoff. "Comets, Meteors, and Eclipses: Art and Science in Early Renaissance Italy." *Meteoritics and Planetary Science* 37 (2002): 1563–78.

Pallen, Thomas A. *Vasari on Theatre.* Carbondale: Southern Illinois University Press, 1999.

Palmieri, Andrea. "Ancient Mechanical 'Oriuoli.'" In *The Line of the Sun: Great Sundials in Florence,* edited by Filippo Camerota, 21–27. Translated by Catherine Frost. Florence, Italy: Edizioni della Meridiana, 2007.

Paoletti, John T., and Gary M. Radke. *Art in Renaissance Italy.* 3rd ed. Upper Saddle River, NJ: Pearson Education, 2005.

Park, Katherine. *Doctors and Medicine in Early Renaissance Florence.* Princeton, NJ: Princeton University Press, 1985.

Piccini, Attilio. *Spedale degli Innocenti: The Foundling Hospital and Its Museum.* Florence, Italy: Lo Studiolo, Amici dei Musei Fiorentini, 1977.

Piccolpasso, Cipriano. *The Three Books of the Potter's Art.* Translated by Ronald Lightbrown and Alan Caiger-Smith. London, UK: Scolar Press, 1980.

Pieraccini, Massimiliano, Daniele Mecatti, Guido Luzi, Maurizio Seracini, Gianpaolo Pinelli, and Carlo Atzeni. "Non-Contact Intrawall Penetrating Radar for Heritage Survey: The Search of *The Battle of Anghiari* by Leonardo da Vinci." *NDT & E International* 38 (2005): 151–57.

Prager, Frank D., and Gustina Scaglia. *Brunelleschi: Studies of His Technology and Inventions.* Mineola, NY: Dover Publications, 2004.

Proietti, Noemi, Donatella Capitani, Raffaele Lamanna, Federica Presciutti, Enrico Rossi, and Anna L. Segre. "Fresco Paintings Studied by Unilateral NMR." *Journal of Magnetic Resonance* 177 (2005): 111–17.

Ptolemy, Claudius. *The Almagest.* Translated by G. J. Toomer. New York: Springer-Verlag, 1984.

Richardson, Brian. *Print Culture in Renaissance Italy: The Editor and the Vernacular Text, 1470–1600.* Cambridge: Cambridge University Press, 2004.

———. *Printing, Writers, and Readers in Renaissance Italy.* Cambridge: Cambridge University Press, 1999.

Ronan, Colin A. *Science: Its History and Development among the World's Cultures.* New York: Facts on File, Inc., 1982.

Saalman, Howard. *Filippo Brunelleschi: The Buildings.* University Park: Pennsylvania State University Press, 1993.

———. *Filippo Brunelleschi: The Cupola of Santa Maria del Fiore.* London: A. Zwemmer Ltd., 1980.

Salemi, Maria C. *Palazzo Vecchio a Firenze.* Florence, Italy: Nardini Editore, 2001.

Salgado, Tomás Garcia. "Geometric Interpretation of the Albertian Model." *Leonardo* 31 (1998): 119–23.

Samelson, Hans. "Letter to the Editor." *Journal of the Society of Architectural Historians* 57 (1998): 360–61.

Saslow, James M. *The Medici Wedding of 1589: Florentine Festival as Theatrum Mundi.* New Haven, CT: Yale University Press, 1996.

Schevill, Ferdinand. *History of Florence from the Founding of the City through the Renaissance.* New York: Harcourt, Brace, 1936.

Schmid, Stefan M., Mervyn S. Paterson, and Jim N. Boland. "High Temperature Flow and Dynamic Recrystallization in Carrara Marble." *Tectonophysics* 65 (1980): 245–80.

Schnabel, Luise. "The Sundial in Piazza dei Giudici." In *The Line of the Sun: Great Sundials in Florence*, edited by Filippo Camerota, 108–10. Translated by Catherine Frost. Florence, Italy: Edizioni della Meridiana, 2007.

Sendova, Mariana, Valentin Zhelyaskov, M. Scalera, and Colwyn Gulliford. "Micro-Raman Spectroscopy Characterization of Della Robbia Glazes." *Archaeometry* 49 (2007): 655–64.

Settle, Thomas B. "Egnazio Danti as a Builder of Gnomons: An Introduction." In *MUSA MUSAEI: Studies on Scientific Instruments and Collections in Honour of Mara Miniati*, edited by Marco Beretta, Paolo Galluzzi, and Carlo Triarico, 93–115. Florence, Italy: Leo S. Olschki Editore, 2003.

———. "The Gnomons and Astronomical Instruments of Santa Maria Novella." In *The Line of the Sun: Great Sundials in Florence*, edited by Filippo Camerota, 71–92. Translated by Catherine Frost. Florence, Italy: Edizioni della Meridiana, 2007.

———, Curator. "Traces of Science and Engineering and Their Protagonists in Florence and Nearby Tuscany." Updated December 15, 2012. http://www.imss.fi.it/~tsettle/traces.html (accessed June 25, 2016).

Siraisi, Nancy G. *Medieval and Early Renaissance Medicine: An Introduction to Knowledge and Practice.* Chicago: University of Chicago Press, 1990.

Smart, Alastair. "Taddeo Gaddi, Orcagna, and the Eclipses of 1333 and 1339." In *Studies in Medieval and Renaissance Art in Honor of Millard Meiss*, Vol. 1, edited by Irving Lavin and John Plummer, 403–14. New York: New York University Press, 1977.

Smith, Graham. "Cosimo I and the Joseph Tapestries for the Palazzo Vecchio." *Renaissance and Reformation* 18 (1982): 183–96.

Sobel, Dava. *Galileo's Daughter: A Historical Memoir of Science, Faith, and Love*. New York: Walker Publishing Company, Inc., 1999.

Spallanzani, Marco. "Medici Porcelain in the Collection of the Last Grand-Duke." *Burlington Magazine* 132 (1990): 316–20.

Stokstad, Marilyn. *Art History: Fourteenth to Seventeenth Century Art*. Upper Saddle River, NJ: Pearson Prentice Hall, 2009.

Stone, Richard E. "Antico and the Development of Bronze Casting in Italy in the End of the Quattrocento." *Metropolitan Museum Journal* 16 (1981): 87–116.

Strano, Giorgio. "The Origins of Gnomonics." In *The Line of the Sun: Great Sundials in Florence*, edited by Filippo Camerota, 10–12. Translated by Catherine Frost. Florence, Italy: Edizioni della Meridiana, 2007.

Strocchia, Sharon T. "The Nun Apothecaries of Renaissance Florence: Marketing Medicines in the Convent." *Renaissance Studies* 25 (2011): 627–47.

———. *Nuns and Nunneries in Renaissance Florence*. Baltimore: Johns Hopkins University Press, 2009.

Strong, Roy. *Art and Power: Renaissance Festivals, 1450–1650*. Berkeley: University of California Press, 1984.

Suh, H. Anna, ed. *Leonardo's Notebooks*. New York: Black Dog and Leventhal Publishers, 2005.

Tite, Michael S. "The Production Technology of Italian Maiolica: A Reassessment." *Journal of Archaeological Science* 36 (2009): 2065–80.

Tobin, Richard. "Ancient Perspective and Euclid's Optics." *Journal of the Warburg and Courtauld Institutes* 53 (1990): 14–41.

Tognetti, Sergio. "The Development of the Florentine Silk Industry: A Positive Response to the Crisis of the Fourteenth Century." *Journal of Medieval History* 31 (2005): 55–69.

Trachtenberg, Marvin. "What Brunelleschi Saw: Monument and Site at the Palazzo Vecchio in Florence." *Journal of the Society of Architectural Historians* 47 (1983): 14–44.

Vasari, Giorgio. *The Lives of the Artists*. Vol. 1. Translated by George Bull. London: Penguin Books, 1987.

———. *Lives of the Most Eminent Painters, Sculptors, and Architects*. Vol. 1. Translated by Gaston du C. de Vere. New York: Abrams, 1979.

Walker, Paul Robert. *The Feud That Sparked the Renaissance*. New York: HarperCollins, 2002.

Wallace, William E. *Michelangelo at San Lorenzo: The Genius as Entrepreneur*. Cambridge: Cambridge University Press, 1994.

Wassell, Stephen R. "Commentary on Elements of Painting." In *The Mathematical Works of Leon Battista Alberti*, edited by Kim Williams, Lionel March, and Stephen R. Wassell, 153–68. Basel, Switzerland: Birkhäuser, 2010.

Wheelis, Mark. "Biological Warfare at the 1346 Siege of Caffa." *Emerging Infectious Diseases* 8 (2002): 971–75.

Wilford, John Noble. *The Mapmakers*. New York: Vintage Books, 1981.

Wirtz, Rolf C. *Art and Architecture, Florence*. Potsdam, Brandenburg, Germany: Konemann, 2005.

Wray, Shona Kelly. "Boccaccio and the Doctors: Medicine and Compassion in the Face of Plague." *Journal of Medieval History* 30 (2004): 301–22.

Wu, Xiaohong, Chi Zhang, Paul Goldberg, David Cohen, Yan Pan, Trina Arpin, and Ofer Bar-Yosef. "Early Pottery at 20,000 Years Ago in Xianrendong Cave, China." *Science* 336 (2012): 1696–1700.

Zucchiatti, Alessandro, Anne Bouquillon, Ildiko Katona, and Alessandra D'Alessandro. "The 'Della Robbia Blue': A Case Study for the Use of Cobalt Pigments in Ceramics during the Italian Renaissance." *Archaeometry* 48 (2006): 131–52.

Zucchiatti, Alessandro, Anne Bouquillon, Jacques Castaing, and Jean-René Gaborit. "Elemental Analyses of a Group of Glazed Terracotta Angels from the Italian Renaissance, as a Tool for the Reconstruction of a Complex Conservation History." *Archaeometry* 45 (2003): 391–404.

About the Authors

Susan Puett has a BA from Duke University in history and education, holds a certificate in the teaching of kindergarten from Belmont University, and is a graduate of the University of Miami Chemical Dependency Training Institute. She has worked throughout her career as a teacher, group facilitator, and advocate for young people and currently is devoting her time to professional writing. She is the author of one historical book, as well as numerous poetic works that have appeared in various journals. In recent years her passion for art and history, as well as her Italian heritage, has coalesced in a desire to immerse herself in the study of the Italian Renaissance, and most particularly Florence.

J. David Puett has a BS and MS in physics and earned his PhD in biochemistry from the University of North Carolina at Chapel Hill. He has held faculty and administrative positions in biochemistry and molecular biology at the Vanderbilt University School of Medicine, the University of Miami School of Medicine, and the University of Georgia, where he served as department head for fourteen years. He is currently Regents Professor and department head emeritus at the University of Georgia and adjunct professor of biochemistry and biophysics at the University of North Carolina School of Medicine at Chapel Hill. His science teaching has focused on human, medical, and physical biochemistry, as well as topics in Renaissance Florence (honors and first-year seminars). He has authored hundreds of scientific publications including research articles, reviews, and books.

Susan and David, who now reside in Chapel Hill, North Carolina, have been traveling to Italy for many years and together have led groups of University of Georgia honors students to Florence on travel-abroad programs focused on art and science in the Renaissance. They have also accompanied study-abroad students to Florence from the University of Georgia Cortona campus where David taught for five years. Although this book is their first major collaborative endeavor, the authors have jointly published an article on Florentine Renaissance apothecaries and their role in both medicine and art.